solid state
devices and
applications

solid state devices and applications

Frederick F. Driscoll
Robert F. Coughlin

Wentworth Institute
Boston, Massachusetts

PRENTICE-HALL, INC., ENGLEWOOD CLIFFS, NEW JERSEY

Library of Congress Cataloging in Publication Data

Driscoll, Frederick F.   1943–
  Solid state devices and applications.

  1. Semiconductors.   2. Transistor circuits.
I. Coughlin, Robert F., joint author.   II. Title.
TK7871.85.D73   621.3815′2   73–17388
ISBN 0–13–822106–5

Printed in the United States of America

10   9   8   7   6   5   4   3   2

Prentice-Hall International, Inc., *London*
Prentice-Hall of Australia, Pty. Ltd., *Sydney*
Prentice-Hall of Canada, Ltd., *Toronto*
Prentice-Hall of India Private Limited, *New Delhi*
Prentice-Hall of Japan, Inc., *Tokyo*

To

JEAN and BARBARA

contents

chapter three
field-effect transistors   97

chapter four
device limitations   138

chapter five
basic dc power supplies   175

chapter six
feedback voltage regulators
using integrated circuits   216

chapter seven
triggering devices   255

chapter eight
silicon controlled rectifier 305

chapter nine
the triac 336

chapter ten
light sources 353

chapter eleven
photodetectors 384

appendices 429

bibliography 462

index 465

preface

Solid State Devices and Applications developed from need to learn how to use a variety of newer semiconductor devices together with the standard diodes, bipolar and field-effect transistors. The newer devices were selected if they

1. are commonly available at reasonable cost
2. are used in everyday consumer or industrial-type applications
3. are growing in importance as indicated by trade literature
4. simplify understanding or operation of relatively complex circuits
5. perform better and at less expense than other devices.

This book is divided into four sections, each of which can stand alone.

SECTION 1—CHAPTERS 1 THROUGH 4

Section 1 introduces the diode, bipolar junction, and field-effect transistor together with heat and voltage limitations. Basic ideas of voltage gain, input resistance, graphical analysis, and circuit models are set forth for use in later sections.

SECTION 2—CHAPTERS 5 AND 6

Section 2 shows how to analyze or design half and full-wave rectifiers, basic zener regulators, series regulators, and feedback regulators. An integrated circuit op-amp is used to reduce design and analysis of a basic feedback regulator to three simple equations. The basic feedback regulator is treated as a system to which we add options of remote programming, current limiting, variable load voltage, and/or remote voltage sensing. This system can also be used as a constant current generator or audio power amplifier and demonstrates what goes on inside any of the host of modern IC regulators.

SECTION 3—CHAPTERS 7 THROUGH 9

Theory, measurements, characteristics, and applications of semiconductor triggering and power switching devices are presented in this section. The unijunction transistor is employed to learn about timing, relaxation oscillators, voltage level sensing, and pulse formation. A two-transistor switch is studied to simplify understanding of the programmable unijunction transistor (PUT), silicon unitateral switch (SUS), silicon bilateral switch (SBS), asymmetrical bilateral switch (ASBS), diac, triac, and silicon controlled rectifier (SCR).

SECTION 4—CHAPTERS 10 AND 11

Section 4 is concerned with optoelectric devices. Theory, measurements, characteristics, and applications of light sources and photo detectors are introduced. These include incandescent tungsten and neon glow lamps, light-emitting diodes (LEDs), photoconductive cells, photovoltaic cells, photodiodes, phototransistors, light-activated-silicon-controlled rectifiers (LASCR), and optical or photo-couplers.

All applications have been tested and may be used in the laboratory or for demonstrations. The simplest and most economical test circuit has been selected for each device to test its most important electrical characteristic. Sweep circuits for testing are shown, where practical, because they give maximum information in a minimum of time. There is sufficient material for a two semester course, and a knowledge of basic algebra would be helpful.

We hereby thank Dean Charles M. Thomson for his encouragement, support, and guidance. For skillful preparation of the manuscript we sincerely thank Mrs. Jean Driscoll, Mrs. Mary Ellen Hatfield, and Mrs. Pauline Campbell. Finally we are grateful to our students for their refreshing insistence on truth, understanding, relevance, simplicity, and the elimination of trivia. We also gratefully acknowledge the influence of Professors Campbell L. Searle and Bruce Wedlock.

solid state
devices and
applications

chapter one
two-terminal
semiconductor devices

1-0 INTRODUCTION

Electrical and chemical properties of an element of matter depend primarily on the number of electrons in the outer shell of the atom. These outer electrons are called *valence* electrons, and atoms with four valence electrons make up the *tetravalent* group of elements that can act as a semiconductor. As its name implies, the semiconductor has a resistance to the flow of current midway between a conductor and an insulator. Silicon and germanium are the most common semiconductor materials. In order to understand how current flows through the semiconductor, we must study it as an environment within which the movement of positive or negative charges constitutes a current.

1

1-1 THE SEMICONDUCTOR ENVIRONMENT

If we could shrink to the size of an electron, maintain all of our faculties, and enter a sample of extremely pure or *intrinsic* semiconductor material, we would see mostly empty space. However, order would be apparent in an endless cubic symmetrical structure in which every atom shared one of its valence electrons with one from each of its four neighbors to form four *covalent bonds*. Each atom thus has a full outer shell with eight electrons (four of its own plus four shared) and forms a very stable crystalline arrangement known as a *cubic lattice*. A two-dimensional view is shown in Fig. 1-1. We should visualize the electrons as moving continuously within the outer shell of each atom, and from outer shell to outer shell of neighboring atoms.

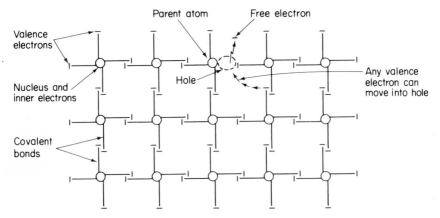

Figure 1-1

Two-dimensional illustration of a semiconductor crystal.

The intrinsic semiconductor crystal receives energy from its surroundings, usually in the form of heat. This heat is conducted to the atoms in order to increase their kinetic energy, and consequently the velocity of some of the valence electrons. A few valence electrons absorb enough energy to escape from their parent atoms and become free to wander through the empty space in the semiconductor environment. There are approximately 5×10^{22} atoms in a cubic centimeter of silicon. At room temperature approximately 1.5×10^{10} electrons have escaped from their parent atoms, or one electron has been freed from every 3.3×10^{12} silicon atoms. As an analogy, one free electron is like the last of the Mohicans with respect to all the people ever born.

The semiconductor structure is not altered because of the freed electrons, but it is important to analyze the consequences of freeing electrons. For each freed electron, one covalent bond has been broken, leaving an absence of one electron or *hole* in the valence shell of the parent atom. The resulting hole can be considered as a free particle with a positive charge equal to that of the electron. For example, in Fig. 1-1 a valence electron from a neighboring atom moves from right to left into the hole. The hole has apparently moved from left to right. Thus the broken covalent bond has generated two free charge carriers: (1) an electron to travel in the space of the crystal, and (2) a hole to travel in the valence shells of the atoms. Each atom that has a hole is missing one electron and has one less negative charge to balance positive charges in the nucleus. The atom thus has a net positive charge and is called a positive ion. (Note: A semiconductor ion is not a hole. The ion cannot move; it is locked in the lattice.)

1-2 CURRENT FLOW IN THE SEMICONDUCTOR

Within the semiconductor, current magnitude depends on the number of available charge carriers and how fast they can move. Hole motion is influenced by the atom's nucleus and is slower than the free electron. Semiconductor devices that depend on electrons rather than holes as current carriers perform better at higher frequencies. The number of current carriers in a pure semiconductor depends on the number of broken covalent bonds, which in turn depend on the amount of energy received. For example, if heat, light, or electromagnetic energy is applied: (1) the atoms vibrate faster to (2) rupture more covalent bonds, (3) releasing more free holes and electrons to (4) raise the total number of free charges. Of course, with more free electrons we increase the chance that a free electron will return to an atom to occupy a hole and reestablish a covalent bond through a process of *recombination*. Once the rate of recombination has increased to equal the higher rate of generation, there is a dynamic equilibrium, with more electrons and holes available for current flow. Resistance of the semiconductor goes *down* with temperature *rise* because the number of carriers increases indirectly with an increase in temperature. This property is utilized in a temperature-sensitive resistor known as the *thermistor* that will be studied in section 1-3.

Once an electron has been freed it will collide with vibrating atoms and rebound with either more or less speed (energy) in an unpredictable direction. We employ this observation to explain diffusion current. For example, we could create a local concentration of holes and electrons by heating one end of a semiconductor rod. The electrons and holes must move with random speeds and directions. Eventually electrons and holes would distribute themselves evenly throughout the sample. This type of hole and electron movement is called *diffusion current*. It is the same principle by which perfume molecules spread from the source throughout a room.

Connecting a battery across a pure semiconductor slightly modifies random motion of holes and electrons. Electrons are attracted toward the positive

battery terminal. The net movement of holes and electrons within the semiconductor is called *drift current*. We conclude that current flow in a pure semiconductor depends on both temperature and applied voltage.

1-3 TEMPERATURE MEASUREMENTS WITH THERMISTORS

"Thermal resistors" or *thermistors* are made by heating under pressure
(sintering) semiconductor ceramic materials made from mixtures of metallic
oxides. Thermistors may be obtained in the shape of beads, rods, disks, probes,
or as required for special applications. Depending on the temperature range
of your application, select a thermistor according to the approximate classification in Table 1.1.

Table 1.1

Temperature Range of Application	Type of Thermistor	Range of Resistance Values at 25°C
−100°F to 150°F	low resistance	100Ω to 2kΩ
150°F to 300°F	intermediate resistance	2kΩ to 75kΩ
300°F to 600°F	high resistance	75kΩ to 500kΩ

In Fig. 1-2 a low-resistance disc thermistor is connected to an ohmmeter.
Temperature of the air ambient is varied and measured by an ordinary mercury
thermometer. At each temperature, resistance is read on the ohmmeter and
corresponding values of resistance-temperature (R-T) are plotted to draw curve
A. The meter needle points to $2.8 \times 100 = 280\Omega$ when the thermometer and
thermistor are at body temperature or $37°C \cong 98°F$. Curve B is the R-T characteristic of a different type of disc thermistor. Figure 1-2 suggests an immediate
application, in that we could buy a thermistor in the shape of a probe and,
during the R-T test, calibrate the ohmmeter scale in terms of temperature. For
example, 37°C (body temperature) would be marked above the needle in Fig.
1-2, and the ohmmeter-thermistor combination can now serve as a thermometer.
Resistance-temperature plots are often furnished by the manufacturer, or the
same information can be given either in table form or as a percentage change in
resistance per degree centigrade over a limited range of temperatures. For
example, thermistor A in Fig. 1-2 could be described by Table 1.2.

Table 1.2

Resistance in ohms	940	800	400	250	100
Temperature in °C	10	15	30	40	70

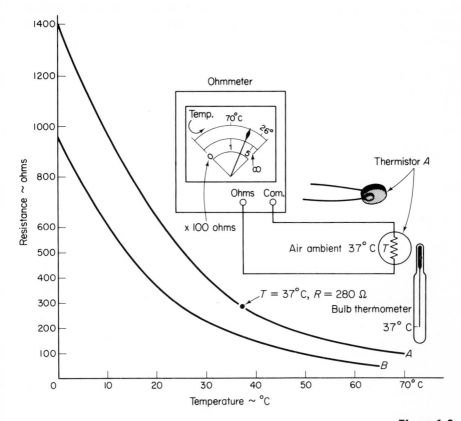

Figure 1-2

Measurement of the resistance-temperature characteristic of a thermistor.

Since *R-T* curve *A* is reasonably straight in the vicinity of 15°C, the thermistor could also be described by the slope of the curve between 10°C and 30°C as

$$\frac{\Delta R}{\Delta T} = \frac{(940 - 400)\Omega}{(10 - 30)°C} = -\frac{540\Omega}{20°C} = -27 \text{ ohms per } °C \qquad (1\text{-}1)$$

Temperature Coefficient

Usually resistance change ΔR is specified as a percentage of the thermistor's resistance at some reference temperature within the temperature range (15°C in this example). The result is called *temperature coefficient of resistance*, α. Evaluating α around 15°C from curve *A*

$$\alpha = \frac{\Delta R/\Delta T}{R \text{ at } 15°C} = \frac{-27 \text{ ohms/}°C}{800 \text{ ohms}} \cong \frac{0.034}{°C} = -3.4\%/°C \qquad (1\text{-}2)$$

The minus signs in Eqs. (1-1) and (1-2) mean that resistance decreases when temperature increases. Therefore, thermistors have a *negative* temperature coefficient.

EXAMPLE 1-1:

A thermistor has a temperature coefficient of resistance equal to $-3.4\%/°C$ and a resistance of 800Ω at $15°C$. What is its resistance at $25°C$?

SOLUTION:

From Eq. (1-2)

$$\Delta R = \alpha(R \text{ at } 15°C)(\Delta T) = (-0.034)(800)(25 - 15) = -272\Omega$$

and

$$R \text{ at } 25°C = R \text{ at } 15°C + \Delta R = 800\Omega - 272\Omega = 518\Omega$$

Check R-T curve A in Fig. 1-2 for verification at $T = 25°C$.

Temperature-to-Voltage

The circuit of Fig. 1-3 illustrates another application of a thermistor. This circuit converts a temperature change of $0°C$ to $50°C$ into a voltage change of 0 to $5V$. At $0°C$ output voltage V_o equals zero if

$$R_B = R_A + 1,400\Omega \tag{1-3}$$

To make $V_o = +5V$ at $50°C$ we want 15 volts to be dropped across R_B out of the total supply voltage of $10V - (-10) = 20V$. See Fig. 1-3b. Using the voltage division law

$$\frac{R_B}{R_B + R_A + 180\Omega}(20V) = 15V \tag{1-4}$$

Solving Eqs. (1-3) and (1-4) for R_A and R_B yields $R_A = 430\Omega$ and $R_B = 1830\Omega$.

There is a disadvantage to this simple design. Since thermistor resistance varies nonlinearly with temperature, output voltage will also be nonlinear. For example, at $25°C$ we would like V_o to be $2.5V$, but instead, since $R_T = 518\Omega$ at $25°C$ (from Ex. 1-1), V_o will actually be

$$V_o = -10V + V_{RB} = -10V + \frac{1830\Omega \times 20V}{1830\Omega + 430\Omega + 518\Omega}$$

$$= -10V + 13.2V = 3.2V$$

Voltage versus temperature is plotted in Fig. 1-3c.

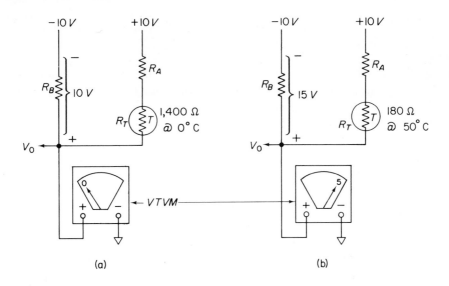

(a)   (b)

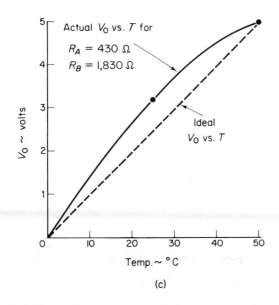

(c)

Figure 1-3

Temperature is converted to voltage by the circuit in (a) and (b)
as shown in (c).

Bridge Improvement

Improved methods of temperature measurements are illustrated in Fig. 1-4. In Fig. 1-4a two thermistors are mounted in one probe to measure the temperature. This technique doubles the sensitivity involved in Fig. 1-3. A 50°C temperature change gives an output voltage of $10V$. Another circuit for measuring temperature is shown in Fig. 1-4b, where V_o is adjusted to zero with matched thermistors at the same temperature. The measuring thermistor is then placed in a different ambient and V_o is proportional to the *difference* in temperature.

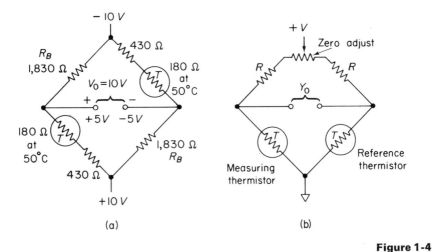

(a)   (b)

Figure 1-4

Bridge arrangements of thermistors give increased sensitivity in (a) and differential temperature measurement in (b).

Thermal Time Constant

We have not considered the possibility of thermistor self-heating. Current passing through a thermistor generates heat power equal to I^2R_T. As long as self-heating is negligible, the thermistor assumes the temperature of its ambient. If temperature is suddenly raised, for example from T_1 to T_2, the thermistor temperature also rises toward T_2. The rate of temperature rise is determined by a *thermal time constant* which is the time interval required for the thermistor to change temperature by 63% of $(T_2 - T_1)$. Time constant magnitude depends mostly on mass and packaging of the thermistor and is on the order of 0.1 to 10 seconds. Self-heating must usually be avoided in applications involving temperature measurements. However, the self-heating property of thermistors is employed in the next section to measure motion or density of liquids or gases.

1-4 FLOW RATE MEASUREMENT WITH THERMISTORS

In Fig. 1-5 a test circuit is employed to show that thermistor current must be limited to 10mA or less to avoid self-heating. Above 10mA the thermistor begins to self-heat and its resistance decreases. For example, at $I = 50$mA, voltage across the thermistor is $9V$ with a resistance of 180Ω and its temperature has stabilized at 50°C or 24°C above the ambient of 26°C. The thermistor will indicate its own temperature, not that of the environment, and cannot be used to measure temperature when self-heating occurs. Thermistor temperature now depends on how fast its heat flows into the environment. If any changes occur in the ambient, such as movement (how fast), composition (what kind), pressure or level (how much), heat will be conducted away from the thermistor at a different rate. Thermistor temperature must change because of environment changes, thus causing resistance changes. Operation in this self-heating mode is summarized by saying that *thermistor thermal conductivity* is controlled by the environment. Thus the thermistor can act as a *transducer* to convert, for example, a flow rate to a voltage reading.

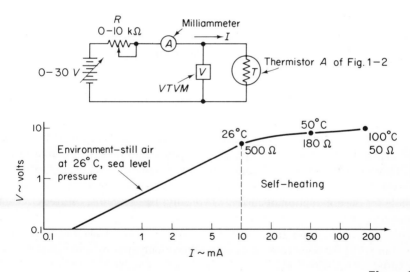

Figure 1-5

Test circuit and current-voltage characteristic of a low-resistance, disc-type thermistor.

Measurement of air flow is undertaken in Fig. 1-6a by two matched thermistors with measurements like those of Fig. 1-5. Both are connected into the bridge circuit of Fig. 1-6b. Resistors R are chosen to give a self-heating current of 200mA. With *no* air flow, the 50Ω balancing resistor is adjusted for zero output voltage. As air is pumped through the line, heat is removed faster from *TH*1 while heat transfer from *TH*2 is unchanged in the still air of the reservoir. Resistance of *TH*1 increases, unbalancing the bridge, to generate an output voltage. V_o may be calibrated in cubic feet per minute.

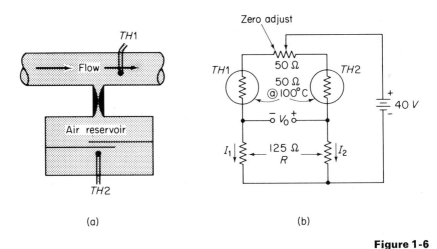

(a)   (b)

Figure 1-6

Flow measurement in (a) by the bridge circuit of (b). $T_1 = T_2 = 200$ mA
at ambient of 26° C in still air.

1-5 DOPING AND THE *pn* JUNCTION

Resistance of pure semiconductor materials is much too high to conduct currents in the range of milliamperes to amperes, and must be lowered by *doping*. Doping is the process of adding precise, small amounts of impurity (non-tetravalent) atoms so that each impurity atom can contribute one free electron or hole for the conduction of current. Typically, 10^{15} free holes or free electrons, depending on the type of dopant, are added to each cubic inch of a semiconductor that previously contained only 1.5×10^{10} free electrons and holes. Thus the number of free electrons or holes has been increased by a factor of roughly $10^{15}/10^{10} = 100,000$. Semiconductor resistance is reduced by the same factor of 100,000.

P-material

There are two types of doping elements—*p*-type, which contribute holes, and *n*-type, which contribute electrons. A *p*-type atom has only three-valence electrons (trivalent). These trivalent atoms form covalent bonds with three

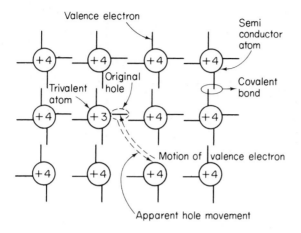

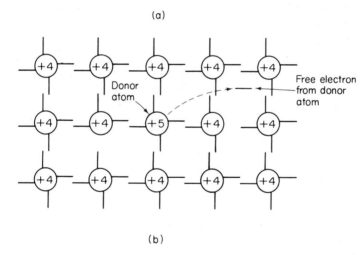

Figure 1-7

Positive charge carriers in (a) and negative charge carriers in (b) are added by doping to form *p*-type and *n*-type semiconductor samples, respectively.

neighboring semiconductor atoms, as in Fig. 1-7a. Absence of the fourth electron in the *p*-type atom constitutes a *hole*. A hole moves because a valence electron from another simiconductor atom moves to occupy the hole. Effectively the hole has moved from the *p*-type atom to a semiconductor atom, causing: (1) a flow of current and (2) the trivalent atom to become a negative ion. To summarize:

1. The resulting *p*-type semiconductor sample is electrically neutral because the total number of positive charges in the nucleus equals the total number of electrons.

2. A few free electrons are present due to thermally broken bonds.

3. A large number of free holes are present because of doping.

4. The majority of current conduction is due to free hole motion, so holes are the *majority* current carriers.

5. Only a small value of current flows due to electrons, so electrons are the *minority* current carriers.

6. All of the doping atoms are ionized and cannot move.

7. Trivalent doping atoms are called *acceptor atoms* because each accepts a valence electron from a tetravalent atom.

N-material

N-type semiconductors are made by doping pure semiconductor material with pentavalent atoms (five-valence electrons), as in Fig. 1-7b. Four of these five-valence electrons form covalent bonds with neighboring semiconductor atoms, and the fifth is free to wander as a current carrier through the lattice. *N*-type material is electrically neutral because only neutral pentavalent atoms have been added to neutral semiconductor atoms. In *n*-type semiconductors, *majority* current carriers are *free electrons*. However, *free holes* are still present due to broken covalent bonds and are the *minority* current carriers in an *n*-type material. All of the *n*-type doping atoms are positive ions and each is locked in the crystal. Pentavalent atoms are often called *donor* atoms because each donates a free electron to the semiconductor sample.

PN Junction

Assume a *p*-type semiconductor sample joined abruptly to an *n*-type sample. Free electrons from the *n*-type sample would initially try to diffuse into the *p*-type sample in an attempt to spread evenly throughout both samples, as in Fig. 1-8a. However, free electrons from donor atoms near the *n*-side of the *pn* junction leave positive charged donor ions locked in the lattice to form a wall of stationary charges, as shown in Fig. 1-8b. These free electrons crossing into the *p*-side occupy acceptor atoms near the junction, setting up a wall of locked negative ions in Fig. 1-8b. The locked ions set up an electric field just like a capacitor to repel majority electrons diffusing into it from the *n*-side and holes from the *p*-side. Therefore, very few free charges will be found in that region surrounding the junction bounded by the locked ions. This small region is appropriately designated as *depletion region*—depleted of free charges—or *space charge* region—location of space containing stationary charges.

The stationary charges form an electric field and consequently a potential difference, similar to voltage across a charged capacitor, exists across the space charge region. Minority carriers that diffuse into this region are accelerated across to the other side. Some majority carriers will attain sufficient energy from heat, lattice collision, or by other means, to overcome the electric field and cross the space charge region to become minority carriers on the other side. A dynamic equilibrium is thus attained.

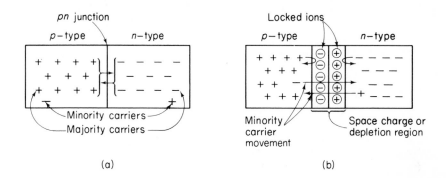

Figure 1-8

A *pn* junction in (a) causes a space charge region in (b) that opposes diffusion of majority carriers across the junction.

Summary

The *pn* junction sets up a space charge region which (1) blocks passage of all but high energy majority carriers, (2) accelerates minority carriers to aid their passage, and (3) exhibits properties of a capacitor.

1-6 BIASING THE *pn* JUNCTION

In Fig. 1-9a the positive battery terminal is connected to the *n*-side of a *pn* junction and the negative battery terminal is connected to the *p*-side. The battery terminals attract majority carriers away from the junction to: (1) uncover more ions, which (2) widens the space charge region to (3) decrease space charge capacitance (separating capacitor plates), and (4) increases the electric field across the space charge region. This condition is known as *reverse bias*. (Current flows in the wires of the external circuit by the net motion of electrons only. Within the semiconductor current flows by the net motion of both holes and electrons. It is the job of the ohmic contacts in Fig. 1-9 to exchange holes for electrons.)

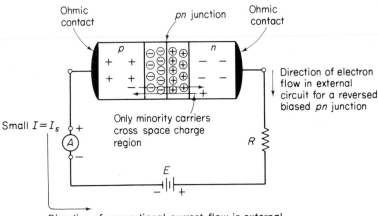

Direction of conventional current flow in external
circuit for a reversed biased *pn* junction

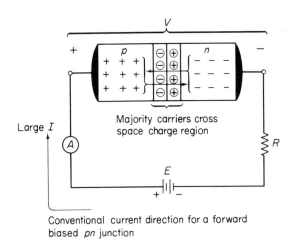

Conventional current direction for a forward
biased *pn* junction

Figure 1-9

Reverse bias in (a) and forward bias in (b) of a *pn* junction.

Reverse Bias

The magnitude of reverse-bias current is small and fixed by the number of minority carriers available. All of the minority carriers flow across the junction, and the resulting current flow is called *saturation current* I_S. Since the number of broken covalent bonds and the number of minority carriers depends on temperature, I_S is highly temperature-dependent. Typical values of I_S at room temperature and temperature dependence are given in Table 1.3.

Table 1.3

	I_S at 25°C	I_S Doubles for Each Temperature Increase of:
Germanium	$1\mu A$	approximately 10°C
Silicon	$10^{-3}\mu A$(or 1nA)	approximately 6°C

Forward Bias

If the battery connections are reversed as shown in Fig. 1-9b, the *pn* junction is *forward biased*. Since body or *bulk resistance* of *n*-type and *p*-type semiconductor material is low, almost all of voltage V is developed across the space charge region. V acts to reduce the electric field at the junction, allowing majority carriers to cross the junction in large numbers, and reduction of the space charge region to increase capacitance (closing capacitor plates). The forward bias voltage must be kept small, in the order of $0.3V$ for germanium junctions and $0.6V$ for silicon junctions, or excessive current will flow. Very briefly we can say that the *pn* junction senses polarity of an applied voltage to conduct a small constant current with reverse bias and a large current with forward bias.

1-7 ELECTRICAL MEASUREMENTS OF THE JUNCTION DIODE

The simplest application of a *pn* junction is the diode-rectifier of Fig. 1-10. Therefore a diode is a two-terminal device which conducts current easily when forward biased, but allows only a small current to flow when reverse biased. The arrowhead points toward *n*-type material, shows the direction of *conventional current flow* under forward bias, and represents the *anode* terminal. To use the diode effectively we should know all possible combinations of current and voltage. This information is best displayed by a graph called the diode's *current-voltage characteristic*, or *I-V* curve.

I-V **Curve**

The circuits of Fig. 1-10d show the diode reverse biased and forward biased. In both circuits, the battery voltage E is varied so that we may observe how I varies. An IN1693 silicon diode is used.

Corresponding values of voltage across the diode, V, and current through the diode, I, are measured and plotted. Reverse saturation current is plotted in the third quadrant of Fig. 1-10d, and is constant and too small to be distinguished from zero on a scale of 1 mA per division. Forward current/versus volt-

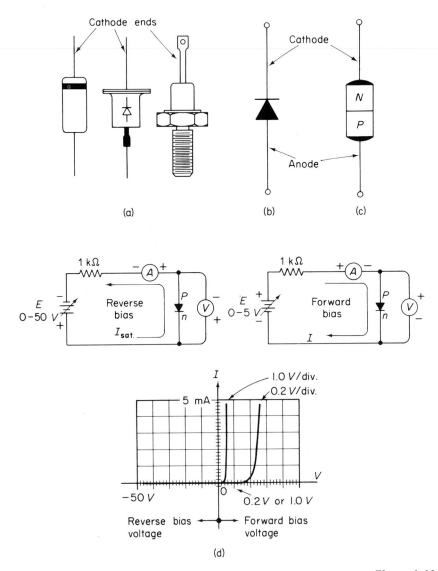

Figure 1-10

Typical diode constructions in (a) compare with the diode symbol in (b) and *pn* junction in (c). Test circuits in (d) give data to plot forward and reverse *I-V* characteristics.

age combination points are plotted in the first quadrant, first on a horizontal scale of 1 *V*/div. and again at 0.2 *V*/div. to show how a choice of scale can hide or emphasize the exponential nature of the diode's *I-V* characteristic.

Sweep-Measurement Techniques

Point-by-point measurement of a diode characteristic is time-consuming. *Sweep-measurement* techniques are faster, just as accurate, and introduce an application of diode D_B as a rectifier in Fig. 1-11. Here we plot the *I-V* characteristic of test diode D_T on a Cathode-ray Oscilloscope (CRO).

An ordinary filament transformer, rectifying diode D_B, and a 3.3 kΩ resistor provide a continuously varying voltage from $0V$ to $8.4V$ and back to

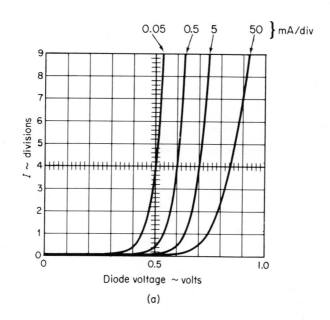

(a)

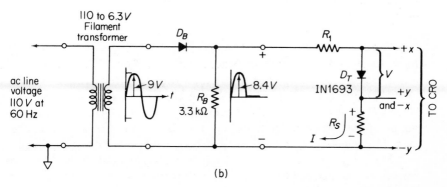

(b)

Figure 1-11

I-V characteristic curves for four forward current ranges are drawn on
a CRO in (a) with the sweep circuit in (b).

$0V$. Controls of the CRO are set for external sweep and direct coupling, with the spot zeroed in the lower left-hand corner of the screen. Test diode voltage is wired to the horizontal x-amplifier to deflect the spot further right with increasing voltage. Sensitivity of the x-amplifier is $0.1V$ per division. The vertical amplifier displays current on the vertical axis of the CRO with a scale dependent on both current sensing resistor R_S and setting of the y-amplifier sensitivity. Four separate characteristic curves are superimposed in Fig. 1-11a to display the I-V characteristic of an IN1693 diode over a wide current range of $50\mu A$ per division to 50 mA per division. Resistor R_1 limits maximum current and is changed as shown below to get full scale deflection. Test conditions are given in Table 1.4.

Table 1.4

R_1	y-amplifier Sensitivity	$\div$	R_S	$=$	Current Scale Sensitivity
18kΩ	10mV/div.	$\div$	200Ω	$=$	0.050mA/div.
1.8kΩ	10mV/div.	$\div$	20Ω	$=$	0.500mA/div.
180Ω	100mV/div.	$\div$	20Ω	$=$	5.0mA/div.
18Ω	1V/div.	$\div$	20Ω	$=$	50.0mA/div.

Although the y-amplifier of the CRO reads voltage drop across R_S, the y-axis is calibrated directly in milliamperes. Resistor R_B insures that D_B is reverse biased on negative half-cycles to prevent glitches (unwanted voltage transients) along the negative x-axis.

EXAMPLE 1-2:

See Fig. 1-12a. The CRO x- and y-amplifiers are set for 0.1/div. and 10mV/div. respectively. What is the value of R_S to calibrate the CRO's vertical axis at 1mA/div.? Assuming the diode's characteristic is identical with Fig. 1-11a, what will be displayed on the CRO if switch, Sw, is connected to (a) point 1, (b) point 2?

SOLUTION:

$R_S = 10mV \div 1mA = 10\Omega$. (a) Since the diode is forward biased, assume it is a short-circuit (dropping 0 volts) and calculate current from $I = E/R_1 = 10V/2k\Omega = 5mA$. Enter Fig. 1-11a at the first division for the curve labeled 5mA/div. and read the diode voltage, V, as $0.64V$. Now our assumption may be corrected to include diode drop and I can be recalculated from $I = (E - V)/R_1 = (10 - 0.64)V/2k\Omega = 4.68mA$. Now check again with Fig. 1-11a to see that V remains approximately at $0.64V$, when I is more correctly calculated at 4.68mA. This operating point is point Q in Fig. 1-12b. (b) With Sw on point 2, the total voltage varies between peak values of $10V + 2V = 12V$ and $10V - 2V = 8V$. Calculate operating point location at $12V$ by assuming a diode drop of $0.64V$ to approximate I by $I = (12 -$

$0.64)V/2k\Omega = 5.67$mA. From Fig. 1-11a, check $V = 0.65V$, and plot point A at $0.65V$, 5.67mA in Fig. 1-12b. Using the same procedure we find $I = 3.68$mA, $V = 0.63V$, at peak source voltage of $8V$, and locate point B in Fig. 1-12b. We conclude, as E_g varies from 0 to $2V$ to $-2V$, the spot moves from Q to A back to Q and on to B along the diode characteristic.

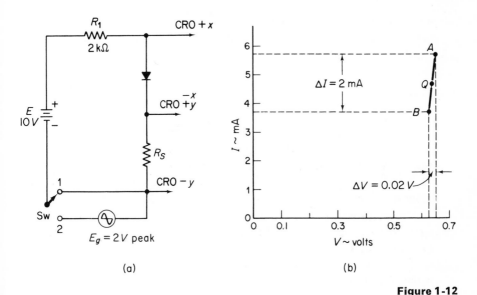

(a)   (b)

Figure 1-12

Circuit for Example 1-2.

Ac Diode Resistance

Line AB in Fig. 1-12b is a *dynamic operating path*, since it shows all possible diode operating points for all possible values of E_g. Several fundamental concepts on the analysis of semiconductor circuits are learned from this figure. First, the Q point is a dc operating point and depends on E, R_1 (R_S is negligible), and a diode voltage drop for silicon diodes that is reasonably constant at $0.6V$ for diode currents of 1 to 10mA. Thus we can solve any dc problem separately from the ac problem. Second, the slope of operating path AB is the reciprocal of *ac diode resistance r_d*. Admittedly a reading of ΔV is inaccurate, but we can estimate r_d from

$$r_d = \frac{\Delta V}{\Delta I} \qquad (1\text{-}5)$$

For Fig. 1-12b

$$r_d = \frac{(0.65 - 0.63)V}{(5.67 - 3.68)\text{mA}} \cong \frac{0.02V}{2\text{mA}} = 10\Omega$$

Third, an ac current will be drawn from E_g that depends primarily on circuit resistance R_1, and not the low value of r_d. Thus E_g sees a zero resistance in battery E, $2k\Omega$ in R_1, and negligible resistance in r_d and R_S, so that

$$\Delta I = \frac{E_g \text{ (peak-to-peak)}}{R_1} = \frac{4V}{2k\Omega} = 2mA$$

and $\Delta V = \Delta I \times r_d = (2mA)(10\Omega) = 0.02V$.

We summarize these ideas into general circuit principles that apply for a forward biased diode.

1. If any ac signal voltage source is present in the circuit, replace the source with a resistance equal to its internal resistance.

2. Temporarily ignore the ac voltage and treat the diode as a $0.6V$ drop if silicon or $0.2V$ drop if germanium. Find dc current I through the diode from dividing total dc voltage minus diode drop by total series resistance. Locate operating point Q at the intersection of I and the diode's I-V curve. (Rarely is it necessary to refine calculation of I from the actual operating point diode voltage.) The dc problem is now solved.

3. Now treat the ac problem as a separate problem. Replace dc voltage sources with their internal resistances (usually zero ohms). Find ac diode resistance by estimating the slope of the I-V curve at the Q point. Find ac current ΔI from dividing ac signal voltage E_g by total series resistance. The ac problem is now solved.

The above principles are really an application of superposition and duplicate actual operation of a device. First we turn the device on, that is, we establish a dc operating point or bias it. Then we inject an ac signal current and superimpose it on top of the existing dc current. There is no reason why we cannot simplify calculation by separating the circuit into a dc problem and an ac problem.

Junction Resistance

There is an easier and perhaps more accurate way of finding ac diode resistance r_d without using the I-V curve. Ac diode resistance r_d is composed of two components which are (1) body or bulk resistance r_b and (2) junction resistance r_j. Junction resistance predominates at diode currents below approximately 25mA and is found simply, although approximately, from the equation

$$r_j = \frac{25mV}{I} \tag{1-6}$$

where r_j is in ohms and I is in milliamperes. In Fig. 1-13 r_j is seen to predominate at low current levels. Above 50mA, bulk resistance is almost constant because the diode characteristic becomes a straight line and r_j is negligible. We evaluate r_d graphically at point Q to find $r_d = \Delta V/\Delta I = 0.12V/400\text{mA} = 0.3\Omega$, and r_j to be $25\text{mV}/300\text{mA} = 0.08\Omega$. Thus $r_d \cong r_b$ because r_j is negligible.

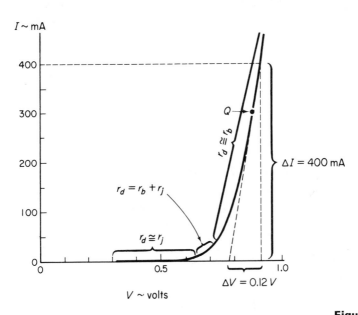

Figure 1-13

AC junction and bulk resistance of a diode.

EXAMPLE 1-3:

Evaluate r_d: (a) at point Q in Fig. 1-12, (b) at $I = 50\ \mu A$, $V = 0.44V$ in Fig. 1-11a, and (c) at $I = 350\text{mA}$, $V = 0.9V$ in Fig. 1-11a.

SOLUTION:

(a) From Eq. (1-6), $r_d \cong r_j \cong 25\text{mV}/4.6\text{mA} \cong 6\Omega$. Compare with results of Eq. (1-5).

(b) From Eq. (1-6)

$$r_d \cong r_j = \frac{25\text{mV}}{50\mu A} = 500\Omega$$

(c) From Eq. (1-6)

$$r_j = \frac{25\text{mV}}{350\text{mA}} = 0.07\Omega$$

Graphically measuring slope at $I = 350\text{mA}$, $r_d = 0.12V/350\text{mA} = 0.34\Omega$. Thus diode ac resistance is due primarily to bulk resistance.

1-8 APPLICATIONS OF JUNCTION DIODES

Polarity Selection Switch

Many devices and circuits are easily destroyed if the polarity of a power supply is incorrectly connected to them. To provide protection against such a happening, the circuit of Fig. 1-14a can be used. If the polarity switch, Sw, is

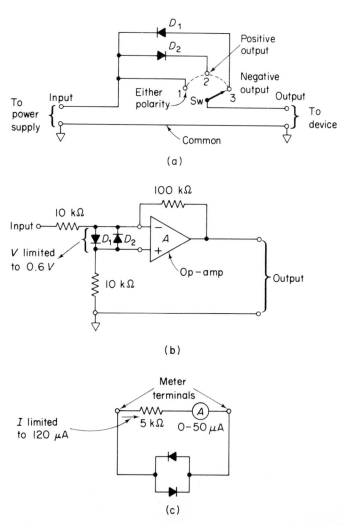

(a)

(b)

(c)

Figure 1-14

Protection circuits: polarity-selecting switch in (a), overvoltage in (b), and overcurrent in (c).

set at position 2, only positive input voltages are connected to the device or circuit. Negative input voltages are blocked by D_2. If switch, Sw, is set at position 3, only negative inputs are connected to the device, while positive input voltages are blocked by D_1.

Overvoltage Protection

An example of *overvoltage protection* by diodes is given in Fig. 1-14b. In normal operation, voltages never exceed a few millivolts between the $+$ and $-$ input terminals of integrated circuit amplifier A. The maximum allowable voltage between these two terminals is often below a few volts. Diodes D_1 and D_2 limit the input voltage to $0.6V$ between these terminals. They do not affect normal operation of the amplifier but do protect against an incorrect wiring connection.

Overcurrent Protection

Overcurrent protection is provided by diodes in Fig. 1-14c. Normally the meter movement reads full scale when conducting $50\mu A$ through its winding resistance of $5k\Omega$. Voltage across the meter at full scale will be $5k\Omega \times 50\mu A = 0.25V$ and is insufficient to forward bias a silicon diode. When excessive current is forced into the meter circuit, the diode holds the voltage across the meter constant at about $0.6V$. Meter current is then limited to $0.6V/5k\Omega = 120\mu A$, and the remaining excess current is conducted by the diode.

Voltage Regulator

We can use the fact that a forward-biased diode's voltage is almost constant and design a simple voltage regulator. For example, from the 50mA per division curve of Fig. 1-11a the diode's voltage stays reasonably constant at $0.8V \pm 0.15V$ for forward current variations between 10mA and 500mA. This characteristic can be applied to illustrate how to power either one or two 3-volt (2-cell) flashlights from an automobile battery. Essentially the diodes are used to step the battery voltage down to form a regulated 3-volt supply. Consider the following example.

EXAMPLE 1-4:

A PR-4 flashlight bulb is rated to draw approximately 1/4A from two series 1.5-volt batteries. Therefore the hot resistance of the bulb is $3V/0.25A = 12\Omega$. The four series diodes in Fig. 1-15a are forward biased so that each will drop approximately $0.8V$, or a total voltage across the four diodes of approximately 3 volts.

In Fig. 1-15c both lights are on and draw 500mA through resistor R. Assuming 10mA of current flows through the diodes to keep them forward biased, $I = 510$mA. (a) Find the required value of R and its power rating. (b) What is the current value that flows through the diodes for each of the three circuits?

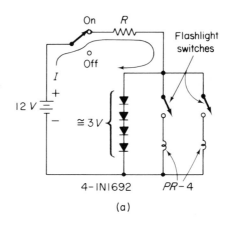

(a)

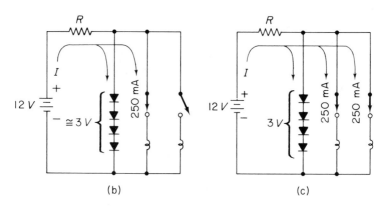

(b)   (c)

Figure 1-15

Diode application of a simple dc voltage regulator.

SOLUTION:

(a) In Fig. 1-15c, the voltage on one side of R is $12V$; on the other side it is $3V$ and R carries 510mA, therefore its value is

$$R = \frac{(12-3)V}{510\text{mA}} \approx 18\Omega$$

The power dissipated by R is

$$P = VI = 9V \times 510\text{mA} \approx 5\text{W}$$

(b) If $R = 18\Omega$ then in Fig. 1-15a

$$I = \frac{(12 - 3)V}{18\Omega} = 510\text{mA}$$

and all of I flows through the diodes.

In Fig. 1-15b I is again 510mA but now one lamp conducts $3V/12\Omega$ $= 250\text{mA}$. The current through the diodes is $510\text{mA} - 250\text{mA} = 260\text{mA}$.

In Fig. 1-15c each lamp conducts 250mA and the current through the diodes is $510\text{mA} - 2(250\text{mA}) = 10\text{mA}$. The value of R must be such that the diodes are always forward biased. In all three circuits we have kept the voltage across the diodes constant. (In practice, however, the voltage increases slightly with an increase in current.)

Clamping Circuits

Consider the circuit of Fig. 1-16a. Capacitor C charges to $10V$, with the indicated polarity, during the first half of the E_i positive half-cycle (from 0 to $\pi/2$ rad). Voltage drop across the diode is negligible, so V_o remains essentially

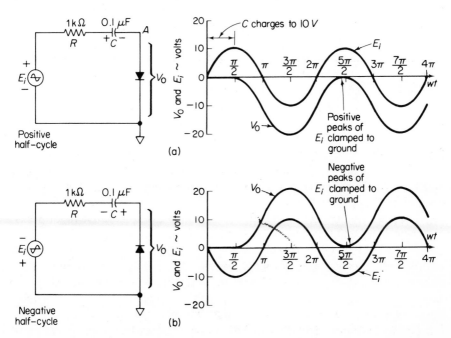

(a)

(b)

Figure 1-16

Diode clamping circuits.

at ground during this interval. As the top terminal of E_i drops below $10V$ (with respect to ground) during $\pi/2$ to π rad, E_i cannot discharge because of the diode. Therefore, V_o will equal the difference between E_i and dc capacitor voltage at each instant. For example, when $E_i = 0V$, the left-hand terminal of C is referenced to ground. Point A goes negative by $10V$ (with respect to ground), so V_o goes negative. At $3\pi/2$, the top terminal of E_i is at $-10V$ and is in series with capacitor voltage to put V_o at $-20V$. We conclude that positive peaks of V_o will be clamped to ground and V_o will have a dc component equal to capacitor voltage. Thus a clamping circuit adds a fixed voltage (positive or negative) to the input waveform.

In Fig. 1-16b, E_i charges to a voltage equal to the peak value of the first negative half-cycle, and cannot discharge because of the diode. V_o is the sum of the instantaneous value of E_i and dc capacitor voltage at each instant. Thus negative peaks of E_i are clamped to ground.

Clipping Circuits

Clipping circuits receive an input voltage and clip the top and/or bottom from the voltage. In Fig. 1-17a the diode is reverse biased for all input voltages that make input terminal A negative with respect to ground. Also, terminal A must go positive by $0.6V$ before the diode is forward biased. The diode is an open circuit both for all negative input voltages and for positive voltages less than $0.6V$. So $V_o = E_i$ for these inputs. But if E_i exceeds $0.6V$, voltage drop across the diode, and therefore V_o, is steady at $0.6V$. Thus Fig. 1-17a is a positive peak clipper. Figure 1-17b is a negative peak clipper. The diode is reverse biased (and therefore an open circuit) for all input voltages that make point A positive or less than $-0.6V$ with respect to ground.

In Fig. 1-17c a battery is added to reverse-bias the diode. Now point A must go positive to $5.6V$ before the diode turns on. For all voltages at point A equal to or greater than $5.6V$, the diode conducts and clamps output terminal B to $5.6V$. For all negative voltages at A and positive voltages less than $5.6V$, the diode is reverse biased. When reverse biased, the diode acts like an open circuit and $V_o = E_i$. Thus Fig. 1-17c is an adjustable positive peak clipper that clips all positive peaks greater than the battery voltage. Fig. 1-17d is an adjustable negative peak clipper because the diode is installed to clip negative input peaks, and the battery adjusts clipping level to approximately $-5V$ (actually $-4.4V$). The circuit of Fig. 1-17e actually clamps the output voltage to $-5.6V$ for all positive voltages and negative voltage less than $-5.6V$. For input voltages more negative than $-5.6V$, $V_o = E_i$. A symmetrical clipper is shown in Fig. 1-17f. D_1 is reverse biased for all positive inputs and negative inputs less than $-0.6V$. D_2 is reverse biased for all negative inputs and positive inputs less than $0.6V$. For all positive inputs greater than $0.6V$, D_2 is forward biased, clamping point B at $0.6V$. For all negative inputs greater than $-0.6V$, D_1 is forward biased, clamping B at $-0.6V$. Thus $V_o = E_i$ over the narrow range of input voltages

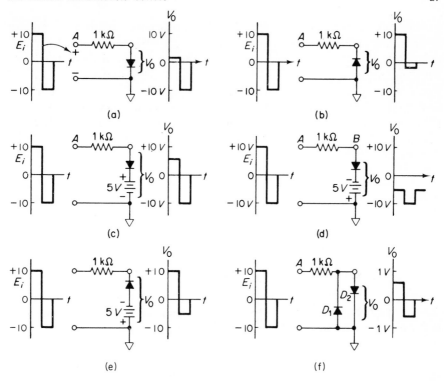

Figure 1-17

Diode clipping circuits.

that lie between $-0.6V$ and $0.6V$. This circuit is sometimes used to convert a sine-wave input signal to a reasonably square-wave output signal. Peak sine-wave voltage should be at least $10V$ to insure a steep rise between 0 and $0.6V$.

Gate Circuits

Figure 1-18 shows a diode AND gate. When both inputs A and B are at a low positive potential, such as ground in Fig. 1-18a, both diodes are forward biased. Output C is close to ground potential and is essentially equal to the input. In Fig. 1-18c both diodes are still forward biased and output C is clamped to about $5V$ (actually $5.6V$). In Fig. 1-18b, inputs A and B are at different potentials. To determine the potential at point C, first assume, for no special reason, that diode B is on. Now check the assumption. If diode B is on, point C is at $+5V$ and diode A must be on. Thus the battery is shorted through diodes A and B and our assumption must be wrong. The only remaining assumption is that diode A is on. If diode A is on, point C is at $0.6V$ and diode B is reverse biased. Our second assumption checks out as valid, so we conclude if either A or B input is at a lower positive potential than the other input, the output will approximately equal the lower potential. In computer logic circuits a low voltage or ground is called zero and a positive voltage is called one. Logically the circuit of Fig. 1-18 gives a one output only when input *A* is one *and* input *B* is one. For example, if an oil burner pump were pumping oil, a pressure sensor could put $+5$ on input *A*, then ground when the pump was off. A stack thermostat could put $+5$ on *B* when the stack is cold and ground when the stack is hot. Oil pumping *and* the stack being cold means an ignition failure. Our AND gate would put out $+5V$ under this one condition to operate a relay which would disconnect all power.

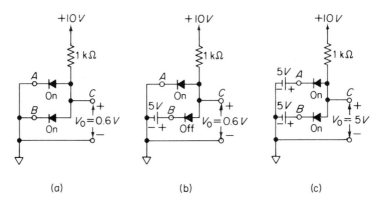

Figure 1-18

Diode and gate.

To see why Fig. 1-18 is called a gate, assume input *B* is grounded and signals of $+5V$ and ground are applied to input *A*. Output *C* remains at ground and the gate is closed to signals on *A*. Now put $+5V$ on gate control *B*. Output *C* goes to $+5V$ when *A* is at $+5V$, and *C* goes to ground when *A* is at ground. Effectively, signals on input *A* are passed through to *C* and the gate is open. Thus $+5V$ on *B* opens the gate and ground on *B* closes the gate. Other diodes could be wired to point *C* to make more inputs. The same AND relationship would still hold. All inputs must be high simultaneously for the output to be high.

Reversing both diodes and supply battery gives the diode OR gate of Fig. 1-19. In both Figs. 1-19a and c, both diodes are on and output voltage V_o approximately equals input voltage. In Fig. 1-19b assume (incorrectly) that diode *A* is on. Point *C* is then at $-0.6V$ and diode *B* must be on (because its anode is more positive than its cathode). If *B* is one, point *C* is at $+5V$. Since *C* cannot be at both $+5$ and $-0.6V$, the assumption was wrong. Now assume *B* is on, to put point *C* at $+5V$. If *C* is at $+5V$, diode *A* is off and the assumption is valid. We conclude that if *A* is one ($+5V$) *or* if *B* is one ($+5V$), output *C* will be one. Now if we wanted to turn power off to an oil burner motor (a) when the oil drum was empty (+5V) *or* (b) when stack temperature was too high (+5V) we would use an OR gate with output C to disconnect power.

Half-Wave Rectifier

A diode half-wave rectifier is shown in Fig. 1-20. During the first positive input voltage the diode conducts and charges filter capacitor C. At the maximum positive input voltage of Fig. 1-20a, C is charged to peak input voltage of about +100V. We can neglect the small diode drop. As E_i drops below +100V, the diode becomes reverse biased and disconnects E_i from C and R_L. Current is now furnished to R_L by the discharge of C. When E_i has reversed and gone to its peak negative voltage, as in Fig. 1-20b, E_i and V_C reverse-bias the diode by 195V. We assume C is large enough to hold most of its charge or voltage and still be able to furnish load current to R_L when the diode is off. When E_i exceeds V_C on the next positive half-cycle, the diode turns on and connects E_i to recharge C.

If R_L is a high-resistance dc voltmeter, C would not discharge very much between positive peaks of E_i. Then the dc voltmeter would read only positive peaks of the ac voltage. Thus Fig. 1-20 is also a diode application for a positive peak reading voltmeter. Reversing the diode would yield a negative peak reading voltmeter. This circuit will be studied in further detail in Chapter 5 because it is the basis of power supply and rectifier applications.

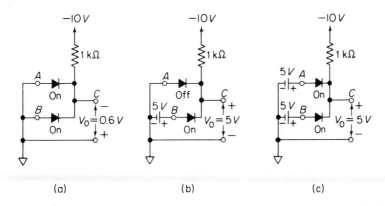

(a)   (b)   (c)

Figure 1-19

Diode or gate.

At $3\pi/2$ rad the diode in Fig. 1-20 is reversed biased by 195V, as shown in Fig. 1-20c. Therefore the diode must be capable of withstanding this large voltage and not be destroyed. A diode chosen for this particular application must have a rating called the *peak inverse voltage* (PIV) capable of withstanding 195V, or approximately 2 E_m. We may now ask what happens if the diode cannot withstand a large reverse voltage? The answer is that the diode breaks down and may be destroyed. Junction breakdown and the mechanisms that cause it are discussed in section 1-9.

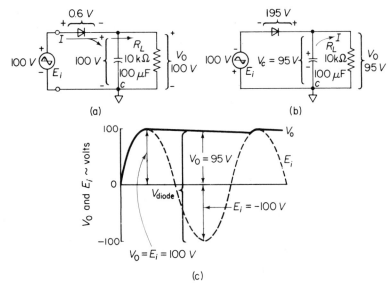

(a)   (b)

(c)

Figure 1-20

Half-wave rectifier or positive-peak reading voltmeter.

Full-Wave Rectifier

A common diode application is in the use of four diodes to build a diode bridge circuit. This important application is a fundamental building block for dc power supplies and will be covered in Chapter 5.

1-9 JUNCTION BREAKDOWN

All diodes have a limit of maximum reverse voltage beyond which the diode has a large reverse current. This voltage is called *breakdown voltage*. The reverse characteristics along with forward characteristics are shown in Fig. 1-21. There are two mechanisms which may break a diode down—*avalanche breakdown* and *zener breakdown*.

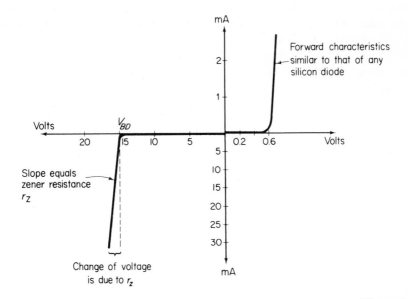

Figure 1-21

Forward and reverse characteristics of a zener diode.

Avalanche Breakdown

Figure 1-9a illustrated that when a diode is reverse biased, a saturation current flows. Remember that saturation current is the result of minority carriers—electrons in the p-side entering the space charge region and being swept into the n-side, and holes in the n-side crossing the junction to the p-side. If the external supply voltage is large enough, it will impart sufficient energy to these minority carriers so that if they collide with an atom, a covalent bond is broken. This broken bond produces a new hole and an electron, each of which can gain enough energy from the external supply to cause another broken bond. The saturation current quickly increases as more bonds are broken, because almost all the carriers—holes and electrons—that are produced contribute to the total current. This type of breakdown is avalanche breakdown.

Zener Breakdown

In section 1-5 we mentioned that typically there are 10^{15} doping atoms per cm³. If the number of doping atoms is increased to approximately 10^{18} atoms/cm³, two things happen at the junction: (1) the width of the space charge region increases; (2) under reverse bias, an extremely large electric field is developed. It is this large electric field that causes zener breakdown. The electric field causes covalent bonds to rupture, producing large quantities of holes and electrons. These holes and electrons cause the rapid increase in saturation current. In comparison, avalanche breakdown is the result of collisions of high-energy minority carriers with atoms to break the covalent bond. Zener breakdown is the result of a strong electric field at the junction to break the covalent bond.

Above 8 volts, diode breakdown is predominantly due to the avalanche mechanism. Diodes which break down below 5 volts are due mostly to zener breakdown. Between 5 and 8 volts, the breakdown mechanism depends upon the amount of impurities and their distribution at the junction.

Diodes which are properly designed are able to be operated in the breakdown region. They are avalanche, breakdown, or zener diodes. Note, however, the term zener diode is most commonly used for any diode which operates in its breakdown region. Section 1-10 and sections 5-6 to 5-8 describe characteristics and applications of zener diodes.

1-10 CHARACTERISTICS AND APPLICATIONS OF ZENER DIODES

Figure 1-21 illustrates typical I-V characteristics of a zener diode. In the forward direction, a zener characteristic is similar to that of an ordinary silicon diode. In the reverse direction, voltage may be increased to the zener voltage breakdown point, V_{BD}, where the pn junction breaks down. Beyond V_{BD}, current increases rapidly for a very small increase in voltage. This current must be limited by the external circuit; otherwise, maximum current capability (I_{Zmax}) of the zener will be exceeded and the device destroyed.

Zener diodes are purchased according to their nominal zener voltage, V_Z, and their maximum power dissipation capabilities, P_{Dmax}. Zener voltage varies with zener current according to

$$V_Z = V_{BD} + I_Z r_Z \qquad (1\text{-}7)$$

Usually $I_Z r_Z \ll V_{BD}$, and therefore

$$V_{BD} \cong V_Z \qquad (1\text{-}8)$$

r_Z is the *zener resistance* measured at test current I_{ZT}, where

$$I_{ZT} = \tfrac{1}{4} I_{Zmax} \qquad (1\text{-}9)$$

and

$$I_{Zmax} = \frac{P_{Dmax}}{V_Z} \qquad (1\text{-}10)$$

EXAMPLE 1-5:

From manufacturer's data for an IN5250 zener diode, $V_Z = 20V$ and $P_{Dmax} = 500\text{mW}$, calculate: (a) I_{Zmax}, and (b) I_{ZT}. (c) If $r_z = 25\Omega$ at I_{ZT}, show Eq. (1-8) is valid.

SOLUTION:

(a) From Eq. (1-10)

$$I_{Zmax} = \frac{500\text{mW}}{20V} = 25\text{mA}$$

(b) From Eq. (1-9)

$$I_{ZT} = \tfrac{1}{4}(25\text{mA}) = 6.2\text{mA}$$

(c) $$I_{ZT}r_Z = (6.2\text{mA})(25) = 0.16V$$

therefore

$$V_{BD} \cong V_Z = 20V$$

EXAMPLE 1-6:

For the circuit of Fig. 1-22, calculate R to limit zener current at I_{ZT}.

SOLUTION:

From Fig. 1-22a

$$R = \frac{E - V_Z}{I_{ZT}} = \frac{(30 - 20)V}{6.2\text{mA}} = 1.6\text{k}\Omega$$

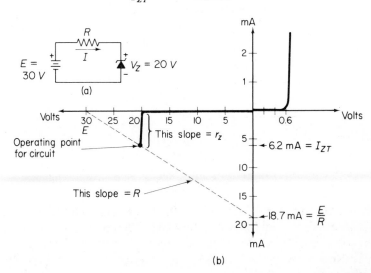

(a)

(b)

Figure 1-22

Operating conditions of circuit (a) are shown on the zener diode's characteristics in (b).

The slope of R is plotted on the zener's characteristics of Fig. 1-22b, and the intersection of this load line and the zener's characteristic is the *operating point* for the circuit.

Zener diodes are available in voltage ranges from $2.4V$ to $200V$ and in power ratings from 250mW to 50W. A principle application for zener diodes is for regulating voltage to a constant value. In Chapters 5 and 6 we will learn how to measure r_z and use the zener in designing regulator sections for power supplies.

Meter Protection with a Zener Diode

Occasionally we may connect voltmeters into circuits before checking to see if the range switch is set at a proper setting. This mistake could result in permanent damage to the meter if it is not adequately protected. In Fig. 1-23 the basic meter movement has a full-scale deflection of $50\mu A$ and an internal resistance of 5kΩ. To convert this meter to a $2.5V$ voltmeter, which is the lowest range, we add a series resistor of

$$R_1 + R_2 = \frac{2.5V - (50\mu A)(5k\Omega)}{50\mu A} = 45k\Omega$$

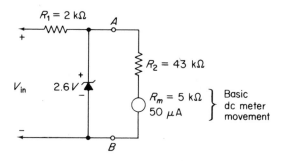

Figure 1-23

Voltmeter protection using a zener diode.

Let $R_1 = 2k\Omega$ and $R_2 = 43k\Omega$. When $50\mu A$ flows through the meter, the voltage between points A and B is $(50\mu A)(43k\Omega + 5k\Omega) = 2.35V$. To protect this combination, choose an IN702 zener diode with $V_Z = 2.6V$ and $P_{Dmax} = 500mW$. Now if the meter is accidentally inserted into a circuit where $V_{in} = 10V$, current through R_1 is

$$I = \frac{(10 - 2.6)V}{2k\Omega} = 3.7mA$$

Since only $50\mu A$ can flow through the meter, the remaining current (7.5mA $-50\mu A = 7.45$mA) flows through the zener. When the voltmeter is connected into a circuit where voltage is less than $2.5V$, the zener draws negligible leakage current.

If $V_{in} = 200V$, the zener will still be able to protect the meter. Under this condition

$$I = \frac{(200 - 2.6)V}{2k\Omega} = 98.7\text{mA}$$

Practically all of I flows through the zener. Therefore I_{Zmax} must be greater than 100mA. From Eq. (1-10), with a 500mW zener,

$$I_{Zmax} = \frac{500\text{mW}}{2.6V} = 192\text{mA}$$

and the zener protects the basic meter movement.

The zener may also replace the diode or diode-battery combinations in the clamping and clipping applications of Figs. 1-16 and 1-17a to e.

PROBLEMS

1. Determine the temperature of coefficient, α, for thermistor A in Fig. 1-2 at 40°C. Compare results with Eq. (1-2).

2. A thermistor has a temperature coefficient of $-2.8\%/°C$. What is its resistance at 20°C if its resistance at 16°C is 500Ω?

3. For thermistor B in Fig. 1-2, determine: (a) $\Delta R/\Delta T$ between 5°C and 15°C and between 35° and 45°; (b) temperature coefficient, α, at 10°C and 40°C.

4. From the results of problem 3, is α increasing or decreasing as the resistance decreases and temperature increases?

5. If thermistor B of Fig. 1-2 is used in the circuit of Fig. 1-3 (in place of thermistor A), calculate R_A and R_B.

6. In Fig. 1-6b, if $TH1$ is thermistor A of Fig. 1-2 and $TH2$ is thermistor B, will the 50Ω variable resistor be adequate for zero adjustment? Consider R remains at 125Ω.

7. (a) A diode conducts $1\mu A$ with $50V$ across it. Is the diode forward or reverse biased? (b) If ambient temperature increases from 25°C to 35°C and the diode current doubles, is the diode silicon or germanium?

8. Show a forward-biased diode, label p and n parts, and indicate direction of conventional current flow.

9. When a diode is reverse biased, is the space charge region increased or decreased?

10. The ambient temperature increases from 25°C to 55°C. Calculate the resulting diode current for: (a) a germanium diode, $I_s = 10\mu A$ at 25°C; (b) a silicon diode, $I_s = 10nA$ at 25°C.

11. From Fig. 1-11a calculate the ac diode resistance r_d at an operating point of 25mA.

12. Using Fig. 1-11a, calculate junction resistance r_j at 25mA. Compare answer with problem 11.

13. Referring to Fig. 1-15, if one of the diodes should short-circuit, is the 18Ω, 5W resistor calculated in Example 1-4 sufficient for R, or does R have to have a larger power rating?

14. Draw the voltage waveform V_o for each of the circuits of Fig. 1-24.

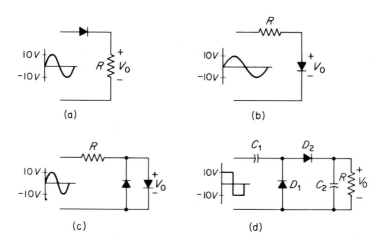

Figure 1-24

Circuit diagrams for Example 1-10.

15. Repeat Example 1-5 for P_{Dmax} equal to 250mW and 1W. Compare answers with Example 1-5.

16. (a) What is the minimum value of R in Fig. 1-22a so that the zener is not destroyed? (b) What is the maximum value of R to keep the zener in regulation? $I_{Zmin} = 5mA$.

17. If the zener in Fig. 1-22 is reversed and $R = 1.6k\Omega$, will the zener be destroyed?

chapter two
bipolar
junction transistors

2-0 INTRODUCTION

There are two major classifications of transistors, *field-effect* transistors (FETs) and *bipolar junction* transistors (BJTs). Both are names for their operating principles. In field-effect transistors it is an electric field of a *pn* junction that controls the resistivity of a conducting channel. FETs depend on the transport of the majority carriers in the conducting channel, either holes or electrons, but not both. For this reason the FET is classified as a *unipolar* device. Chapter 3 treats the FET. In this chapter we discuss only the BJT. Bipolar junction transistors depend on the motion of both charge carriers—holes and electrons—and the action of two *pn* junctions. BJTs are formed by many processes, but each has the same general configuration: a thin base region of doped semiconductor material is sandwiched between two regions of oppositely doped semiconductor material. An *n*-type layer sandwiched between two *p*-type layers is designated as a *pnp* transistor. A *p*-type layer sandwiched between two *n*-type layers is designated as an *npn* transistor. The middle layer is the *base*, while one outside layer is the *emitter* and the other outside layer is the *collector*.

In Fig. 2-1 the BJTs have two *pn* junctions, one between the emitter and base, the other between the collector and base. As we discussed in Chapter 1, every *pn* junction has a space charge region associated with it. The width of the space charge region depends on whether the junction is forward or reverse biased. In this text we are primarily concerned with operating the transistor as an amplifier. To do this the emitter-base junction must be forward biased and the collector-base junction must be reverse biased.

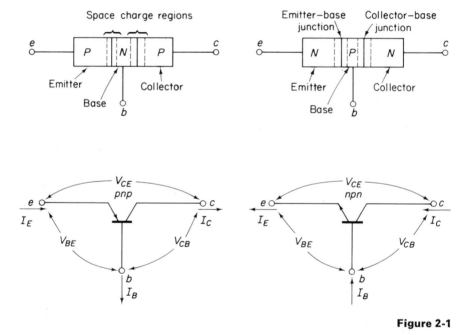

Figure 2-1

Construction and circuit symbols for bipolar junction transistors.

The pictorial representations in Fig. 2-1a are not drawn to scale. The base region is actually very thin. Figure 2-1b shows the circuit symbols for a *pnp* and *npn* transistor. The arrowhead identifies the emitter and points in the direction of conventional current flow and toward the *n*-type material. Therefore, for a *pnp* transistor the arrowhead points into the base, and for an *npn* transistor the arrowhead points toward the emitter.

2-1 OPERATING MODES AND CIRCUIT CONFIGURATIONS

Since there are two junctions, each of which may be either forward or reverse biased for a particular application, there are four possible modes of operation, as defined in Table 2.1.

Table 2.1

Operating Mode	Bias on the Emitter Junction	Bias on the Collector Junction
Active	forward	reverse
Saturated	forward	forward
Cutoff	reverse	reverse
Inverse	reverse	forward

Switching circuits operate in either the *saturated* or *cutoff* mode. That is, the collector-emitter of the transistor acts as a short circuit or as an open circuit when the transistor is saturated or cut off, respectively. *Amplifying* circuits require operation in the *active* mode. The inverse mode is not commonly found.

In this text we wish to use a transistor as an amplifier, that is, a device that accepts a signal at two input terminals and delivers an amplified reproduction to a load at two output terminals. Since the transistor is only a three-terminal device, one terminal must be *common* to both input and output. There are three standard circuit configurations, as shown in Fig. 2-2: common-emitter (*CE*), common-collector (*CC*), and common-base (*CB*).

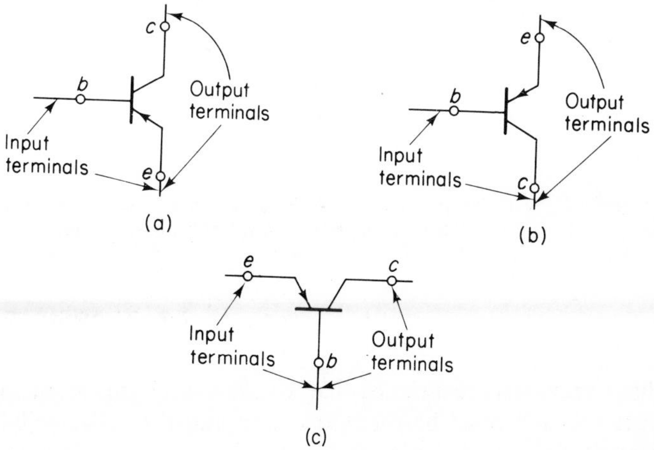

Figure 2-2

Standard circuit configurations: (a) common-emitter, (b) common-collector, (c) common-base.

The simplest rule for classifying a circuit configuration is to (1) identify the input terminal and (2) identify the output terminal. (3) The remaining terminal is common and names the classification. Although the configurations of Fig. 2-2 are drawn with *pnp* transistors, they are the same terminals for *npn* transistors.

The proper junction biasing of a transistor is established by external power supplies. Basic supply potentials are connected in Fig. 2-3 to show the required polarity to bias a *pnp* transistor in the active mode for all three configurations. For an *npn* transistor, reverse all the supply polarities and all the current directions. Resistor R_L in all three configurations is the load. Resistor R_B in the common-emitter and common-collector configurations limits the base current I_B. Resistor R_E in the common base configuration limits the emitter current.

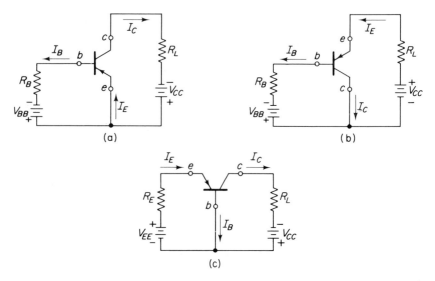

Figure 2-3

Supply-voltage polarities for biasing a *pnp* transistor in the active mode:
(a) common-emitter, (b) common-collector, (c) common-base.

2-2 TRANSISTOR ACTION

We have previously mentioned that, to use a transistor as an amplifier, the emitter-base junction must be forward biased and the collector-base junction reverse biased. To investigate transistor action, let us consider the common-base *pnp* transistor circuit of Fig. 2-4. Note the polarities of the external batteries, V_{EE} and V_{CC}, insure the proper biasing of the junctions.

The forward bias of V_{EE} reduces the space charge region at the emitter-base junction and allows holes to be injected from the heavily doped emitter into the

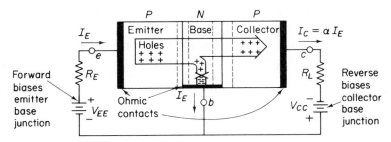

Figure 2-4

Common-base action for a *pnp* transistor operating in the active region.

lightly doped base. Since the base is an *n*-type material, these injected holes are minority carriers in the base. Practically all of these holes diffuse through the base toward the collector junction. Remember from Section 1-6 that, for reverse-biased junctions, majority carriers are repelled but minority carriers are swept through the space charge region. The same principle applies at the reverse-biased collector-base junction. As holes reach the collector-base space charge region, they are swept across the junction into the collector. It is the external supply V_{CC} that reverse-biases the collector-base junction. Not all of the holes that are injected from the emitter reach the collector; a small percentage (less than 2 percent) recombine with electrons in the base. To maintain a state of equilibrium within the transistor, a small number of electrons must enter from the base contact. These entering electrons resupply the electrons in the base that have been lost due to recombinations.

There is another component of current through the reverse-biased collector-base junction, consisting of base minority holes and collector minority electrons. This minority current is temperature-sensitive because electrons in the collector are thermally generated by broken covalent bonds. At room temperature this minority current is in the order of microamperes for germanium and nanoamperes for silicon. In this text we will restrict our circuits to silicon transistors, and thus this minority current becomes unimportant and need not be considered even if the ambient temperature increases.

Common-base Configuration

The emitter current in Fig. 2-4 is I_E, that fraction α_F (alpha) of the emitter current which reaches the collector is $\alpha_F I_E$. In equation form

$$I_C = \alpha_F I_E \qquad (2\text{-}1)$$

where I_C is the collector current, I_E the emitter current, and α_F is called the *forward current transfer ratio*. Typical values for α_F are 0.98 to 0.99, which means that from 98 to 99 percent of the emitter current reaches the collector.

We have previously mentioned that a state of equilibrium must exist within the transistor. In terms of current this means that Kirchhoff's current law cannot be violated—currents entering a transistor equal currents leaving. For a *pnp* transistor I_E enters and I_B and I_C leave. In equation form

$$I_E = I_B + I_C \qquad (2\text{-}2)$$

The base current needed for recombinations is the difference between emitter current and collector current, or

$$I_B = (1 - \alpha_F)I_E \qquad (2\text{-}3)$$

Equation (2-1) is the relationship between output current (collector current) and input current (emitter current) for the common-base configuration. A relationship may also be obtained for common-emitter and common-collector configurations.

Common-emitter Configuration

The output current in a common-emitter circuit is collector current, and the input current is base current. Dividing Eq. (2-1) by Eq. (2-3) yields

$$\beta_F \equiv \frac{I_C}{I_B} = \frac{\alpha_F I_E}{(1 - \alpha_F)I_E} = \frac{\alpha_F}{1 - \alpha_F} \qquad (2\text{-}4)$$

β_F ranges from 50 to 500 and is the *dc current gain for the common-emitter configuration*. β_F is also the ratio of collector current to base current. Rearranging Eq. (2-4) for α_F in terms of β_F

$$\alpha_F = \frac{\beta_F}{1 + \beta_F} \qquad (2\text{-}5)$$

Common-collector Configuration

The output current in a common-collector circuit is emitter current, and the input current is base current. Therefore the ratio I_E/I_B is our current gain

$$\frac{I_E}{I_B} = \frac{I_E}{I_C} \times \frac{I_C}{I_B}$$

$$\frac{I_E}{I_B} = \frac{1}{\alpha_F} \times \beta_F$$

Substituting Eq. (2-5) for α_F yields

$$\frac{I_E}{I_B} = \frac{1}{\beta_F/(1 + \beta_F)} \times \beta_F = 1 + \beta_F \qquad (2\text{-}6)$$

Equation (2-6) shows that the current gain is largest for a common-collector circuit.

EXAMPLE 2-1:

From manufacturer's data $\alpha_F = 0.98$, calculate current gain if the transistor is connected in (a) common-emitter configuration and (b) common-collector configuration.

SOLUTION:

(a) From Eq. (2-4)

$$\beta_F = \frac{0.98}{1 - 0.98} = 49$$

(b) From Eq. (2-6), $\beta_F + 1 = 50$.

2-3 COMMON-EMITTER CHARACTERISTIC CURVES

One of the most typical and useful ways in which transistor information is displayed is by a set of characteristic curves. These curves are a plot of output parameters—current versus voltage—for a fixed value of input current. Since the common-emitter configuration is the one most often used for amplifying circuits, its characteristics are most important. For common-emitter characteristics, the vertical axis is collector current I_C, the horizontal axis is collector-emitter voltage V_{CE}, and the input current is base current I_B. Manufacturer's data sheets often give a set of typical common-emitter characteristics. To obtain a set of characteristics for a particular transistor we may use commercially available curve tracers or measure them ourselves with the sweep circuit of Fig. 2-5.

Sweep Measurement Techniques

Figure 2-5 illustrates how common-emitter characteristics are obtained using an ordinary cathode-ray oscilloscope (CRO). The procedure involves (1) establishing a constant input current, I_B, and (2) varying output voltage to measure corresponding values of output current. The result is plotted by the CRO as a graph of output current versus output voltage, and yields one characteristic curve. We then establish a different constant input current in step (1) and repeat step (2) to obtain another characteristic curve. Repeat steps (1) and (2) until a family of output characteristics is obtained, as in Fig. 2-6.

Referring to Fig. 2-5, the constant input current is established by varying either V_{BB} or the 50kΩ potentiometer, or both. The sweep automatically

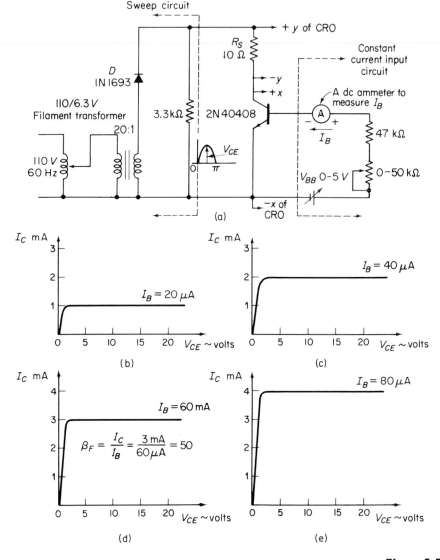

Figure 2-5

The sweep circuit in (a) displays V_{CE} on the x-axis and I_C on the Y-axis, for each value of I_B in (b) through (e).

varies the collector-emitter voltage. Diode D insures that the collector is positive with respect to the emitter. V_{CE} is displayed on the x-axis of the CRO. R_S samples collector current I_C and connections to the y-amplifier display output collector current on the vertical axis. Dial a vertical sensitivity of 10mV/cm for $10mV/10\Omega = 1mA/cm$. Both x and y CRO amplifiers should be direct coupled. For best results, the spot of the CRO should be zeroed at the lower left corner of the screen. Figures 2-5b, c, d, and e display a typical characteristic curve for different base currents. By plotting these curves on one set of axes, a family of common-emitter characteristics is obtained, as shown by Fig. 2-6.

Note on the curves of Fig. 2-6 that once the collector junction is reverse biased (V_{CE} larger than approximately $0.6V$), the magnitude of V_{CE} has practically no effect on I_C. That region of the graph where I_C does not change appreciably with changes in V_{CE} is called the *active region*. If the transistor is operated at any point within this region, its emitter-base junction is forward biased and its collector-base junction is reverse biased.

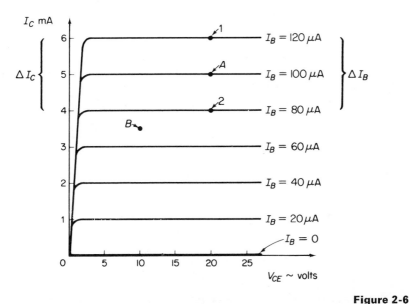

Figure 2-6

Typical collector characteristics of small-signal transistors.

2-4 MEASUREMENT OF COMMON-EMITTER CURRENT GAIN

Equation (2-4) defined the ratio of dc collector current to dc base current for a common emitter configuration as dc current gain β_F.

$$\beta_F = \frac{I_C}{I_B}$$

One of the primary uses of collector characteristic curves is to measure β_F. For example, in Fig. 2-6 the value of β_F is calculated for point A to be

$$\beta_F = \frac{I_C}{I_B} = \frac{5\text{mA}}{100\mu\text{A}} = 50$$

If we desire β_F at point B, which is between two known base current curves, we must approximate the value of I_B. For example, point B is approximately half-way between $I_B = 60\mu\text{A}$ and $I_B = 80\mu\text{A}$, thus we estimate $I_B = 70\mu\text{A}$ at point B. The corresponding value of collector current is $I_C = 3.5\text{mA}$, thus

$$\beta_F = \frac{I_C}{I_B} = \frac{3.5\text{mA}}{70\mu\text{A}} = 50$$

Small-Signal Current Gain

Small-signal current gain β_0 can also be measured from the characteristic curves. Suppose I_B is changed by a small signal ΔI_B to vary between $80\mu\text{A}$ and $100\mu\text{A}$, as shown by points 1 and 2 in Fig. 2-6. V_{CE} is held constant at $20V$ during the change. We want to know the small-signal current gain β_0 that relates ΔI_B to the *resulting* change in collector current ΔI_C. In equation form

$$\beta_0 = \frac{\Delta I_C}{\Delta I_B}\bigg|_{V_{CE}=\text{constant}} \tag{2-7}$$

For a $20\mu\text{A}$ peak-to-peak swing between points 1 and 2 on Fig. 2-6, $\Delta I_C = 6\text{mA} - 4\text{mA} = 2\text{mA}$, so that

$$\beta_0 = \frac{\Delta I_C}{\Delta I_B}\bigg|_{V_{CE}=\text{constant}} = \frac{2\text{mA}}{20\mu\text{A}} = 100$$

From the above analysis, we conclude that if the characteristic curves are equally spaced and parallel, then $\beta_0 = \beta_F$. This is usually the situation for small-signal transistors (1 watt or less). For power transistors, however, the characteristic curves may be neither equally spaced nor parallel, as shown in Fig. 2-7. Table 2.2 summarizes calculations of β_F for different points on Fig. 2-7. We will not calculate a β_0 for power transistors because these transistors are used with large changes in base current. Therefore, if a β_0 for a power transistor under signal conditions is needed, we will calculate the worst possible β_F. That is a β_F at the largest value of collector and base current. (Note that the smallest value of V_{CE} occurs at this time.) It should be noted that a commonly accepted symbol for β_F is h_{FE}, and one for β_0 is h_{fe}.

Table 2.2

| I_B(mA) | $V_{CE} = 5\text{V}$ | | $V_{CE} = 15\text{V}$ | |
	I_C(A)	β_F	I_C(A)	β_F
10	0.78	78	0.78	78
20	1.48	74	1.50	75
30	2.00	67	2.05	68
45	2.60	58	2.70	60

2-5 GRAPHICAL ANALYSIS

Graphical analysis of transistor circuits gives a pictorial representation of the interaction of a transistor and its external circuit. From this type of analysis we can: (1) visualize the dependence of operating point on both transistor charac-

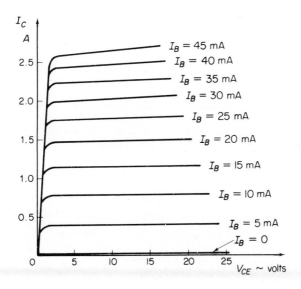

Figure 2-7

Typical collector characteristics of power transistors.

teristics and external circuit elements; (2) predict operating point measurements for the purpose of verification and interpretation; (3) learn the concept of amplification; and (4) learn the concept of maximum possible output voltage swing.

Load Lines

A graphical procedure which superimposes circuit characteristics over device characteristics is known as drawing a *load line*. In Fig. 2-8 V_{CC} and R_L are the external circuit elements which determine the load line. (Note V_{BB} and R_B determine the base current I_B and are not involved with drawing the *load line for the output loop*.) Resistors are the simplest loads, and since many other

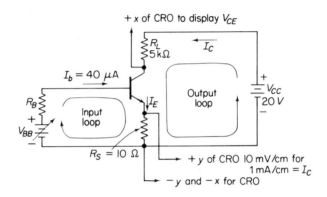

Figure 2-8

CE circuit for graphic analysis and display of operating points on a CRO (CRO amplifiers are direct coupled).

types of loads can often be represented by a resistor we use a resistor load to introduce graphical analysis. Begin the analysis by applying Kirchhoff's voltage law to the output loop to obtain the load line equation

$$V_{CC} = V_{CE} + I_C R_L \qquad (2\text{-}8)$$

To plot this expression on a set of characteristic curves, we need to know any two points on the curve. The easiest points to find are the current and voltage axis intercepts.

1. To find the *current axis intercept* assume the device is a short circuit and set $V_{CE} = 0$. Then $V_{CC} = 0 + I_C R_L$ or $I_C = V_{CC}/R_L$. This intercept is point A on Fig. 2-9.
2. To find the *voltage-axis intercept* assume the device is an open circuit and set $I_C = 0$. Then $V_{CC} = V_{CE} + (0)R_L$ or $V_{CE} = V_{CC}$. This intercept is point B on Fig. 2-9.
3. The load line is a straight line drawn between points A and B as shown in Fig. 2-9. Now that the procedure for drawing load lines has been developed, they will henceforth be drawn directly on the transistor's curves. Figure 2-10 is the load line and the characteristic curves for the circuit of Fig. 2-8.

Operating Point

Remember the curves are a characteristic of the transistor and the load line is a characteristic of the external circuit. When both the transistor and circuit are connected together, *an equilibrium or operating condition* must exist so that *Eq. (2-8) is satisfied.* The *intersection* of the load line and the characteristic curves locate all possible combinations of operating conditions. However, if a particular base current is specified, then the intersection of the load line and this base current curve can locate only one *operating point.* By connecting a CRO to display V_{CE} (neglecting the small drop across R_S) and $I_C \approx I_E$ as in Fig. 2-8, we can display an operating point. Varying V_{BB} will vary I_B and trace out a load line on the CRO. For example, the circuit of Fig. 2-8 has a base current $I_B = 40\mu$A.

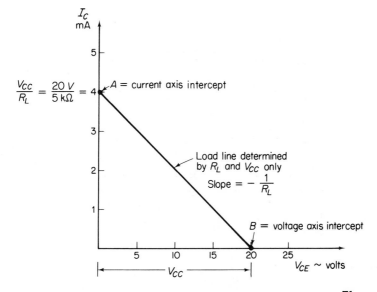

Figure 2-9

Diagram illustrating the procedure for plotting the load line equation (2-8).

The characteristic curves and circuit load line are drawn together in Fig. 2-10. The intersection of $I_B = 40\mu$A and the load line is the operating point for this transistor circuit. The circuit in Fig. 2-10 is drawn to show what dc voltage appears across the transistor and what dc voltage appears across the load. Note that operating point Q identifies one specific combination of base current I_B, collector current I_C, and collector-emitter voltage V_{CE} allowed by a transistor circuit. From Fig. 2-10, $V_{CE} = 10V$, $I_C = 2$mA, and $I_B = 40\mu$A. The voltage across R_L, V_{R_L}, is the difference between V_{CC} and V_{CE}. For our example, $V_{R_L} = 20V - 10V = 10V$. V_{R_L} may also be found by calculating the product of I_C and R_L, or $I_C R_L = (2\text{mA})(5\text{k}\Omega) = 10V$.

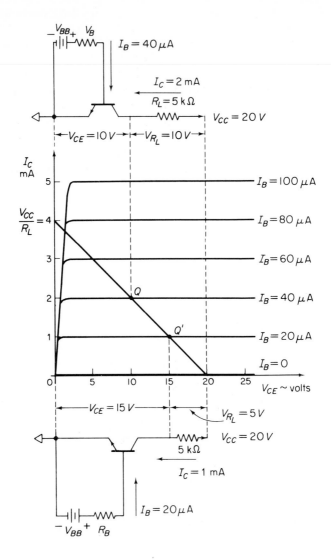

Figure 2-10

Operating points Q and Q' describe two separate equilibrium conditions for the same circuit.

Another example of matching the graphical solution to the circuit condition is shown in the bottom circuit of Fig. 2-10. For this circuit I_B has been changed to $20\mu A$ and the operating point is Q'. At Q', $V_{CE} = 15V$ and $I_C = 1mA$. The voltage across R_L is $V_{R_L} = 20V - 15V = 5V$ or $I_C R_L = (1mA)(5k\Omega) = 5V$. Section 2-6 shows how to design an input circuit to deliver the proper value of base current to a transistor and thus establish an operating point. In conclusion,

V_{CC} and R_L establish the load line, and the base current I_B establishes where on the load line the transistor circuit will operate. The following two examples examine the effect of varying circuit elements V_{CC} and R_L on the operating point.

EXAMPLE 2-2:

The circuit of Fig. 2-8 has a base current $I_B = 40\mu A$. (a) If V_{CC} is changed from $20V$ to $15V$, find the new operating point and load line. (b) Calculate β_F at both operating points.

SOLUTION:

(a) The new voltage axis intercept is $V_{CE} = V_{CC} = 15V$. The current axis intercept is $V_{CC}/R_L = 15V/(5k\Omega) = 3mA$. As shown in Fig. 2-11, if V_{CC} decreases and R_L is not changed, the load line shifts to the left, with the same slope. With $I_B = 40\mu A$ then the new operating point Q' is at $V_{CE} = 5V$, and $I_C = 2mA$.

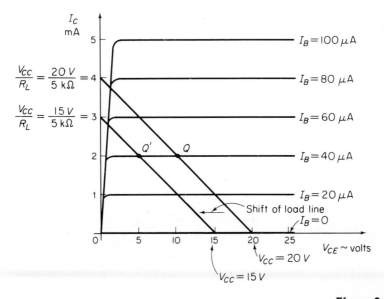

Figure 2-11

Solution to Example 2-2.

(b) From Fig. 2-11, the value of I_C at Q and at Q' is the same (2mA), therefore the value of β_F at both Q and Q' is

$$\beta_F = \frac{I_C}{I_B} = \frac{2mA}{40\mu A} = 50$$

EXAMPLE 2-3:

Again refer to the circuit of Fig. 2-8. This time V_{CC} is held constant at $20V$ and R_L is increased from 5kΩ to 8kΩ. (a) Show the new load line and operating point, and (b) calculate β_F at the new operating point.

SOLUTION:

(a) The current axis intercept is $V_{CC}/R_L = 20V/8kΩ = 2.5mA$. As shown in Fig. 2-12, the I_C intercept moves down the I_C axis. Note with V_{CC} held constant the load lines are pivoted about the horizontal or voltage axis intercept at $I_C = 0$, $V_{CE} = V_{CC}$. With I_B still $40\mu A$, operating point Q' is at $V_{CE} = 4V$ and $I_C = 2mA$.

(b) From Fig. 2-12, the value of I_C at Q and Q' is the same (2mA), therefore the value of β_F at both Q and Q' is

$$\beta_F = \frac{I_C}{I_B} = \frac{2mA}{40\mu A} = 50$$

Once an operating point is chosen, collector current, base current, and collector-emitter voltage are known, and a bias network is designed to deliver that point. Designing the bias network is our next topic.

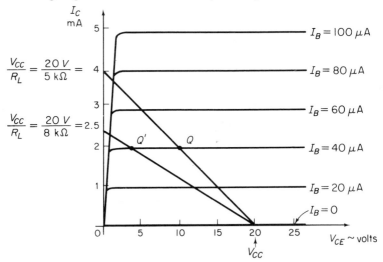

Figure 2-12

Solution to Example 2-3.

2-6 BIASING

The first step in designing an amplifier is to select an operating point on the characteristic curves. As we have seen in section 2-5, for a common-emitter circuit the operating point is specified by a particular value of base current I_B, collector current I_C, and collector-to-emitter voltage V_{CE}. *Biasing* is the procedure that establishes an operating point in a transistor circuit. Remember that to operate the transistor as an amplifier the base-emitter junction must be forward biased while the collector-base junction is reverse biased. As long as we operate in the active region, this condition is met. Since bias arrangements are often selected experimentally, however, we should have some knowledge of design and analysis. In order to keep the operating point from drastically changing with either environmental changes or transistor replacement, the circuit must be stabilized. The final choice of a bias arrangement will be made with the idea of stability in mind.

Our biasing procedure will generally follow the practice of locating the operating point midway between saturation and cutoff to allow for maximum output swing. This corresponds to a location halfway up the dc load line. The reason for this decision will be more fully understood in section 2-7. We shall develop here bias arrangements for common-emitter circuits. The biasing procedure is identical for common-base and common-collector configurations.

The circuit of Fig. 2-13a is a basic bias arrangement. Although its analysis is easy to understand, its disadvantages, soon to be pointed out, limit its extensive use. One advantage is that only a single resistor and supply, V_{CC}, is needed to furnish both I_B and I_C. Figure 2-8 required two supplies—V_{BB} and V_{CC}.

Referring to Fig. 2-13a and writing the loop equation for I_C yields

$$V_{CC} = I_C R_L + V_{CE} \tag{2-9}$$

Writing the loop equation for I_B yields

$$V_{CC} = I_B R_B + V_{BE} \tag{2-10}$$

where $V_{BE} \approx 0.6V$.

EXAMPLE 2-4:

Consider the circuit of Fig. 2-13a. The transistor is made of silicon and its characteristic curves are given in Fig. 2-13b. For operating point Q, calculate (a) R_L, (b) R_B, (c) β_F. Let $V_{CE} = \frac{1}{2}V_{CC}$.

SOLUTION:

From the operating point, $I_B = 20\mu A$, $I_C = 2mA$, and $V_{CE} = 10V$. Since $V_{CE} = \frac{1}{2}V_{CC}$, then $V_{CC} = 2V_{CE} = 2(10V) = 20V$. Rearranging Eq. (2-9)

(a)
$$R_L = \frac{V_{CC} - V_{CE}}{I_C} = \frac{(20 - 10)V}{2mA} = 5k\Omega$$

(b) For a silicon transistor $V_{BE} \approx 0.6V$, rearranging Eq. (2-10)

$$R_B = \frac{V_{CC} - V_{BE}}{I_B} = \frac{(20 - 0.6)V}{20\mu A} \simeq 1M\Omega$$

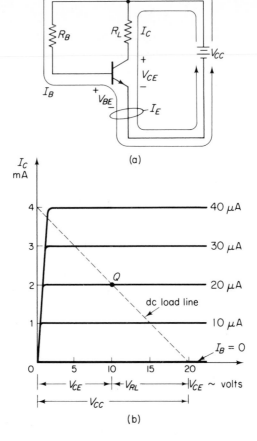

Figure 2-13

Operating point Q in (b) is established by selection of R_B in (a) for
Example 2-4.

(c) Using Eq. (2-4)

$$\beta_F = \frac{I_C}{I_B} = \frac{2\text{mA}}{20\mu\text{A}} = 100$$

EXAMPLE 2-5:

If a transistor with $\beta_F = 150$ is substituted into the solution circuit of
Example 2-4, base current will remain at $I_B = 20\mu\text{A}$. Calculate (a) I_C and
(b) V_{CE}.

SOLUTION:

(a) From Eq. (2-4)

$$I_C = \beta_F I_B = (150)(20\mu\text{A}) = 3\text{mA}$$

(b) Rearranging Eq. (2-9)

$$V_{CE} = V_{CC} - I_C R_L = 20V - (3mA)(5k\Omega) = 5V$$

Figure 2-14 shows the location of the operating point Q.

We conclude from Examples 2-4 and 2-5 that because the operating point changed as β_F changed, the circuit of Fig. 2-13a is unable to compensate for changes in β_F. Changes in β_F could be due either to a temperature change or a new transistor being used in the circuit. The circuit of Fig. 2-13a therefore has an unstable operating point.

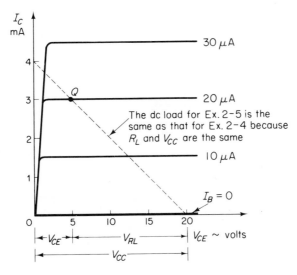

Figure 2-14

Solution to Example 2-5.

A modification of Fig. 2-13a is shown in Fig. 2-15. In this circuit, an emitter resistor R_E is connected between emitter terminal and ground. Writing the loop equation for I_C yields

$$V_{CC} = I_C R_L + V_{CE} + I_E R_E \qquad (2\text{-}11)$$

Since $I_E \approx I_C$, then

$$V_{CC} = I_C(R_L + R_E) + V_{CE} \qquad (2\text{-}12)$$

Writing the loop equation for I_B yields

$$V_{CC} = I_B R_B + V_{BE} + I_E R_E \qquad (2\text{-}13)$$

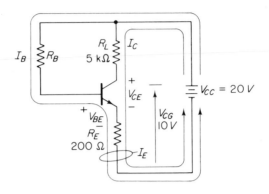

Figure 2-15

Circuit for Examples 2-6 and 2-7.

Substituting Eq. (2-6) into Eq. (2-13)

$$V_{CC} = V_{BE} + I_B[R_B + (1 + \beta_F)R_E] \qquad (2\text{-}14)$$

EXAMPLE 2-6:
For the circuit of Fig. 2-15 calculate (a) I_C, (b) I_B, (c) V_{CE}, and (d) R_B. The transistor is made of silicon and has a $\beta_F = 100$.

SOLUTION:
(a) Since V_{CG}, collector-to-ground voltage, is $10V$, then the voltage across R_L is $V_{R_L} = V_{CC} - V_{CG} = 20V - 10V = 10V$. The collector current is $I_C = 10V/5k\Omega = 2mA$.
(b) Using Eq. (2-4)

$$I_B = \frac{I_C}{\beta_F} = \frac{2mA}{100} = 20\mu A$$

(c) Rearranging Eq. (2-12) we obtain (remember $I_E \approx I_C$)

$$V_{CE} = V_{CC} - I_C(R_L + R_E)$$
$$V_{CE} = 20V - (2mA)(5.2k\Omega) = 9.6V$$

(d) Rearranging Eq. (2-13)

$$R_B = \frac{V_{CC} - V_{BE} - I_E R_E}{I_B}$$

$I_E \approx I_C = 2mA$.

$$R_B = \frac{20V - 0.6V - (2mA)(200\Omega)}{20\mu A} = \frac{19V}{20\mu A} = 0.95M\Omega$$

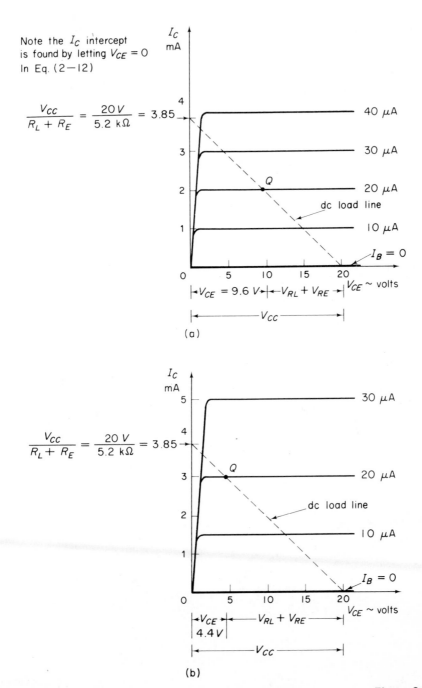

Note the I_C intercept is found by letting $V_{CE} = 0$ In Eq. (2-12)

$$\frac{V_{CC}}{R_L + R_E} = \frac{20\,V}{5.2\,k\Omega} = 3.85$$

$$\frac{V_{CC}}{R_L + R_E} = \frac{20\,V}{5.2\,k\Omega} = 3.85$$

Figure 2-16

Solutions to Examples 2-6 and 2-7.

EXAMPLE 2-7:

Let a transistor with $\beta_F = 200$ be substituted in the solution circuit of Ex. 2-6. Find (a) I_B, (b) I_C, and (c) V_{CE}.

SOLUTION:

(a) From Eq. (2-14)

$$I_B = \frac{V_{CC} - V_{BE}}{R_B + (\beta_F + 1)R_E} = \frac{19.4V}{950\text{k}\Omega + (201)(200\Omega)}$$

$$= \frac{19.4}{990\text{k}\Omega} = 19.7\mu A$$

(b) I_B did not change significantly and remains at about $20\mu A$. From Eq. (2-4)

$$I_C = \beta_F I_B = (150)(20\mu A) = 3\text{mA}$$

(c) Rearranging Eq. (2-12)

$$V_{CE} = V_{CC} - I_C(R_L + R_E)$$
$$V_{CE} = 20V - (3\text{mA})(5.2\text{k}\Omega) = 4.4V$$

Figure 2-16a shows the location of operating point Q for Example 2-6, while Fig. 2-16b locates operating point Q for Example 2-7. Note that for both examples the load line is the same. However, as β_F increases (Example 2-7) the operating point has moved up the load line, causing I_C to increase and V_{CE} to decrease. Although this circuit does have advantages over the circuit of Fig. 2-13a, operating point stability is not one of them.

Therefore, to keep the operating point I_C from moving when β_F changes, the circuit must automatically change I_B. However, the circuits of both Fig. 2-13a and Fig. 2-15 are unable to do this because I_B is established mainly by V_{CC}

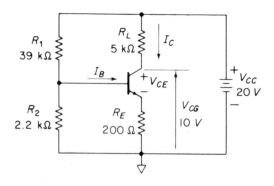

Figure 2-17

Biasing for operating point stability.

and R_B. The only way in which we could decrease I_B in these circuits for an increase in β_F is to manually increase R_B, and this is completely impractical.

Another commonly encountered biasing arrangement is shown in the circuit of Fig. 2-17. (The circuit values, although typical, have been inserted primarily for the following examples.) This circuit, like Figs. 2-13a and 2-15, has a single biasing supply, V_{CC}. Its major advantage is the circuit's ability to maintain a stable operating point. The best way to analyze this circuit is to redraw it with two biasing supplies—one for the output circuit and one for the input circuit—similar to Fig. 2-8. In Fig. 2-18, V_{BB} is in series with R_B, forming a Thévenin voltage source in series with a Thévenin resistor. This series combination is the definition of a Thévenin equivalent circuit. A procedure for obtaining an equivalent V_{BB} and R_B from the biasing resistors R_1 and R_2 is:

1. Remove the transistor and measure the voltage V_{BB} between points AA, as in Fig. 2-18a. Note in Fig. 2-18a that V_{BB} is the voltage across R_2; using the voltage division law

$$V_{BB} = \frac{R_2}{R_1 + R_2} V_{CC} \tag{2-15}$$

2. With the transistor still removed, measure the Thévenin equivalent resistance, R_B, between points AA. Note that V_{CC} must be replaced by its internal resistance (usually zero ohms), as in Fig. 2-18b. From Fig. 2-18b, R_1 is in parallel with R_2, then

$$R_B = \frac{R_1 R_2}{R_1 + R_2} \tag{2-16}$$

The Thévenin equivalent circuit is shown in Fig. 2-18c. Both of the above tests may be performed in a laboratory in order to verify calculated values.

The output loop in circuit Fig. 2-17 or Fig. 2-18c is the same as that of Fig. 2-15, and therefore the output loop equations must be the same.

$$V_{CC} = I_C R_L + V_{CE} + I_E R_E \tag{2-17}$$

Again letting $I_E \approx I_C$, then

$$V_{CC} = V_{CE} + I_C(R_L + R_E) \tag{2-18}$$

Equation (2-18) gives the axis intercepts

$$V_{CE} \text{ intercept} = V_{CC}$$

$$I_C \text{ intercept} = V_{CC}/(R_L + R_E)$$

The input loop equation for the Thévenin equivalent circuit of Fig. 2-18c is

$$V_{BB} = I_B R_B + V_{BE} + I_E R_E \tag{2-19}$$

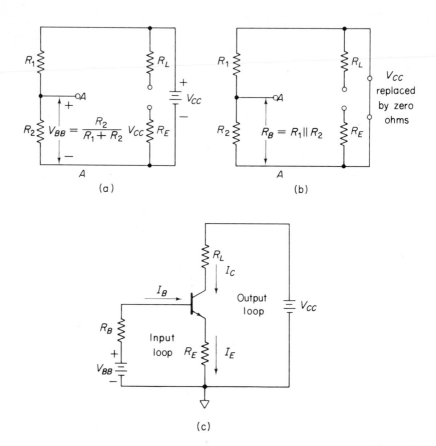

Figure 2-18

Thévenin equivalent input circuit for Figure 2-17 is found from (a) and (b) and shown in (c).

Since $I_E = I_B(\beta_F + 1)$, then Eq. (2-19) may be rewritten in terms of base current only:

$$V_{BB} = I_B R_B + V_{BE} + I_B(\beta_F + 1)R_E \qquad (2\text{-}20)$$

Solving for I_B we obtain

$$I_B = \frac{V_{BB} - V_{BE}}{R_B + (\beta_F + 1)R_E} \qquad (2\text{-}21)$$

The superiority of Fig. 2-17 over Fig. 2-13a and Fig. 2-15 is demonstrated in the following examples.

EXAMPLE 2-8:

For the circuit of Fig. 2-17 calculate (a) I_C, (b) V_{CE}, and (c) I_B for $\beta_F = 100$.

(Note V_{CG} is the voltage from collector to ground, not voltage V_{CE} from collector to emitter.)

SOLUTION:

(a) Since V_{CG} is the voltage from collector to ground ($V_{CG} = V_{CE} + V_{RE}$), then the voltage across R_L is $V_{R_L} = V_{CC} - V_{CG} = 20V - 10V = 10V$ and

$$I_C = \frac{V_{R_L}}{R_L} = \frac{10V}{5k\Omega} = 2mA$$

(b) Letting $I_E \approx I_C = 2mA$, then

$$V_{R_E} \approx I_C R_E = (2mA)(200\Omega) = 0.4V$$

and

$$V_{CE} = V_{CG} - V_{R_E} = 10V - 0.4 = 9.6V$$

(c) Using Eq. (2-4)

$$I_B = \frac{I_C}{\beta_F} = \frac{2mA}{100} = 20\mu A$$

or from Eq. (2-21)

$$I_B = \frac{V_{BB} - V_{BE}}{R_B + (\beta_F + 1)R_E}$$

Solving for V_{BB} from Eq. (2-15)

$$V_{BB} = \frac{R_2}{R_1 + R_2}V_{CC} = \frac{2.2k\Omega}{39k\Omega + 2.2k\Omega}(20V) \cong 1.05V$$

Solving for R_B from Eq. (2-16)

$$R_B = \frac{R_1 R_2}{R_1 + R_2} = \frac{(39k\Omega)(2.2k\Omega)}{39k\Omega + 2.2k\Omega} = 2.08k\Omega$$

Therefore

$$I_B = \frac{1.05V - 0.6V}{2.08k\Omega + (101)(200\Omega)} \cong 20\mu A$$

The real advantage of this circuit and Eq. (2-21) is seen in the following example, when β_F increases by 50 percent *without* a corresponding 50 percent increase in I_C.

EXAMPLE 2-9:

Again consider the circuit of Fig. 2-17. If all resistor values and V_{CC} remain the same but β_F increases to 150, calculate the new operating point values— I_B, I_C, and V_{CE}.

SOLUTION:

To find I_B use Eq. (2-21). From Example 2-8, $V_{BB} = 1.05V$ and $R_B = 2.08k\Omega$, and now $\beta_F = 150$, therefore

$$I_B = \frac{1.05V - 0.6V}{2.08k\Omega + (151)(200\Omega)} = 13.9\mu A$$

From Eq. (2-4)

$$I_C = \beta_F I_B = (150)(13.9\mu A) = 2.09mA$$

Rearranging Eq. (2-18) and letting $I_E = I_C = 2mA$

$$V_{CE} = V - I_C(R_L + R_E)$$
$$V_{CE} = 20V - (2.09mA)(5.2k\Omega) = 9.2V$$

Let us compare the operating point of Example 2-8 with the operating point of this example.

EXAMPLE 2-8:

$$I_B = 20\mu A, \ I_C = 2mA, \ V_{CE} = 9.6V$$

EXAMPLE 2-9:

$$I_B = 13.9\mu A, \ I_C = 2.09mA, \ V_{CE} = 9.2V$$

What the circuit of Fig. 2-17 has been able to do that the circuits of neither Fig. 2-13a nor Fig. 2-15 could do is maintain approximately the same I_C and V_{CE} for changes in β_F. Remember the changes in β_F may be due to either environmental changes or transistor replacement. The circuit of Fig. 2-17 is able to maintain a stable operating point by automatically decreasing base current I_B as β_F increases. If β_F should decrease, I_B will increase, trying to keep a constant dc collector current.

To investigate further how I_B decreases as β_F increases (maintaining the product $I_C = \beta_F I_B$ constant), refer to Fig. 2-19. Notation used in Fig. 2-19: An increase is represented by $\uparrow$ and a decrease by $\downarrow$. V_B is the measured voltage from base to ground, which is the voltage across R_2 and also equals $V_{BE} + V_{RE}$. Start by assuming β_F increases (either because of a temperature increase or a transistor replacement), which in turn leads to the following:

1. I_C tends to increase.
2. As I_C increases, V_{RE} increases.
3. V_B increases because $V_B = V_{BE} + V_{RE}$.
4. V_B is also the voltage across R_2, thus I_2 must increase.

5. As V_B increases and V_{CC} remains constant, then V_{R_1} must decrease because $V_{R_1} = V_{CC} - V_B$.

6. For a decrease in V_{R_1}, I_1 must decrease.

7. Since $I_1 = I_B + I_2$ or $I_B = I_1 - I_2$, and I_1 is decreasing while I_2 is increasing, then I_B must decrease.

8. As I_B decreases and because $I_C = \beta_F I_B$, then I_C tends to decrease, offsetting its original increase in 1.

9. I_C controls the voltage drops in the output loop, and since I_C tends to remain constant, then V_{CE} cannot change very much.

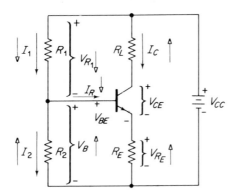

Figure 2-19

Current and voltage changes for an increase in β_F

Equations (2-15) and (2-21) depend on R_B. In order for a circuit similar to Fig. 2-17 to stabilize best the operating point, R_B should be chosen 10 to 15 times R_E. (The choice of values for both R_2 and R_E are discussed in section 2-7.) To meet other specifications, however, the 10 to 15 times R_E rule may have to be relaxed, and R_B may have to be chosen as high as 25 times R_E. In circuits where $R_B = 25R_E$ and β_F increases, the operating point moves further up the dc load line than it would if $R_B = 10R_E$. Thus the designer is faced with a compromise of operating point stability versus other specifications, such as input resistance. In Examples 2-8 and 2-9, $R_B = 2.08\text{k}\Omega$ and $R_E = 200\Omega$, therefore $R_B = 10.4R_E$. Example 2-9 showed that I_C and V_{CE} varied very little, and yet β_F increased from 100 to 150.

2-7 AC ANALYSIS

In this book we shall be concerned primarily with low-frequency amplifiers. Low frequency is a range of frequencies in which circuit coupling and by-pass capacitors act as short circuits to ac currents, and internal transistor capacitances act as open circuits. This frequency range extends approximately from 100 to 10,000Hz.

Hybrid π Model

At this time we wish to develop an *ac model* for the transistor. This model will help us to analyze ac problems of transistor circuits and will contain no dc voltages or currents. There are many types of models in the literature, and most are based on measurements. One model, based on the physics of BJTs, is the *hybrid pi* model in Fig. 2-20. This model is simple, allows easy calculations of ac parameters—voltage gain, input resistance, and output resistance—and requires a minimum of data sheet information.

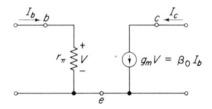

Figure 2-20
AC, small-signal circuit model of a BJT.

Resistance r_π is the approximate ac input base-to-emitter resistance of the transistor. The current generator, $g_m V$ or $\beta_0 I_b$, is a *dependent* current generator. That is, ac collector current, I_c, depends on the ac signal voltage across r_π. The term g_m is a *transconductance* parameter; it is the ac output collector current divided by the ac input voltage.

Let us consider the physical significance of each element and its role in the model.

Resistance r_π

In section 2-2 we discussed the fact that forward biasing a base-emitter junction decreases the potential barrier at the junction and allows carriers (holes for *pnp* transistors and electrons for *npn* transistors) to be injected into the base. If this voltage is increased, more carriers are injected into the base, and therefore more recombinations occur in the base. To maintain equilibrium the base current must increase to feed these recombinations. The ratio of the *increase* in base-emitter voltage to *increase* in base current is r_π.

The above paragraph can be written with the words decrease and less substituted for increase and more. Therefore, if the base-emitter voltage alternately increases and decreases, the base current must follow these variations. This alternate increase and decrease is an ac signal, so resistance r_π is an ac parameter, and not dc. For most practical considerations r_π is the ac resistance between base and emitter, and corresponds to the h_{ie} given on manufacturer's data sheets.

$$r_\pi = \frac{V}{I_b} \cong h_{ie} \qquad (2\text{-}22)$$

Dependent-current Generator, $g_m V(\beta_0 I_b)$

Again consider the condition in which the base-emitter voltage is increased by a signal. The increase in carriers that reach the collector cause an increase in collector current. Alternatively, if input voltage decreases, the collector current decreases; the ratio of this changing base-emitter voltage to changing collector current is transconductance, g_m. Note it is the ac voltage, V, across r_π that causes g_m and ac collector current, $I_c = g_m V$.

Therefore, *rms* collector current I_c is proportional to the *rms* base-emitter junction voltage V times the transconductance g_m, or

$$I_c = g_m V \qquad (2\text{-}23)$$

Since V is the voltage across r_π and I_b flows through r_π, then

$$V = I_b r_\pi \qquad (2\text{-}24)$$

Substituting Eq. (2-24) into Eq. (2-23) yields

$$I_c = g_m r_\pi I_b \qquad (2\text{-}25)$$

Express Eq. (2-25) in terms of incremental currents and compare it with Eq. (2-7):

Eq. (2-25)   Eq. (2-7)

$$g_m r_\pi = \frac{\Delta I_C}{\Delta I_B} \quad \text{and} \quad \frac{\Delta I_C}{\Delta I_B} = \beta_0$$

Therefore the small signal current gain β_0 is related to g_m by r_π as

$$\beta_0 = g_m r_\pi \qquad (2\text{-}26)$$

We know how to find β_0 from the collector characteristics, and if we also know how to determine g_m, then r_π may be calculated.

Although g_m is an ac parameter, it may be approximated quite simply from

$$g_m = \frac{I_C}{25\text{mV}} \qquad (2\text{-}27)$$

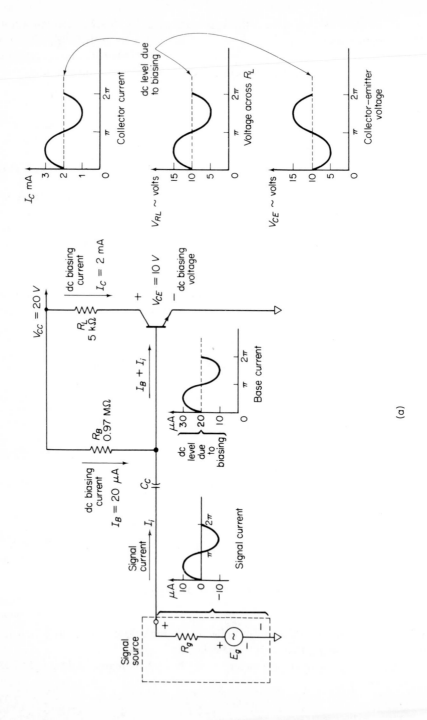

(a)

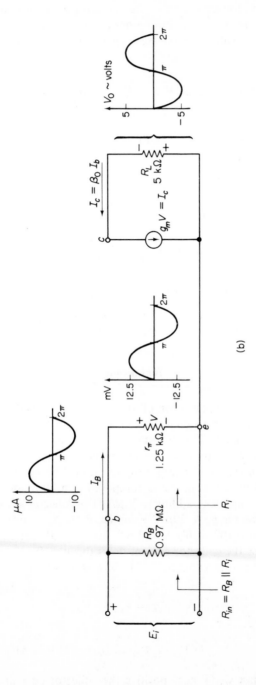

Figure 2-21

Ac performance of the BJT in (a) is analyzed from its ac circuit model in (b).

where I_C is the dc collector current (the operating point current). Equation (2-27) is the transition from dc operating point conditions to the ac model. Equation (2-27) gives best results up to *dc collector current* of 10mA.

EXAMPLE 2-10:

For the circuit of Fig. 2-13a calculate the base-to-emitter resistance.

SOLUTION:

From Ex. 2-4, $I_C = 2\text{mA}$, and from Fig. 2-13b

$$\beta_0 = \frac{\Delta I_C}{\Delta I_B}\bigg|_{V_{CE}=\text{const.}} = \frac{(3-1)\text{mA}}{(30-10)\mu\text{A}} = 100$$

Applying Eq. (2-27)

$$g_m = \frac{I_C}{25\text{mV}} = \frac{2\text{mA}}{25\text{mV}} = 0.08\text{mho}.$$

Rearranging Eq. (2-26)

$$r_\pi = \frac{\beta_0}{g_m} = \frac{100}{0.08\mho} = 1.25\text{k}\Omega$$

Instantaneous Current and Voltage Waveforms

To illustrate the application of an ac signal, reconsider the circuit of Fig. 2-13a. This circuit with a signal source is drawn in Fig. 2-21a with a $10\mu\text{A}$ peak signal current applied. The instantaneous waveforms are shown in Fig. 2-21. Note the base current, collector current, and collector-emitter voltage waveforms are composed of (1) the dc component and (2) the ac wave riding on the dc.

At $\pi/2$ rad the base current is composed of a $20\mu\text{A}$ dc component plus a $10\mu\text{A}$ peak ac component, adding to $30\mu\text{A}$. The collector current corresponding to this base current is $\beta_0 \times 30\mu\text{A}$. From Ex. 2-10, $\beta_0 = 100$, therefore the peak collector current is $100 \times 30\mu\text{A} = 3\text{mA}$. See the collector current waveform of Fig. 2-21a. Since the collector current flows through R_L, the voltage across the load at $\pi/2$ rad is $(3\text{mA})(5\text{k}\Omega) = 15V$. Kirchhoff's voltage law applied to the output loop is

$$V_{CC} = V_{CE} + V_{RL}$$

then

$$V_{CE} = V_{CC} - V_{RL} = 20V - 15V = 5V$$

Note as the input signal goes positive the collector current increases and the collector-emitter voltage decreases.

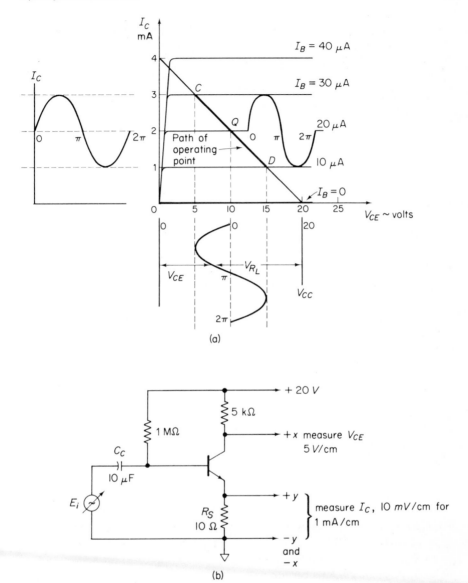

Figure 2-22

Current-voltage wave shapes in (a) for Figure 2-21a. The load line may be
displayed by the sweep circuit in (b).

This same information can be obtained from the collector characteristics.
Figure 2-22a shows the characteristic curves of the transistor and the load line
determined from the external circuit—V_{CC} and R_L. The signal source varies the
base current, causing the operating point to move up and down the load line.

For this example the dc base current is $20\mu A$ and the peak ac base current is $10\mu A$; therefore, the operating point moves from $20\mu A$ to $20\mu A + 10\mu A = 30\mu A$, and then back again. When the signal source goes negative (π to 2π), the operating point goes from $20\mu A$ to $20\mu A - 10\mu A = 10\mu A$, and then back again. This variation in base current along operating path CD causes both collector current and collector-emitter voltage to vary. From Fig. 2-22a an increase in I_B causes (1) an increase in collector current and (2) a decrease in collector-emitter voltage. A decrease in I_B causes (1) a decrease in collector current and (2) an increase in collector-emitter voltage. The entire operation can be seen on a CRO connected as in Fig. 2-22b.

The basic common-emitter amplifier shown in Fig. 2-21 is driven by an ideal voltage source E_i. The voltage E_g is the open-circuit voltage of a practical supply, and therefore is larger than E_i. Ideal voltage E_i may be achieved in practice by first connecting the amplifier to the source and then adjusting the value of input voltage E_i. E_i should be monitored so that it may be kept constant. The ac circuit model is drawn in Fig. 2-21b. Capacitor, C_C, and dc voltage supply, V_{CC}, are replaced by short circuits because their internal impedance is assumed negligible at midfrequency (1,000Hz). Since Fig. 2-21b is an ac model, dc levels of the waveforms of Fig. 2-21a are *not* included. Note $V = I_b r_\pi$, and it is this voltage that increases and decreases base-emitter voltage and thus controls the ac output current $I_c = \beta_0 I_b = g_m V$.

The ac model allows us to determine two important amplifier parameters— *input resistance* and *voltage gain*.

Input Resistance

Input resistance to a transistor from base to ground is represented by R_i in Fig. 2-21b. Input resistance for the stage which includes the biasing resistor R_B is R_{in} where

$$R_{in} = R_B \| R_i \tag{2-28}$$

For the circuit of Fig. 2-21b, $R_i = r_\pi = 1.25\text{k}\Omega$ (calculated from Ex. 2-10) and $R_{in} = 0.97\text{M}\Omega \| 1.25\text{k}\Omega \approx 1.25\text{k}\Omega$. For this circuit $R_i \ll R_B$, then $R_{in} \approx R_i = r_\pi$. When a design stipulates that the input resistance of an amplifier must equal or be greater than a certain value, it is R_{in} that must meet this condition. Thus we see another disadvantage of this circuit. R_{in} is controlled by the low resistance r_π, which in turn is controlled by the dc operating point current I_C.

Voltage Gain

An extremely useful method of determining voltage gain of a circuit is

$$\text{voltage gain} = A_V = \text{current gain} \times \frac{\text{load resistance}}{\text{input resistance}} \tag{2-29}$$

The current gain for the transistor of a common-emitter amplifier is β_0. The load resistance is R_L. The input resistance *for the transistor* is R_i. For the circuit of Fig. 2-21, $R_i = r_\pi$ and

$$A_V = \frac{-\beta_0 R_L}{r_\pi} = -g_m R_L = \frac{V_0}{E_i} \tag{2-30}$$

For the values in Fig. 2-21

$$A_V = \frac{(100)(5\text{k}\Omega)}{1.25\text{k}\Omega} = 400$$

Let us analyze the same circuit for a different β_F and β_0. The minus sign signifies that V_0 goes negative when E_i goes positive for a 180° phase reversal. R_B does not affect A_V because R_B does not affect I_b.

EXAMPLE 2-11:

In Fig. 2-21a let dc base bias current $I_B = 20\mu\text{A}$, and $\beta_F = \beta_0 = 150$. Calculate (a) input resistance R_i and R_{in} and (b) voltage gain.

SOLUTION:

(a) To calculate R_i we need to know g_m, which means finding the dc collector current.

$$I_C = \beta_F I_B = (150)(20\mu\text{A}) = 3\text{mA}$$

Applying Eq. (2-27)

$$g_m = \frac{I_C}{25\text{mV}} = \frac{3\text{mA}}{25\text{mV}} = 0.12\mho$$

Rearranging Eq. (2-26)

$$r_\pi = \frac{\beta_0}{g_m} = \frac{150}{0.12\mho} = 1.25\text{k}\Omega$$

Then

$$R_i = r_\pi = 1.25\text{k}\Omega$$

and

$$R_{in} = R_B \| R_i = 0.97\text{M}\Omega \| 1.25\text{k}\Omega \cong 1.25\text{k}\Omega$$

(b) Applying Eq. (2-29)

$$A_V = 150 \times \frac{5\text{k}\Omega}{1.25\text{k}\Omega} = 600$$

We may conclude, for the circuit of Fig. 2-21: (1) if $\beta_F = \beta_0$, then r_π is independent of β because I_C and g_m decrease in the same proportion that β increases; (2) voltage gain is directly proportional to β_0, or g_m and consequently dc collector current I_C.

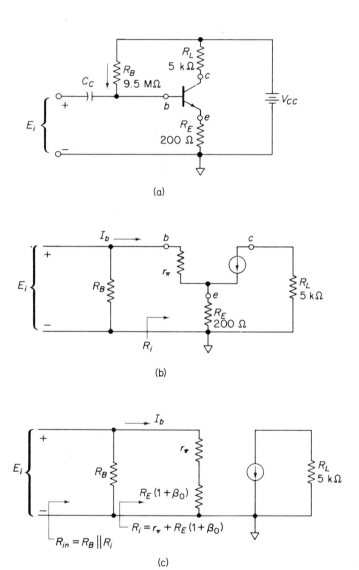

(a)

(b)

(c)

Figure 2-23

Gain is stabilized and ac input resistance increased by addition of an emitter resistor.

Let us consider the circuit of Fig. 2-21, this time with an added emitter resistor of 200Ω (see Fig. 2-23). In section 2-6 we analyzed this circuit for biasing and showed that the emitter resistor, R_E, unfortunately did not hold the dc operating point constant for variations in β_F. Under ac conditions, however, the emitter resistor increases the input resistance of the transistor and stabilizes voltage gain. That is, voltage gain remains fairly constant for variations in β_0.

Input Resistance with an Emitter Resistor

To determine input resistance of the transistor we need the ratio of E_i/I_b. Applying Kirchhoff's voltage law to the input loop in Fig. 2-23b, E_i equals the voltage drop across r_π ($V = I_b r_\pi$) plus the voltage drop across R_E ($I_e R_E$), or

$$E_i = I_b r_\pi + I_e R_E \qquad (2\text{-}31)$$

Since

$$I_e = I_b + I_c$$

and

$$I_c = \beta_0 I_b$$

then

$$I_e = I_b + \beta_0 I_b = I_b(\beta_0 + 1) \qquad (2\text{-}32)$$

Substituting Eq. (2-32) into Eq. (2-31) yields

$$E_i = I_b r_\pi + I_b(\beta_0 + 1)R_E$$

Therefore

$$R_i = \frac{E_i}{I_b} = r_\pi + (\beta_0 + 1)R_E \qquad (2\text{-}33)$$

Equation (2-33) shows that a *resistor connected in the emitter lead is reflected into the base lead by multiplying it by* ($\beta_0 + 1$). The circuit of Fig. 2-23c illustrates that R_i is now the series combination of r_π and $R_E(\beta_0 + 1)$. Note that the load resistor R_L does not affect the input resistance.

For a dc collector current of 2mA we calculated r_π from Ex. 2-10 to be 1.25 kΩ. If $\beta_0 = 100$, then R_i is

$$R_i = 1.25\text{k}\Omega + (100 + 1)\,200\Omega = 21.4\text{k}\Omega$$

Note that R_i is approximately ($\beta_0 + 1)R_E$ because r_π is small compared to it. If $\beta_0 = 150$, then $R_i = 1.25\text{k}\Omega + (150 + 1)200\Omega = 31.4\text{k}\Omega$.

The input resistance *to the stage* R_{in} is now no longer dependent on r_π but rather on the parallel combination of R_B and R_i.

For $\beta_0 = 100$

$$R_{in} = R_B \| R_i = 0.95\text{M}\Omega \| 21.4\text{k}\Omega \approx 21\text{k}\Omega$$

For $\beta_0 = 150$

$$R_{in} = R_B \| R_i = 0.95\text{M}\Omega \| 31.4\text{k}\Omega \approx 31\text{k}\Omega$$

Voltage Gain with an Emitter Resistor

Using Eq. (2-29) and $\beta_0 = 100$

$$A_V = \frac{V_0}{E_i} = \frac{\beta_0 R_L}{R_i} = \frac{(100)(5\text{k}\Omega)}{21.4\text{k}\Omega} = 23.4$$

For $\beta_0 = 150$

$$A_V = \frac{V_0}{E_i} = \frac{\beta_0 R_L}{R_i} = \frac{(150)(5\text{k}\Omega)}{31.4\text{k}\Omega} = 23.9$$

Note that although β_0 increased (from 100 to 150), the variation in voltage gain for the circuit of Fig. 2-23 is small. Therefore we conclude that a common-emitter amplifier with an emitter resistor stabilizes voltage gain and raises input resistance. This circuit still has the disadvantage of not being able to keep the dc operating point from changing as β_F changes. However, the circuit of Fig. 2-24 does stabilize the dc operating point, as we saw in Section 2-6, and having an emitter resistor benefits ac considerations. Thus the circuit of Fig. 2-24 should be the best choice for a common-emitter amplifier. Figures 2-24b and c show the ac model.

The hybrid π model of Fig. 2-24a is drawn in Fig. 2-24b. Figure 2-24b is similar to Fig. 2-23b except that R_B is the parallel combination of R_1 and R_2.

Input Resistance and Operating Point Stability

Input resistance from base to ground is the same as that of Fig. 2-23:

$$R_i = r_\pi + (\beta_0 + 1)R_E \qquad (2\text{-}34)$$

Input resistance of the stage is the parallel combination of R_B and R_i

$$R_{in} = \frac{R_B R_i}{R_B + R_i} \qquad (2\text{-}35)$$

where $R_B = R_1 \| R_2$ as given by Eq. (2-16). For the circuit values given in Fig. 2-24

$$R_i = 1.25\text{k}\Omega + (101)(200\Omega) = 21.4\text{k}\Omega$$

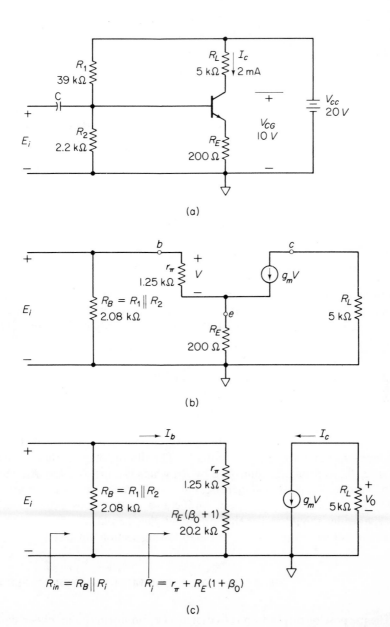

Figure 2-24

Common-emitter circuit with stable dc operating point and voltage gain.

This checks with R_i of Fig. 2-23 because both circuits have the same values of I_C, β_0 and R_E.

From Eq. (2-16)

$$R_B = \frac{(39\text{k}\Omega)(2.2\text{k}\Omega)}{39\text{k}\Omega + 2.2\text{k}\Omega} = 2.08\text{k}\Omega$$

and from Eq. (2-35)

$$R_{in} = \frac{(2.08\text{k}\Omega)(2.45\text{k}\Omega)}{2.08\text{k}\Omega + 21.45\text{k}\Omega} = 1.9\text{k}\Omega$$

Voltage Gain Approximation

Using Eq. (2-29), then

$$A_V = \beta_0 \frac{R_L}{r_\pi + (\beta_0 + 1)R_E} \tag{2-36}$$

If $r_\pi \ll (\beta_0 + 1)R_E$, as is usually the case, then

$$A_V \cong \beta_0 \frac{R_L}{(\beta_0 + 1)R_E} \cong \frac{R_L}{R_E} \tag{2-37}$$

Thus the voltage gain of a common-emitter circuit with an emitter resistor is approximately the ratio of R_L to R_E. This ratio is also applicable to the circuit of Fig. 2-23. To compare Eqs. (2-36) and (2-37), let us apply both expressions to Fig. 2-24.

$$A_V = \frac{\beta_0 R_L}{R_i} = (100)\frac{(5\text{k}\Omega)}{21.4\text{k}\Omega} = 23.4$$

$$A_V \cong \frac{R_L}{R_E} = \frac{5\text{k}\Omega}{200\Omega} = 25$$

The advantages of Fig. 2-24 are operating point stability (when $R_B = 10$ to 15 times R_E) and voltage gain stability. The disadvantage is that the input resistance, R_{in}, is primarily controlled by R_2, which is approximately R_B. Therefore, to increase R_{in}, R_B should increase, but in order to maintain operating point stability R_E would have to increase. An increase in R_E means a lower voltage gain. The reader should realize that no circuit yields an ideal solution; any circuit design will be a compromise, as the following section illustrates.

2-8 DESIGN OF A COMMON-EMITTER AMPLIFIER

The purpose of this section is to design a common-emitter amplifier similar to Fig. 2-17 with the following specifications: $V_{CC} = 24V$, $R_{in} \geqslant 1\text{k}\Omega$, and $A_V = 30$. The transistor used in the design is a silicon 40408.

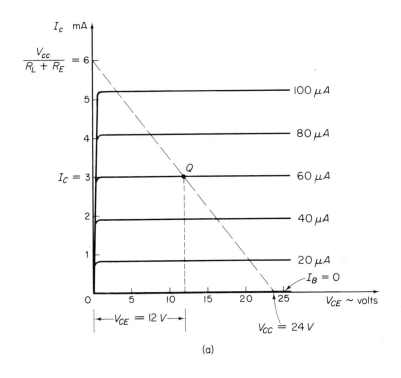

$\dfrac{V_{cc}}{R_L + R_E} = 6$

$I_c = 3$

Q

$\longleftarrow V_{CE} = 12\,V \longrightarrow$

$100\,\mu A$

$80\,\mu A$

$60\,\mu A$

$40\,\mu A$

$20\,\mu A$

$I_B = 0$

$V_{CE} \sim volts$

$V_{CC} = 24\,V$

(a)

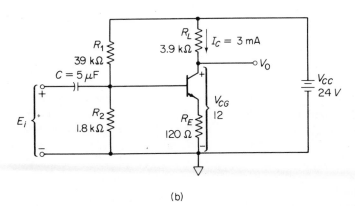

R_1 39 kΩ

$C = 5\,\mu F$

R_2 1.8 kΩ

E_i

R_L 3.9 kΩ

$I_C = 3\,mA$

V_0

V_{CG} 12

R_E 120 Ω

V_{CC} 24 V

(b)

Figure 2-25

Design example for a common-emitter amplifier.

SOLUTION:

We begin the design by first obtaining the characteristic curves and then choosing an operating point. Both the curves and operating point, Q, are

shown in Fig. 2-25a. At Q, $I_B = 60\mu A$, $I_C = 3mA$, and $V_{CE} = 12V$. Using
Eq. (2-4)

$$\beta_F = \frac{I_C}{I_B} = \frac{3mA}{60\mu A} = 50$$

An alternate procedure is to obtain a typical value of β_F from a data sheet.
From Eq. (2-7)

$$\beta_0 = \frac{\Delta I_C}{\Delta I_B}\bigg|_{V_{CE}=12V} = \frac{(4.1 - 1.9)mA}{(80 - 40)\mu A} = 55$$

Given $V_{CC} = 24V$ is the V_{CE} intercept, and choosing the operating point Q,
the load line (also shown in Fig. 2-25a) may be drawn. The I_C intercept is $V_{CC}/(R_L + R_E)$, from the curves

$$\frac{V_{CC}}{R_L + R_E} = 6mA$$

or

$$R_L + R_E = \frac{V_{CC}}{6mA} = \frac{24V}{6mA} = 4k\Omega$$

Rearranging Eq. (2-37)

$$R_L = A_V R_E = 30R_E$$

then

$$30R_E + R_E = 4k\Omega$$
$$R_E = 129\Omega$$

and

$$R_L = 4k\Omega - 129\Omega \approx 3.87k\Omega$$

Using 10 percent resistors and choosing the closest value by color code, then
$R_E = 120\Omega$ and $R_L = 3.9k\Omega$.
From Eq. (2-27)

$$g_m = \frac{I_C}{25mV} = \frac{3mA}{25mV} = 0.12\mho$$

and from Eq. (2-26)

$$r_\pi = \frac{\beta_0}{g_m} = \frac{55}{0.12\mho} = 457\Omega$$

Using $R_E = 120\Omega$, the input resistance from base to ground is

$$R_i = r_\pi + (\beta_0 + 1)R_E$$
$$R_i = 457\Omega + (55 + 1)(120\Omega) \cong 7.18k\Omega$$

Since R_{in} has to be equal to or greater than 1kΩ, let $R_{in} = 1.5$kΩ, so that when the circuit is built with 10 percent resistors we will have a margin of safety. From Eq. (2-35) $R_{in} = R_B \| R_i$; rearranging to solve for R_B yields

$$R_B = \frac{R_i R_{in}}{R_i - R_{in}} = \frac{(7.18\text{k}\Omega)(1.5\text{k}\Omega)}{7.18\text{k}\Omega - 1.5\text{k}\Omega} = 1.9\text{k}\Omega$$

Note the ratio $R_B/R_E = 1.9$kΩ/120Ω $\cong 16$. We conclude that we had to sacrifice very little in the way of operating point stability to meet the input resistance specification. Refer to Fig. 2-17, and writing the input loop equation which is also given by Eq. (2-19)

$$V_{BB} = I_B R_B + V_{BE} + I_E R_E$$

Since $I_E \approx I_C = 3$mA and for a silicon transistor $V_{BE} = 0.6V$, then

$$V_{BB} = (60\mu\text{A})(1.9\text{k}\Omega) + 0.6V + (3\text{mA})(0.20\text{k}\Omega) \approx 1.07V$$

Solving Eqs. (2-15) and (2-16) simultaneously for R_1 and R_2 yields

$$R_1 = \frac{V_{CC}}{V_{BB}} R_B \tag{2-38}$$

and

$$R_2 = \frac{V_{CC}}{V_{CC} - V_{BB}} R_B \tag{2-39}$$

Therefore

$$R_1 = \frac{24V}{1.07V}(1.9\text{k}\Omega) = 42.6\text{k}\Omega$$

and

$$R_2 = \frac{24V}{24V - 1.07V}(1.9\text{k}\Omega) = 2\text{k}\Omega$$

Using 10 percent resistors, a selection for R_1 and R_2 could be $R_1 = 39$kΩ and $R_2 = 1.8$kΩ. A 5μF capacitor should be placed at the input so that the dc biasing current flows into the transistor and not into a low impedance signal source. The completed design is shown in Fig. 2-25b.

Measuring the Operating Point

The first test after the circuit is built should be a dc voltage reading from collector to ground. For this circuit the reading should be approximately $12V$ because V_{RE} is small ($\cong I_C R_E = 0.36V$). If V_{CE} is not $12V$, what must be done? R_L and R_E must not be varied because these resistors control voltage gain. Let's say the dc voltage reading is $15V$. To readjust this voltage to $12V$, I_C and I_B must increase to move the operating point up the load line. To increase I_B, more current must flow into the base of the transistor. R_2 may be increased or R_1 decreased. Either change, however, will slightly affect operating point stability because $R_B = R_1 \| R_2$ and R_B will be increased. If V_{CE} is lower than $12V$, say $10V$, I_B must be decreased. To decrease I_B we could decrease R_2, but R_2 is a major factor controlling input resistance. A better choice is an increase of R_1. Again operating point stability will be slightly affected. Now consider that the voltage gain specification is not met. Since R_L and R_E control voltage gain, either one or both resistors must be changed. The best choice would be a decrease of R_E, since an increase of R_L will greatly affect the load line. Remember, however, that R_i depends on R_E and a large decrease of R_E will introduce other problems. Hopefully, the above discussion illustrates some obstacles and their solutions in designing a common-emitter amplifier. The reader must remember that at best his end design will be a compromised solution.

2-9 THE AC LOAD LINE

Until now the output circuit contained V_{CC} and either R_L or R_L and R_E (see Figs. 2-13a and 2-17). In both circuits the path of the ac current is the same as that of the dc current. However, it is more common to have different dc and ac loads. The dc load establishes a dc load line, and it is this load line and its intersection with a particular base current, I_B, that locates the operating point. The ac load determines the ac load line which gives the ac output voltage and current. For the circuits of Figs. 2-13a and 2-17 the dc and ac loads are the same, and therefore only one load line appears on their characteristic curves. The following discussion analyzes two of the most common circuits which have an ac load different from the dc load.

Emitter By-Pass Capacitor

Capacitor C_E in Fig. 2-26a is called an *emitter by-pass* capacitor and is connected in parallel with R_E. This capacitor does not restrict the path of the dc current because I_E still flows through R_E. Remember that it is this voltage drop $(I_E R_E \cong I_C R_E)$ that gives operating point stability. However, C_E is chosen so that its impedance $(X_{CE} = 1/\omega C_E)$ is less than $0.1 R_E$ at midfrequency (1kHz). Then the parallel combination of R_E and X_{CE} is small and negligible ac voltage drop occurs across it. For this reason R_E is not included in the ac model of Fig. 2-26b. Without R_E the circuit has the advantage of higher voltage gain, as given by Eq. (2-30). The disadvantage once again is the circuit's low input resistance, both R_i and R_{in}. For the circuit of Fig. 2-26

$$A_V = \beta_0 \frac{R_L}{r_\pi} = g_m R_L$$

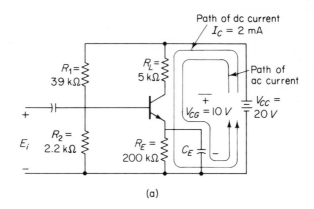

(a)

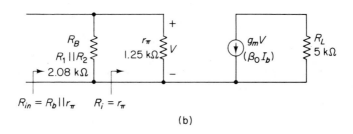

(b)

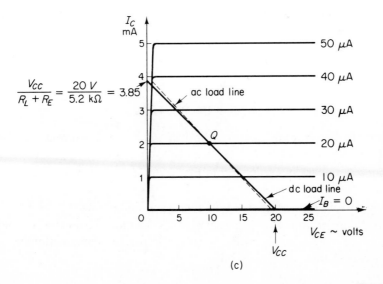

(c)

Figure 2-26

Ac load R_L differs from dc load $R_L + R_E$ because of by-passing by C_E.

where

$$g_m = \frac{I_C}{25\text{mV}} = \frac{2\text{mA}}{25\text{mV}} = 0.08\mho$$

$$A_V = (0.08\mho)(5\text{k}\Omega) = 400$$

Since the curves are the same as Fig. 2-22 $\beta_F = \beta_0 = 100$ and then

$$R_i = r_\pi = \frac{\beta_0}{g_m} = \frac{100}{0.08\mho} = 1.25\text{k}\Omega$$

Compare A_V and R_i with Fig. 2-21. From Fig. 2-26b

$$R_{in} = R_B || r_\pi$$

Since

$$R_B = R_1 || R_2 = 39\text{k}\Omega || 2.2\text{k}\Omega = 2.08\text{k}\Omega$$

then

$$R_{in} = 2.08\text{k}\Omega || 1.25\text{k}\Omega = 770\Omega$$

The circuit designer must be willing to accept this low input resistance for the increase in voltage gain and good dc operating point stability.

Constructing ac Load Lines

A procedure for drawing both the dc and ac load lines on the collector characteristics is:

1. The dc load line intercepts are

$$V_{CE} \text{ intercept} = V_{CC}$$
$$I_C \text{ intercept} = V_{CC}/R_{dc}$$

where R_{dc} is the total resistance through which the dc current flows. For Fig. 2-26a, $R_{dc} = R_L + R_E$, which is the same as that of Fig. 2-17.
2. Choose an operating point. From Fig. 2-26c, $I_B = 20\mu\text{A}$, $I_C = 2\text{mA}$, and $V_{CE} = 9.6V$.
3. The ac load line must pass through the operating point.
4. The V_{CE} intercept is $V_{CE} + I_C R_{ac}$ where R_{ac} is the total ac resistance. For the circuit of Fig. 2-26, $R_{ac} = R_L$, because the parallel combination R_E and X_{CE} is negligible. See Fig. 2-26b.

Note I_C is the dc collector current. The maximum output voltage swing without distortion is $I_C R_{ac}$ or V_{CE}, whichever is smaller. The operating point now moves up and down the ac load line and *not* along the dc load line.

Capacitor Coupled Load

Another circuit commonly encountered is shown in Fig. 2-27a. When there is not enough gain available from one transistor, its output is connected to the input of another transistor. The second transistor stage amplifies the output of

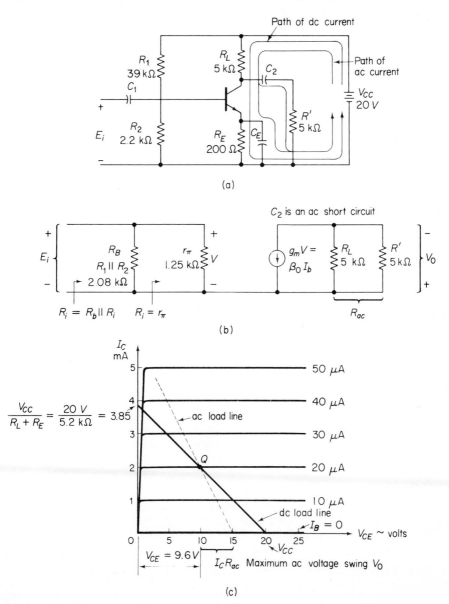

(a)

(b)

(c)

Figure 2-27

CE amplifier with an ac load.

the first so that the overall gain is the product of the individual stages. Therefore, R' in the circuit of Fig. 2-27a represents the input resistance of the second stage. Capacitor C_2 blocks the dc collector current, I_C, from flowing into R', but C_2 appears as a short circuit for ac current. Capacitor C_E performs the same by-pass function as in the previous example. In this circuit the dc resistance in the output loop is $R_{dc} = R_L + R_E$. Therefore the intercepts for the dc load line are

$$V_{CE} \text{ intercept} = V_{CC}$$

$$I_C \text{ intercept} = V_{CC}/R_{dc} = V_{CC}/(R_L + R_E)$$

The operating point is chosen at $I_B = 20\mu A$, $I_C = 2mA$, and $V_{CE} = 9.6V$. The ac resistance is $R_{ac} = R_L \| R' = 5k\Omega \| 5k\Omega = 2.5k\Omega$. The V_{CE} intercept for the ac load line is $V_{CE} + I_C R_{ac} = 9.6V + (2mA)(2.5k\Omega) = 14.6V$. See Fig. 2-27c. The maximum voltage swing without distortion is $I_C R_{ac} = 5V$. With R_E still shunted by C_E, input resistance is not increased, thus $R_i = r_\pi = 1.25k\Omega$ and $R_{in} = R_B \| R_i = 2.08k\Omega \| 1.25k\Omega = 770\Omega$. Voltage gain is decreased because the ac load resistance is decreased.

$$A_V = \beta_0 \frac{R_{ac}}{r_\pi} = g_m R_{ac} = (0.08\mho)(2.5k\Omega) = 200$$

2-10 MEASUREMENT OF AC VALUES

The following procedures illustrate a simplified laboratory method for measuring ac parameters. To measure A_V and g_m refer to Fig. 2-28:

1. Measure dc values of V_{CC}, V_{CG}, and R_L.
2. Use the above values to calculate I_C from $I_C = (V_{CC} - V_{CG})/R_L$.
3. Measure E_i with a CRO or ac VTVM. E_i should not be larger than 10mV. Most audio oscillators do not go down to 10mV, therefore the voltage divider network at the input must be used. From Fig. 2-28

$$E_i = \frac{10\Omega}{10k\Omega + 10\Omega}(10V) = 10mV$$

4. Measure V_0 with the same ac VTVM used for E_i. This is done so errors in different meters will not be encountered.
5. Calculate voltage gain, $A_V = V_0/E_i$, using the measured values of steps 3 and 4.
6. Rearranging Eq. (2-30), $g_m = A_V/R_L$, use the calculated value of A_V from step 5 and the measured value of R_L from step 1. This value of g_m from

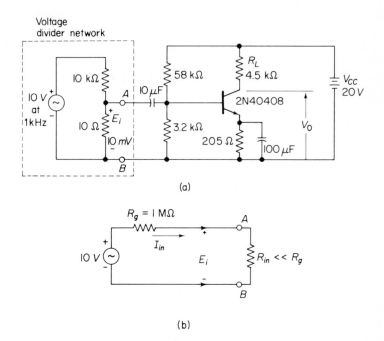

(a)

(b)

Figure 2-28

Circuit to measure ac voltage gain and input resistance.

measured values should be compared to the approximate formula given by Eq. (2-27), $g_m = I_C/25\text{mV}$, where I_C is the dc operating point current measured in step 2.

The circuit values in Fig. 2-28a are measured values and *not* color coded values. Following the above procedure:

1. $\left.\begin{array}{l} V_{CC} = 20V \\ V_{CG} = 10.6V \end{array}\right\}$ dc voltage readings
2. $I_C = (20 - 10.6)V/4.5\text{k}\Omega = 2.09\text{mA}$.
3. $E_i = 10\text{mV}$ measured with an ac VTVM.
4. $V_0 = 3.5V$ measured with the same ac VTVM.
5. $A_V = V_0/E_i = 3.5V/10\text{mV} = 350$.
6. From Eq. (2-30), $g_m = A_V/R_L = 350/4.5\text{k}\Omega = 0.078\mho$.

From the approximation Eq. (2-27)

$$g_m = \frac{I_C}{25\text{mV}} = \frac{2.09\text{mA}}{25\text{mV}} = 0.083$$

We see that Eq. (2-27) does yield results within 10 percent, which in most cases is good enough for commercial design.

Input resistance may be measured by removing the voltage divider network and connecting the network of Fig. 2-28b. With circuit connected, E_i was measured with an ac VTVM to be 5.7mV. Therefore, since R_{in} is small with respect to R_g

$$I_{in} \cong \frac{10V}{R_g} = \frac{10V}{1M\Omega} = 10\mu A$$

The ratio of E_i to I_{in} is the ac input resistance to the stage.

$$R_{in} = \frac{E_i}{I_{in}} = \frac{5.7mV}{10\mu A} = 570\Omega$$

Because R_E is shunted by $100\mu F$, the input resistance from base to ground $R_i = r_\pi$ and $R_{in} = R_B \| r_\pi$. $R_B = R_1 \| R_2 = 58k\Omega \| 3.2k\Omega = 3.03k\Omega$. Solving for r_π

$$r_\pi = \frac{R_B \times R_{in}}{R_B - R_{in}} = \frac{(3.03k\Omega)(570\Omega)}{3.03k\Omega - 570\Omega} \cong 700\Omega$$

The value of β_0 at the operating point is 58. Using the calculated value of g_m and Eq. (2-26)

$$r_\pi = \frac{\beta_0}{g_m} = \frac{58}{0.083\mho} = 700\Omega$$

Again excellent agreement has been obtained between measured and calculated values.

2-11 COMMON-COLLECTOR AMPLIFIER

When an input signal is connected to the base and the output signal is extracted from the emitter, the configuration is a common-collector (CC) or emitter-follower circuit (see Fig. 2-29a). The biasing procedure employed for a common-collector circuit is identical with that employed for a common-emitter circuit. The following discussion therefore deals only with ac parameters—input resistance and voltage gain. Figure 2-29b is the ac model of Fig. 2-29a.

Input Resistance

The input circuit of Fig. 2-29b is the same as that of Fig. 2-24. Therefore, the input resistance of the transistor from base to ground is determined by

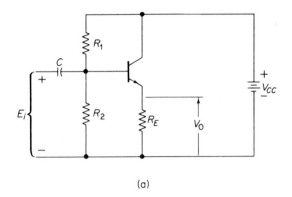

(a)

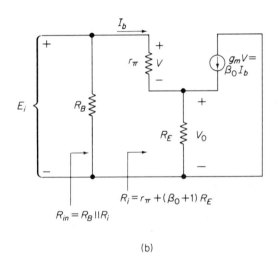

$R_i = r_\pi + (\beta_0 + 1) R_E$

$R_{in} = R_B \| R_i$

(b)

Figure 2-29

Common-collector circuit of (a) along with its hybrid π model in (b).

$$R_i = r_\pi + (\beta_0 + 1)R_E \tag{2-40}$$

Once again the input resistance of the stage is the parallel combination of R_B and R_i

$$R_{in} = \frac{R_B R_i}{R_B + R_i} \tag{2-41}$$

where

$$R_B = R_1 \| R_2$$

Voltage Gain

Apply Eq. (2-29) to calculate voltage gain.

$$\frac{V_0}{E_i} = A_V = \text{current gain} \times \frac{\text{load resistance}}{\text{input resistance}}$$

where current gain for a common-collector circuit is the ratio of emitter current to base current. From Eq. (2-6)

$$\frac{I_e}{I_b} = \beta_0 + 1 = \text{current gain}$$

For a common-collector circuit, output voltage V_0 is developed across R_E, therefore the load resistor is R_E. Input resistance is given by Eq. (2-40). Voltage gain for a common-collector circuit is

$$A_V = \frac{V_0}{E_i} = (\beta_0 + 1)\frac{R_E}{r_\pi + (\beta_0 + 1)R_E} \qquad (2\text{-}42)$$

Usually $r_\pi \ll (\beta_0 + 1)\, R_E$ and thus Eq. (2-42) reduced to

$$A_V \simeq \frac{(\beta_0 + 1)R_E}{(\beta_0 + 1)R_E} = 1 \qquad (2\text{-}43)$$

EXAMPLE 2-12:

Design the common-collector circuit of Fig. 2-29a for the following circuit parameters: $I_C = 2\text{mA}$, $V_{CE} = 10V$, $V_{BE} = 0.6V$, $V_{CC} = 20V$, and $\beta_F = \beta_0 = 100$.

SOLUTION:

Since $V_{CE} = 10V$ and $V_{CC} = 20V$ then the dc voltage across R_E is $V_{CC} - V_{CE} = 20V - 10V = 10V$, and letting $I_E \cong I_C$

$$R_E = \frac{10V}{2\text{mA}} = 5\text{k}\Omega$$

$$I_B = \frac{I_C}{\beta_F} = \frac{2\text{mA}}{100} = 20\mu\text{A}$$

Choose $R_B = 10R_E = 10(5\text{k}\Omega) = 50\text{k}\Omega$. Writing the dc input loop equation

$$V_{BB} = I_B R_B + V_{BE} + I_E R_E$$
$$V_{BB} = (20\mu\text{A})(50\text{k}\Omega) + 0.6V + (2\text{mA})(5\text{k}\Omega) = 11.6V$$

Applying Eqs. (2-38) and (2-39)

$$R_1 = \frac{V_{CC}}{V_{BB}} R_B = \frac{20V}{11.6V}(50k\Omega) = 86.3k\Omega$$

$$R_2 = \frac{V_{CC}}{V_{CC} - V_{BB}} R_B = \frac{20V}{20V - 11.6V}(50k\Omega) = 119k\Omega$$

Thus, the design of a common-collector circuit about a particular operating point is accomplished by the same procedure as that used for a common-emitter amplifier. Using the closest color coded 10 percent resistors, $R_E = 4.7k\Omega$, $R_1 = 82k\Omega$, and $R_2 = 120k\Omega$.

EXAMPLE 2-13:
Using the results of Example 2-12, calculate ac input signal resistance R_i and R_{in}.

SOLUTION:
Applying Eq. (2-27)

$$g_m = \frac{I_C}{25mV} = \frac{2mA}{25mV} = 0.08\mho$$

and

$$r_\pi = \frac{\beta_0}{g_m} = \frac{100}{0.08\mho} = 1.25k\Omega$$

From Eq. (2-34)

$$R_i = r_\pi + (\beta_0 + 1)R_E = 1.25k\Omega + (101)5k\Omega \cong 502k\Omega$$

and from Eq. (2-35)

$$R_{in} = \frac{R_B R_i}{R_B + R_i} = \frac{(50k\Omega)(502k\Omega)}{50k\Omega + 502k\Omega} = 45.5k\Omega$$

Note that R_{in} approximately equals R_B.

EXAMPLE 2-14:
If the ac input signal E_i in Ex. 2-13 is $0.5V$ peak, calculate (a) peak base current, I_{bp}, (b) ac voltage across r_π, V, (c) ac output voltage, V_0.

SOLUTION:
From Fig. 2-29b we see that:

(a) $$I_{bp} = \frac{E_i}{R_i} = \frac{0.5V}{502k\Omega} = 0.99\mu A$$

(b) $\qquad V = I_{bp}r_\pi = (0.99\mu\text{A})(1.25\text{k}\Omega) = 1.24\text{mV}$

(c) $\qquad I_0 = I_e = (\beta_0 + 1)I_{bp} = (101)(0.99\mu\text{A})$

$$\cong 100\mu\text{A} = 0.1\text{mA (peak)}$$

$$V_0 = I_0 R_E = (0.1\text{mA})(5\text{k}\Omega) = 0.5V \text{ (peak)}$$

Note that practically all of E_i is developed across R_E and not across r_π.

2-12 RESISTANCE SEEN BETWEEN EMITTER AND GROUND

In the previous sections we have established that, looking into the base terminal, we see any resistance in the base unchanged, while any resistance in the emitter is reflected into the base by multiplying it by $(\beta_0 + 1)$. We used these properties to calculate R_i for both common-emitter and common-collector configurations. Looking into a collector terminal, we see the very high resistance of current generator $g_m V$ or $\beta_0 I_b$. The collector terminal is the output terminal for both the common-emitter and common-base configurations. Therefore, the output resistance of these configurations is quite high. However, calculating output resistance of common-collector configurations and input resistance of common-base configurations requires transforming any resistance in the base lead to the emitter lead. We shall define *output resistance* as the resistance seen looking into the output terminals *with the load disconnected*. Common-base configuration is studied in section 2-13.

A standard procedure for determining the resistance between any two terminals is to apply a test voltage and measure current that would be drawn from this source. The ratio of test voltage to current is the impedance "seen" between the two terminals. Note that although we may use this procedure to

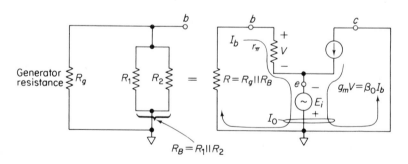

Figure 2-30

Circuit to find ac output resistance $R_0 = E/I_0$.

analyze a circuit with pencil and paper, it may not always be possible in a laboratory. Figure 2-30 shows test voltage E connected between emitter and ground. Practical voltage source E_g (refer to Fig. 2-21a) is replaced by its internal resistance R_g. R_g is in parallel with R_B and is shown as the equivalent resistance R. Test voltage E causes loop current I_b to flow through r_π, thereby causing the current $g_m V(\beta_0 I_b)$ to flow. Both loop currents (I_b and $\beta_0 I_b$) flow through E. If we define R_0 (output resistance) as the resistance seen by E, or, to put it another way, the ratio of test voltage to current I_0 through it, then

$$\frac{E}{I_0} = R_0 = \frac{E}{(\beta_0 + 1)I_b} \tag{2-44a}$$

From the input loop equation $I_b = E/(R + r_\pi)$. Substituting into Eq. (2-44a)

$$R_0 = \frac{R + r_\pi}{\beta_0 + 1} \tag{2-44b}$$

where $R = R_g \| R_B$. We conclude from Eq. (2-44b) that *any resistance in the base lead is divided by $\beta_0 + 1$ when we look into the emitter terminal.* This principle is shown in Ex. 2-15.

2-13 COMMON-BASE AMPLIFIER

The common-base (CB) configuration found extensive use in early applications of transistor circuits because of its operating stability and high-frequency performance. In fact, its application is still used in very high-frequency amplifiers. The biasing arrangement for the common-base circuit of Fig. 2-31a is similar to that for the common-emitter circuit of Fig. 2-17. Differences exist in the application of an ac signal. The input signal is placed between emitter-base and extracted between collector-base. We shall again analyze the ac model to determine input resistance and voltage gain. Capacitor C_B in Fig. 2-31a is large, so that at midfrequency (1000Hz) its reactance is small and places the base at ac ground. Thus the reactance of C_B shorts out R_B, and for this reason R_B is not shown in the ac model of Fig. 2-31c.

Input Resistance

In the circuit of Fig. 2-31b, input voltage, E_i, is directly across R_E. Although R_E is a current drain on E_i it does not affect I_b and consequently does not affect input resistance *of the transistor.* In Fig. 2-31c R_E has been removed to emphasize it is not to be included in the following analysis.

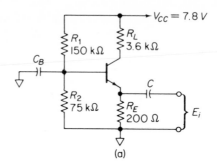

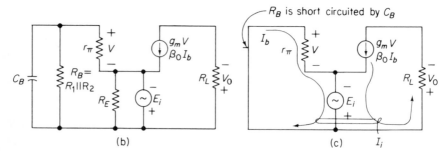

Figure 2-31

Common-base circuit in (a) is analyzed for voltage gain and input resistance in (b) and (c).

Input resistance R_i is the ratio of E_i to I_i, or $R_i = E_i/I_i$. From Fig. 2-31c $I_i = I_b + \beta_0 I_b = I_b(\beta_0 + 1)$, or $I_b = I_i/(\beta_0 + 1)$. Writing the input loop equation

$$E_i = I_b r_\pi$$

$$\therefore \quad E_i = \frac{I_i}{\beta_0 + 1} r_\pi$$

and

$$R_i = \frac{E_i}{I_i} = \frac{r_\pi}{\beta_0 + 1} \tag{2-45}$$

From Eq. (2-45) we conclude that the resistance seen between emitter and ground equals the resistance in the base lead, including r_π *divided by* $(\beta_0 + 1)$. Remember that R_i for CE and CC circuits is the resistance seen between base and ground and is approximately the resistance in the emitter lead *multiplied by* $(\beta_0 + 1)$.

Input resistance for a common-base stage is the parallel combination of R_i and R_E.

$$R_{in} = R_i \| R_E = \frac{r_\pi}{\beta_0 + 1} \middle\| R_E \tag{2-46}$$

Applying Eq. (2-29)

$$A_V = \text{current gain} \times \frac{\text{load resistance}}{\text{input resistance}}$$

where current gain for common-base circuits is the ratio of $I_c/I_e = \alpha_0$. Using Eq. (2-4) to express α_0 in terms of β_0 yields

$$\alpha_0 = \frac{\beta_0}{\beta_0 + 1} \tag{2-47}$$

Load resistance is R_L and input resistance is given by Eq. (2-45). Therefore

$$A_V = \frac{V_0}{E_i} = \frac{\beta_0}{\beta_0 + 1} \times \frac{R_L}{r_\pi/(\beta_0 + 1)} = \frac{\beta_0 R_L}{r_\pi} = g_m R_L \tag{2-48}$$

Equation (2-48) shows that the voltage gain for a common-base circuit is identical with that of a common-emitter circuit without an emitter resistor, or a common-emitter circuit with a by-passed emitter resistor.

EXAMPLE 2-15:
Analyze the common-base amplifier in Fig. 2-31 to find (a) dc operating point I_C, V_{CE}, and I_B, (b) voltage gain A_V, (c) input resistance R_i and R_{in}. $\beta_F = \beta = 50$.

SOLUTION:

(a) $\qquad R_B = R_1 \| R_2$

$$R_B = \frac{(150\text{k}\Omega)(75\text{k}\Omega)}{150\text{k}\Omega + 75\text{k}\Omega} = 50\text{k}\Omega$$

$$V_{BB} = \frac{R_2}{R_1 + R_2}V_{CC} = \frac{75\text{k}\Omega}{150\text{k}\Omega + 75\text{k}\Omega}(7.8V) = 2.6V$$

Rearranging Eq. (2-14)

$$I_B = \frac{V_{BB} - V_{BE}}{R_B + (\beta_F + 1)R_E} = \frac{2.6V - 0.6V}{50\text{k}\Omega + (51)(1\text{k}\Omega)} \cong 20\mu\text{A}$$

$$I_C = \beta_F I_B = (50)(20\mu\text{A}) = 1\text{mA}$$

$$V_{CE} \cong V_{CC} - I_C(R_L + R_E) = 7.8V - (1\text{mA})(3.6\text{k}\Omega + 0.2\text{k}\Omega) = 4V$$

(b) $\qquad\qquad g_m = \frac{I_C}{25\text{mV}} = \frac{1\text{mA}}{25\text{mV}} = 0.04\mho$

$$r_\pi = \frac{\beta_0}{g_m} = \frac{50}{0.04\mho} = 1.25\text{k}\Omega$$

From Eq. (2-48)

$$A_V = \frac{\beta_0 R_L}{r_\pi} = \frac{(50)(3.6\text{k}\Omega)}{1.25\text{k}\Omega} = 144$$

or

$$A_V = g_m R_L = (0.04\text{℧})(3.6\text{k}\Omega) = 144$$

(c) Using Eq. (2-45)

$$R_i = \frac{r_\pi}{\beta_0 + 1} = \frac{1.25\text{k}\Omega}{51} = 24.5\Omega$$

and applying Eq. (2-46)

$$R_{in} = R_i \,\|\, R_E = \frac{(24.5\Omega)(200\Omega)}{24.5\Omega + 200\Omega} = 21.8\Omega$$

This low input resistance is a major disadvantage of the common-base amplifier.

PROBLEMS

1. The forward current-transfer ratio of a common-base configuration is 0.99. Calculate current gain if the transistor is in (a) common-emitter configuration and (b) common-collector configuration.
2. Using Fig. 2-7, calculate the dc current gain at (a) $I_C = 2.5$A and $V_{CE} = 15V$, (b) $I_B = 25$mA and $V_{CE} = 15V$, (c) $I_C = 0.5A$ and $V_{CE} = 15V$.
3. For the three operating points in problem 2, determine the ac current gain. Use a 10mA peak-to-peak base current swing.
4. Consider the circuit of Fig. 2-13a and the collector characteristics of Fig. 2-6. If the operating point of the transistor is at point B, calculate R_L and R_B. Let $V_{CC} = 30V$ and $V_{BE} = 0.6V$.
5. Repeat Example 2-4, but this time let $V_{CE} = 3/4V_{CC}$. Compare results.
6. To keep $V_{CE} = 10V$ in Example 2-6, with $\beta_F = 50$, what value is required for R_B?
7. If $\beta_F = 50$ for the transistor of Fig. 2-17, what are the operating point values—I_B, I_C, and V_{CE}? Compare answers with Examples 2-8 and 2-9.
8. In Fig. 2-17, if $V_{CG} = 15V$, calculate the values for biasing resistors R_1 and R_2. $\beta_F = 100$ and $R_B = 10R_E$.
9. Calculate the ac input resistance for each circuit of Fig. 2-32. Assume $r_\pi = 1\text{k}\Omega$ and $\beta_0 = 99$ for each transistor.

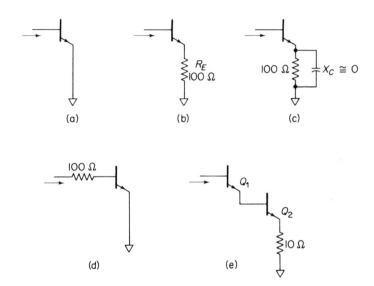

Figure 2-32

Circuits for Problem 2-9.

10. Plot r_π versus I_C for operating point currents between 0.1mA and 10mA for $\beta_0 = 50$ and $\beta_0 = 100$ on the same graph paper.

11. If $R_B = 500$kΩ in Fig. 2-15, calculate (a) g_m, (b) r_π, (c) R_i and R_{in}. Let $\beta_F = \beta_0 = 149$ and $V_{BE} = 0.6V$.

12. In the circuit of Fig. 2-24a, if $R_L = 5$kΩ, $R_E = 200\Omega$, $R_1 = 48.8$kΩ, $R_2 = 2.09$kΩ (answers to problem 8), $V_{CG} = 15V$, $V_{CC} = 20V$, calculate (a) r_π, (b) R_i, and (c) R_{in}. $\beta_F = \beta_0 = 100$.

13. The peak value of base current of Fig. 2-21 is 5μA; calculate (a) E_i, (b) E_g ($R_g = 500\Omega$), and (c) V_0.

14. Consider Fig. 2-21. What is the maximum ac base current, collector current, and output voltage before the circuit is overdriven and the waveforms become clipped? $\beta_F = \beta_0 = 100$.

15. If $E_i = 0.3V$ in Fig. 2-24a, calculate I_b, I_c, and V_0, $\beta_0 = 100$.

16. Design a common-emitter amplifier similar to Fig. 2-17 for the following specifications: $V_{CC} = 30V$, $V_{CG} = 15V$, $A_v = 20$, $I_C = 5$mA, $V_{BE} = 0.6V$, and $\beta_F = \beta_0 = 50$. For stability, let $R_B = 10R_E$.

17. If $E_i = 0.4V_{peak}$ for problem 16, draw the following instantaneous waveforms: (a) base current, (b) collector current, and (c) output voltage.

18. In Fig. 2-27a, R' is changed to 1kΩ. (a) Does dc operating point change? (b) Does input resistance change? (c) Does voltage gain change? (d) Calculate maximum ac voltage swing.

19. For the common-collector circuit of Fig. 2-29, $R_1 = 30$kΩ, $R_2 = 6$kΩ,

$R_E = 1k\Omega$, and $V_{CC} = 20V$. What is the dc operating point, I_B, I_C, V_{CE}? $V_{BE} = 0.6V$, $\beta_F = \beta_0 = 60$.

20. For the circuit values of problem 19, calculate ac input resistance R_i and R_{in}.

21. Repeat Example 2-14 with an ac input voltage $E_i = 0.2V$.

22. For each circuit of Fig. 2-32, what is the resistance seen looking from ground back toward the base? Include emitter resistors. $r_\pi = 1k\Omega$, $\beta_0 = 99$.

23. For the common-base circuit of Fig. 2-31, $R_1 = 60k\Omega$, $R_2 = 20k\Omega$, $R_L = 6k\Omega$, $R_E = 100\Omega$, and $V_{CC} = 12V$. Calculate the dc operating point, I_B, I_C, V_{CE}. $\beta_F = \beta_0 = 80$, $V_{BE} = 0.6V$.

24. For circuit values of Problem 2-23 determine (a) ac input resistance, R_i and R_{in}, and (b) voltage gain.

25. For an input signal, E_i, of 4mV in Fig. 2-31a, calculate I_b, I_c, and V_0. $\beta_0 = 50$.

chapter three
field-effect transistors

3-0 INTRODUCTION

In Chapter 2 we studied bipolar junction transistors (BJTs) and saw that their operation depended on two charge carriers—holes and electrons. Unlike BJTs, operation of *field-effect transistors* (FETs) depends on the transport of one type of carrier, either holes or electrons, but not both. For this reason FETs are classified as unipolar devices. Output current of bipolar transistors depends on controlling two *pn* junctions—emitter-base and base-collector. Field-effect transistors use the electric field of a reverse-biased *pn* junction to control resistance of a channel, and thereby control output current.

There are two types of field-effect transistors: (1) the *junction field-effect transistor* (JFET), and (2) the *insulated gate field-effect transistor* (IGFET), more commonly known as the *metal-oxide semiconductor field-effect transistor* (MOSFET).

Advantages of FETs over BJTs are: (1) high input impedance, (2) lower noise, (3) fewer steps in the manufacture process, (4) more devices able to be packaged into a smaller area for integrated circuit arrays, and (5) no thermal runaway. Disadvantages of FETs are: (1) poor high-frequency performance and (2) low power-handling ability.

3-1 INTRODUCTION TO THE JFET

As shown in Fig. 3-1a, an *n*-channel JFET is made by diffusing *p*-type semiconductor material into an *n*-type channel. Connections to both *p*-type materials are brought out to a single *gate* terminal *G*. Ohmic connections to either end of the channel are brought out to *source S* and *drain D* terminals respectively. Although electrons are conducted between source and drain, we will use conventional current flow in the external circuit, as shown by the direction of I_D (drain current). Figure 3-1b shows typical circuit symbols for an *n*-channel JFET, where arrows point toward the *n*-type material. Packaging and lead location for a 2N5457 JFET in a TO-92 case are given in Fig. 3-1c. Pictorial diagram and a third common symbol type for a *p*-channel JFET appear in Figs. 3-1d and e.

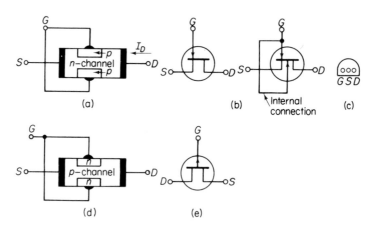

Figure 3-1

Construction, symbols, and packaging for JFETs.

To measure drain or $I_D - V_{DS}$ characteristics of an *n*-type JFET begin by short-circuiting gate to source to insure that gate voltage V_{GS} equals zero. As shown in Fig. 3-2, make the drain more positive than source by first increasing supply voltage V_{DD} from zero volts to about 1 volt. I_D increases directly with an increase in V_{DS}, in a straight line exactly like the *I-V* characteristic of a resistor. Thus the $I_D - V_{DS}$ characteristic curve is said to have an *ohmic region* for small values of V_{DS}. At $V_{GS} = 0$ and near the origin (where $I_D = 0$, $V_{DS} = 0$), the slope of this characteristic has a special symbol $r_{ds(on)}$ and is measured graphically from Fig. 3-2 as

$$r_{ds(on)} = \frac{V_{DS}}{I_D} \text{ at } V_{DS} = 0, V_{GS} = 0 \tag{3-1}$$

$$r_{ds(on)} = \frac{0.5V}{1.0\text{mA}} = 500\Omega$$

As V_{DS} is increased beyond about $1V$, in Fig. 3-2 I_D tends to remain constant. To explain this behavior, assume V_{DS} is $3V$. V_{DS} will divide equally along the n-channel between drain and source. The source end of both pn junctions will be reverse-biased by about $1V$. The drain end will be reverse-biased by about $2V$. The space charge regions of both pn junctions are widened by an amount proportional to the reverse bias. Since no carriers exist in these space charge regions, the conducting channel shrinks and a large portion becomes

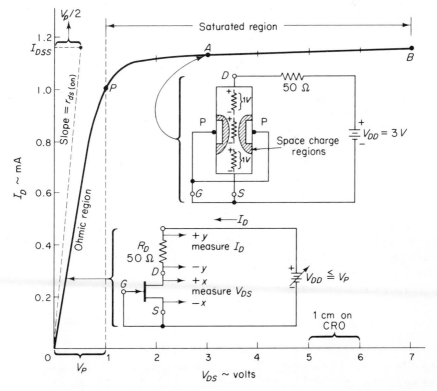

Figure 3-2

Common-source, $I_D - V_{DS}$ characteristic for an n-channel JFET with $V_{GS} = 0$.

depleted of carriers. At point A of Fig. 3-2, drain current is limited to about 1.13 mA with $V_{GS} = 0$ and $V_{DS} = 3V$. At point B drain current increases only slightly to about 1.16 mA when V_{DS} is raised to $7V$. Operation along the relatively flat horizontal portion of the curve is described by the terms (1) *pentode region*, as an analogy for those trained in vacuum tubes, (2) *saturation region*, to indicate the FET is carrying all possible current or (3) *constant current* region.

Locate *pinchoff voltage* V_p in Fig. 3-2 by visually noting the center of the curve portion which is the transition between ohmic and saturation region. V_p is that drain-source voltage where space charge regions almost touch at the drain end of the channel, and for this particular FET, $V_p = 1V$. A second important FET electrical characteristic is the drain-source current at some point well into the saturation region, with $V_{GS} = 0$. This measurement is specified by the symbols I_{DSS} or I_{DO}, and is about 1.15mA in Fig. 3-2.

3-2 MEASUREMENT OF JFET CHARACTERISTICS

A 1 percent lkΩ resistor is added across base and emitter test terminals of a BJT curve tracer in Fig. 3-3a. Collector sweep voltage is set to $+$ (*npn* position) and base step generator is set to $-$ (*pnp* position). The resistor converts the curve tracer's step current generator to a calibrated step-voltage-generator, since the reverse-biased gate draws no gate current. This procedure allows *n*-channel FETs to be tested on the curve tracer. By dialing collector current and collector-emitter volts on vertical and horizontal axes respectively, we get the I_D versus V_{DS} display in Fig. 3-3b. Reverse polarity of the collector sweep (to $-$ or *npn*) and base-step (to $+$ or *pnp*) to test *p*-channel JFETs.

Observe the variation in I_D with V_{DS} for V_{GS} held constant at $0.2V$. In the ohmic region, less I_D flows for the same value of V_{DS} when compared with the curve for $V_{GS} = 0$. This is because the reverse bias of $V_{GS} = 0.2V$ shrinks the channel by widening the depletion or space charge regions. Thus the JFET saturates at a lower value of drain current in the saturation region of the characteristic. Since both V_{GS} and V_{DS} act together to pinch off the channel and limit I_D, it can be shown that

$$V_{GS} + V_{DS} = V_p \qquad (3\text{-}2)$$

As V_{GS} is increased in steps, we see that no drain current flows during the fifth step, where $V_{GS} = 1.0V$. The channel is pinched off at $V_{DS} = 0$ as well as for any other value of V_{DS}, and $V_p = V_{GS} = 1.0V$. Actually, measurement of V_p demands a value judgment, so JFET manufacturers often specify V_p as *that value of gate-source voltage required to reduce I_D to $0.01 \times I_{DSS}$.*

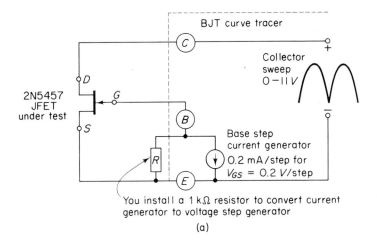

You install a 1 kΩ resistor to convert current generator to voltage step generator

(a)

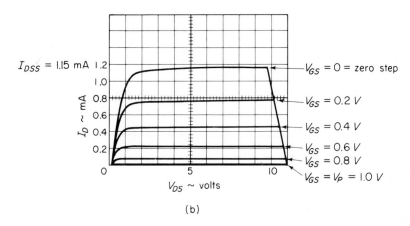

(b)

Figure 3-3

Conversion of BJT curve tracer to measure FET characteristics in (a) yields the drain characteristics in (b).

Forward Transconductance

Since gate voltage determines drain current in the saturation region ($V_{DS} > V_p$), we can measure the output-input relation between them as *forward transconductance* g_{fs}, g_m, or y_{fs}. For simplicity we shall use the symbol g_{fs}, and evaluate it graphically from the drain characteristics of Fig. 3-3b by

$$g_{fs} = \frac{\Delta I_{DS}}{\Delta V_{GS}}, \quad V_{DS} = \text{constant} \tag{3-3}$$

Forward transconductance has a *maximum* value at $V_{GS} = 0$ with the symbol g_{fo}. From Fig. 3-3b evaluate

$$|g_{fo}| \text{ at } V_{DS} = 10V, \text{ is } \frac{(1.15 - 0.75)\text{mA}}{(0 - 0.2)V} = 2000\mu\mho$$

An approximation may be used where

$$500\Omega = r_{ds(on)} \approx \frac{1}{g_{fo}} \; ; \; \frac{1}{500\Omega} = 2000 \times 10^{-6}\mho \qquad (3\text{-}4a)$$

A simpler method of measuring g_{fo} is to dial collector current on the vertical scale and base-emitter voltage on the horizontal scale of a BJT curve plotter to display the $I_D - V_{GS}$ transfer characteristic in Fig. 3-4a. Graphically measure its slope at $V_{GS} = 0$ and note that

$$g_{fs(max)} = g_{fo} = \frac{\Delta I_D}{\Delta V_{GS}} = \frac{I_{DSS}}{V_p/2} = \frac{2I_{DSS}}{V_p}, \text{ at } V_{GS} = 0 \qquad (3\text{-}4b)$$

Evaluating graphically, $g_{fo} = 1.15\text{mA}/0.57V = 2000\mu\mho$. Note also: (a) $V_{GS} = V_p$ where $I_D = 0$; (b) g_{fs} is the slope of the $I_D - V_{GS}$ curve measured at the operating point; and (c) g_{fs} decreases with decreasing I_D and increasing V_{DS}. Economical sweep circuits to display and measure the $I_D - V_{GS}$ and $I_D - V_{DS}$ characteristics with a CRO are shown in Figs. 3-4b and 3-4c respectively.

Using Manufacturer's Data

When characteristic curves are not available, values for I_{DSS} and V_p are obtainable from the manufacturer's data sheet. Drain current I_D for any value of gate voltage V_{GS} can be estimated in the *saturation region* ($V_{DS} \geq 2V_p$) from

$$I_D = I_{DSS}\left(1 - \frac{V_{GS}}{V_p}\right)^2 \qquad (3\text{-}5a)$$

or V_{GS} can be expressed in terms of I_D and I_{DSS} from

$$V_{GS} = V_p\left[1 - \sqrt{\frac{I_D}{I_{DSS}}}\right] \qquad (3\text{-}5b)$$

Transconductance g_{fs} can also be estimated from

$$g_{fs} = g_{fo}\left(1 - \frac{V_{GS}}{V_p}\right) \qquad (3\text{-}6)$$

where g_{fo} is found from Eq. (3-4b) as

$$g_{fo} = \frac{2I_{DSS}}{V_p} \qquad (3\text{-}4b)$$

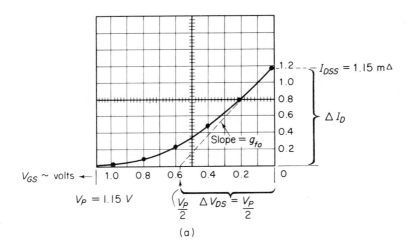

(a)

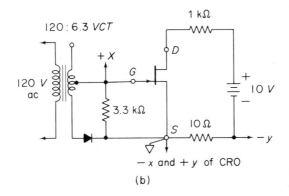

(b)

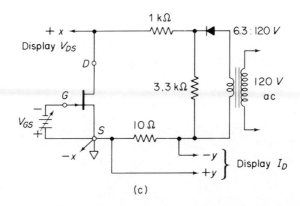

(c)

Figure 3-4

The $I_D - V_{GS}$ characteristic in (a) is measured by the sweep circuit of (b).
$I_D - V_{DS}$ curves are displayed with the sweep circuit of (c).

EXAMPLE 3-1:

Given $I_{DSS} = 1.15\text{mA}$, $V_p = 1.0V$, $g_{f_0} = 2000\mu\mho$. Find (a) I_D and (b) g_{f_s} when $V_{GS} = 0.6V$.

SOLUTION:

(a) From Eq. (3-5), $I_D = 1.15\text{mA}\,(1 - 0.6/1)^2 = 0.185\text{mA}$. Check Fig. 3-3b. The actual value of I_D at $V_{GS} = 0.6V$ is 0.2mA. (b) From Eq. (3-6)

$$g_{f_s} = 2{,}000\left(1 - \frac{0.6}{1}\right) = 800\mu\mho$$

3-3 BIASING THE JFET

The JFET must be turned on or conditioned to process a signal by a procedure known as biasing. Since we will be operating JFETs in the saturation region, we select an operating Q-point from inspection of the circuit's load line, drawn on the JFET's $I_D - V_{DS}$ drain characteristic. Normally V_{DS} should equal or exceed $2V_p$ to insure saturation. Then we modify the circuit to establish a reverse bias voltage V_{GSQ} that will cause the desired operating drain current I_{DQ}. Reverse bias expands the space charge regions to shrink the drain-source channel and *deplete* the channel of carriers. Thus JFETs are said to be biased or operated in the *depletion mode*.

Drain characteristic curves in Fig. 3-5a are shown for the JFET used in the circuit of Fig. 3-5b. (It is the same type as in Fig. 3-3b, to give you an idea of the variations to be expected among transistors.) If any drain current flows it will cause a voltage drop across *source-bias* resistor R_S. Gate resistor R_G is in the circuit to allow the dc voltage drop across R_S to be applied between gate and source with the proper polarity to make the gate negative with respect to the source. Since the gate is reverse-biased and will draw no current, we can write

$$V_{GS} = I_D R_S \tag{3-7}$$

Drawing the Load Line

If R_S equaled zero, V_{GS} would equal zero and the FET must operate along the $I_D - V_{DS}$ curve in Fig. 3-5a, where $V_{GS} = 0$. Now superimpose the circuit load line on the graph which is the loop equation for I_D where

$$V_{DD} = V_{DS} + I_D R_{dc} \tag{3-8a}$$

Supply voltage = FET drop + dc circuit resistance drops where $R_{dc} = R_L + R_S$.

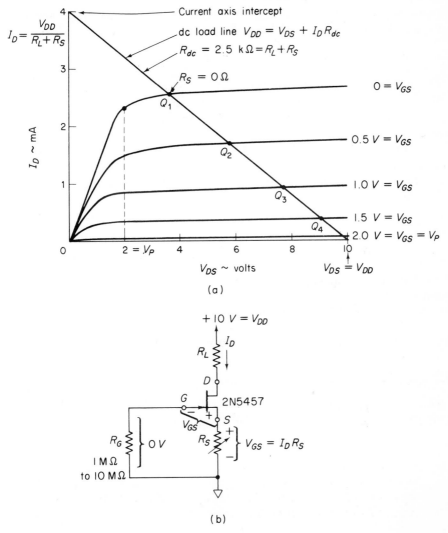

(a)

(b)

Figure 3-5

Operating points in (a) are determined by source bias resistor R_S together with V_{DD} and R_L in (b).

When the FET *acts* as a short circuit or $V_{DS} = 0$, drain current is limited by circuit voltage V_{DD} and circuit resistance R_{dc}. Substitute $V_{DS} = 0$ into Eq. (3-8a) to see that drain current would be

$$I_{Dmax} = \frac{V_{DD}}{R_{dc}}, \quad V_{DS} = 0 \tag{3-8b}$$

This point is plotted as the *current axis intercept* in Fig. 3-5a, corresponding to $V_{DS} = 0$, $I_{Dmax} = 10V/2.5k\Omega = 4mA$. When the FET acts as an *open* circuit or $I_D = 0$, V_{DS} is found from Eq. (3-8a) to be

$$V_{DS} = V_{DD}, I_{DS} = 0 \tag{3-8c}$$

This point is plotted as the *voltage axis intercept* in Fig. 3-5a. For any intermediate value of I_D, the FET must assume an operating point drain current and drain-source voltage located on the *dc load line* drawn between the two intercepts.

Now with $R_S = 0$ and $R_{dc} = R_L = 2.5k\Omega$, the FET must operate simultaneously on the $V_{GS} = 0$ curve and the load line. Their intersection at operating point Q_1 is the only point that will satisfy *both* FET (characteristics) and circuit (V_{DD} and R_{dc}) restrictions. Thus we would expect to measure $V_{DSQ} = 3.6V$ and $I_{DQ} = 2.5mA$ in the circuit of Fig. 3-5b when R_S and consequently $V_{GS} = 0$.

Calibrating Bias Lines

Suppose we wanted to operate at point Q_2 in Fig. 3-5a. Because I_D is held almost constant at 1.7mA for $V_{GS} = 0.5V$, we could choose an R_S to develop 0.5V when it carried 1.7mA from Eq. (3-7) or

$$R_S = \frac{V_{GS}}{I_D} = \frac{0.5V}{1.7mA} \cong 300\Omega$$

Installing $R_S = 300\Omega$ and $R_L = 2200\Omega$ (for $R_{dc} = 2500\Omega$) would give the same dc load line as in Fig. 3-5a, but set up a gate voltage of 0.5V. The FET would operate at point Q_2 with $I_{DQ} = 1.7mA$, $V_{DSQ} = 5.7V$.

EXAMPLE 3-2:

From the drain characteristics in Fig. 3-5a, I_D is practically constant in the saturation region. For example, $I_D \approx 0.9mA$ for $V_{GS} = 1.0V$; $I_D \approx 0.4mA$ for $V_{GS} = 1.5V$; and $I_D \approx 0.5mA$ for $V_{GS} = 2.0V$. Calculate values of bias resistor R_S to operate at each bias voltage.

SOLUTION:

From Eq. (3-7)

V_{GS}		I_D	$R_S = V_{GS}/I_D$
0	V	2.7mA $= I_{DSS}$	0Ω
0.5		1.7	300
1.0		0.9	1,100
1.5		0.4	3,750
2.0 $= V_p$		0.05	40,000

We conclude from Ex. 3-2 that each drain characteristic curve can be cali-brated with a bias resistance value for R_S. Figure 3-6 shows the calibrations and we illustrate their use in the next example.

EXAMPLE 3-3:

Find the dc operating point for the circuit of Fig. 3-6b.

SOLUTION:

Since $R_S = 300\Omega$, V_{GS} must equal $0.5V$ from Fig. 3-6a, and $I_D = 1.7$mA. Draw the dc load line from the dc loop between V_{DD} and ground

$$V_{DD} = I_D R_{dc} + V_{DS}$$

where $R_{dc} = R_L + R_S = $ total dc resistance carrying drain current. The dc load line runs from the I_D axis intercept at $V_{DS} = 0V$, $I_{Dmax} = V_{DD}/R_{dc}$ $= 10V/2.5$k$\Omega = 4$mA to the V_{DS} axis intercept at $I_D = 0$, $V_{DS} = V_{DD}$ $= 10V$.

Operating point Q is located at the intersection of dc load line and gate bias line in Fig. 3-6a.

EXAMPLE 3-4:

Find the ac output voltage swing V_0 and gain in the circuit of Fig. 3-6b.

SOLUTION:

E_i is in series aiding with gate bias voltage V_{DS}. Since voltage *across* C_S will remain constant at $0.5V$, E_i will make the gate negative with respect to the source by $1.0V$ when E_i goes negative to its $0.5V$ peak, as shown in Fig. 3-6b. This lowers drain current to about 0.9mA. But the *change* in drain current develops a change in voltage only across R_L (not R_S), because the current change is shunted around R_S by the low reactance of C_S. This change in voltage is also developed across drain and source. Capacitor C_0 passes the voltage change to cause signal output voltage V_0. The change in E_i from 0 to $-0.5V$ moves the operating point from Q to B along the ac load line, R_L, in Fig. 3-6a.

When E_i goes positive to $0.5V$, V_{GS} goes to zero from point Q to A along the ac load line. Drain current changes from 1.7mA to 2.5mA and V_{DS} goes negative from $5.8V$ to $4.0V$. Voltage gain is measured from points A and B as

$$A_V = \frac{\Delta V_0}{\Delta E_i} = \frac{(4-6.6)V}{[0.5-(-0.5)]V} = \frac{-2.6}{1} = -2.6$$

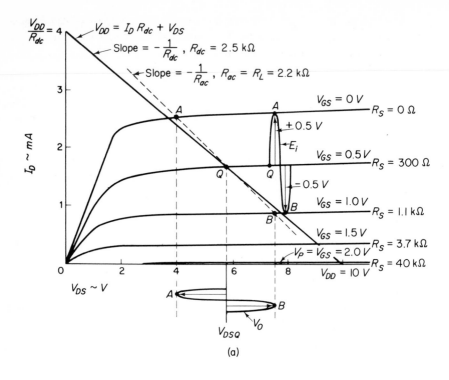

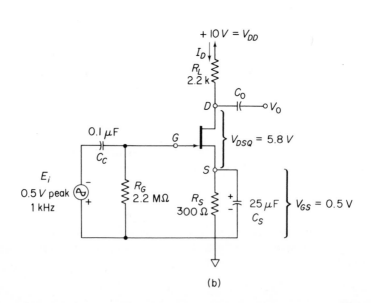

(b)

Figure 3-6

Graphical analysis of the circuit in (b) is performed in (a).

3-4 SIMPLIFIED JFET BIASING

$I_D - V_{GS}$ characteristics are shown in Fig. 3-7 for two of the same type JFETs to illustrate normal differences between transistors. JFETs with large I_{DSS} tend to have higher maximum transconductance. However, a large I_{DSS} requires a larger V_p to pinch off the channel, so transistors of the same number have roughly equal g_{f_o} as predicted by Eq. (3-4b). From the geometry of Fig.

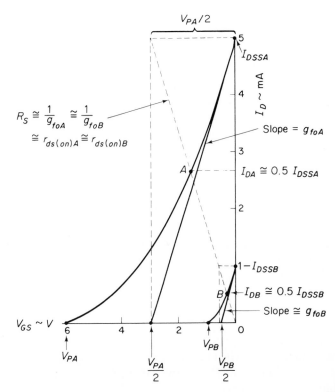

Figure 3-7

For $R_S \cong 1/g_{fo}$, $V_{GS} \cong 0.3V_p$, $I_D \cong 0.5I_{DSS}$, and $g_{fs} \cong 0.7\,g_{fo}$.

3-7, a bias line for $R_S \approx 1/g_{f_o}$ is drawn from origin to the point $V_{PA}/2, I_{DSSA}$. The bias line almost goes through the point $V_{PB}/2, I_{DSSB}$. Significance of this drawing is brought out in the next example.

EXAMPLE 3-5:

When $I_D = I_{DSS}/2$, what is the value of (a) V_{GS} and (b) g_{f_s}? (c) What value of R_S is required to make $I_D \approx I_{DSS}/2$?

SOLUTION:

(a) From Eq. (3-5a), when $I_D = I_{DSS}/2 = I_{DSS}(1 - V_{GS}/V_p)^2$, solve for $V_{GS} = 0.3V_p$, after cancelling I_{DSS}.

(b) When $V_{GS} = 0.3V_p$, $(1 - V_{GS}/V_p) = 0.7$, and from Eq. (3-6), $g_{fs} = 0.7g_{fo}$.

(c) Substituting for $I_D = I_{DSS}/2$ and $V_{GS} = 0.3V_p$ in Eq. (3-7) gives

$$R_S = \frac{0.6V_p}{I_{DSS}}, \text{ but from Eqs. (3-4a) and (3-4b)}$$

$$R_S = \frac{1.2}{g_{fo}} \approx 1.2r_{ds(on)}$$

We conclude from Fig. 3-7 and Example 3-5 that choosing $R_S \approx r_{ds(on)} \cong 1/g_{fo}$ will bias the JFET to a gate-source voltage of about $0.3V_p$, with $I_D = I_{DSS}/2$. Transconductance g_m will equal about 0.7 g_{fo}. We also conclude from Fig. 3-7 that while transconductance may remain fairly constant as we change transistors, drain current will change drastically (from 2.5mA to 0.5mA). A technique of holding drain current reasonably constant is introduced in the next section.

3-5 CONSTANT-CURRENT BIASING FOR JFETS

Transistor Q_A has an I_{DSS} about five times that of Q_B in Fig. 3-8. Suppose we had to restrict drain bias-current change to 0.3mA when transistors are installed, in the same circuit, that have characteristics between the extremes shown in Fig. 3-8a. One solution would be to choose a large $R_S = 10k\Omega$ and operate at low drain currents between limits represented by A' and B'. However, transconductance would be very low at these low drain currents.

Higher drain currents and, consequently, higher transconductance can be obtained through raising the bias line with an external *forward*-bias voltage V_{GG}, as shown in Fig. 3-8b and c. From the gate-source loop equation, V_{GG} changes the bias line equation to

$$V_{GS} = I_D R_S - V_{GG} \tag{3-9}$$

where R_S is found from the geometry of Fig. 3-8a to be

$$R_S = \frac{\Delta V_{GS}}{\Delta I_D} = \frac{(3.3 - 0.3)V}{(1 - 0.7)mA} = \frac{3V}{0.3mA} = 10k\Omega \tag{3-10}$$

V_{GG} is found either from point A and Eq. (3-9) as

$$V_{GG} = I_{DA}R_S - V_{GSA} = (1.0mA)(10k\Omega) - 3.3V = 6.7V$$

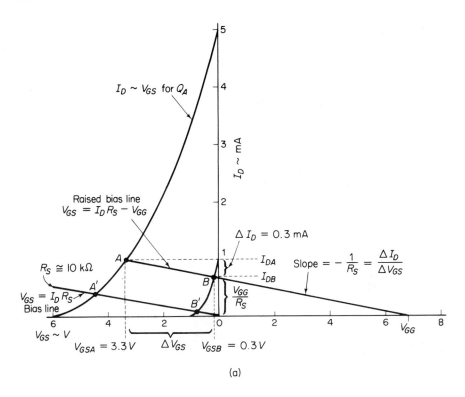

(a)

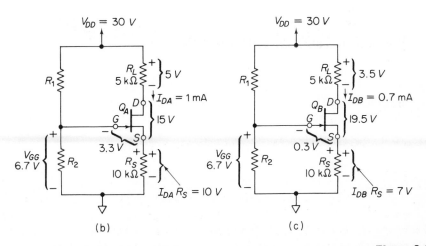

(b)   (c)

Figure 3-8

Resistors R_1 and R_2 in (b) and (c) stabilize drain bias current for different
JFETs as shown in (a).

or from point B as

$$V_{GG} = I_{DB}R_S - V_{GSB} = (0.7\text{mA})(10\text{k}\Omega) - 0.3V = 6.7V$$

EXAMPLE 3-6:

Choose a supply voltage of 20, 25, or 30V for V_{DD} in Fig. 3-8 to insure both JFETs are biased into the saturation region.

SOLUTION:

Voltage drop across R_L is 5V for Q_A in Fig. 3-8b, and 10V for R_S when $I_D = 1\text{mA}$. Allow $2V_{pA} = 2 \times 6V = 12V$ minimum for V_{DSA} to insure saturation region operation. V_{DD} must supply $V_{R_L} = 5V$ plus $V_{R_S} = 10V$ plus $V_{DS} = 12V$ or 27V. Pick $V_{DD} = 30V$. Dc levels are shown for Q_B in Fig. 3-8b, where $V_{DS} = 19.5V$, $2V_{pB} = 2V$.

R_1 and R_2 divide V_{DD} to establish V_{GG} (since the gate draws no current) from

$$V_{GG} = \frac{R_2}{R_1 + R_2} V_{DD} \tag{3-11a}$$

Any ac signal applied to the gate terminal will see R_1 in parallel with R_2 or an effective input resistance R_{GG} where

$$R_{GG} = R_1 \| R_2 \tag{3-11b}$$

EXAMPLE 3-7:

Pick $R_2 = 2\text{M}\Omega$ in Fig. 3-8b to find (a) R_1 and (b) R_{GG}.

SOLUTION:

From Eq. (3-11a)

$$6.7V = \frac{2\text{M}\Omega}{R_1 + 2\text{M}\Omega}(30V)$$

and $R_1 \approx 7\text{M}\Omega$.
(b) From Eq. (3-11b)

$$R_{GG} = 7\text{M}\Omega \| 2\text{M}\Omega \approx 1.5\text{M}\Omega$$

See Fig. 3-9a.

Large-value resistors may change values drastically when exposed to temperature variations, and may also be noisy. Thus, if we needed $R_{GG} = 2\text{M}\Omega$ in Example 3-7, choose $R_1 \approx 9\text{M}\Omega$ and $R_2 \approx 2.8\text{M}\Omega$. A better circuit is shown in Fig. 3-9b, where divider R_1 and R_2 are lower-valued resistors, and series resistor R_G is added to obtain the necessary input resistance of 2$\text{M}\Omega$.

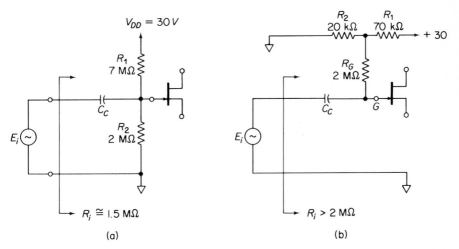

(a)   (b)

Figure 3-9

Resistor R_G in (b) eliminates need for the 7 megohm resistor in (a) and increases ac input resistance.

3-6 VOLTAGE GAINS AND RESISTANCE LEVELS—COMMON-SOURCE AMPLIFIER

In the common-source amplifier of Fig. 3-10a, we assume the JFET is biased properly and dc drain current I_D sets up a dc operating point that insures operation in the saturation region ($V_{DS} < 2V_p$). The JFET is now ready to process an ac signal, represented by E_i. Assume E_i is about 1kHz and reactances of circuit capacitors C_C, C_S, and C_0 are negligible with respect to circuit resistors. That is, treat these coupling (C_C, C_0) and by-pass (C_S) capacitors as ac short circuits. Supply voltage V_{DD} has a low ac impedance and so presents a short circuit to the flow of signal (ac) currents.

Ac Model

Fig. 3-10b shows an ac model that explains how the JFET processes low-frequency signal voltages. Since the gate should never conduct current, the model shows an open circuit between gate and source terminals. When a change in a signal voltage V is applied between gate and source to *reduce* the dc gate bias, increases in drain current will occur. The *increase* in drain current was in response to the change in signal voltage and is considered separately as the ac drain current. Ac drain current is shown as $g_{fs}V$. In Fig. 3-10b, when the gate goes positive, $g_{fs}V$ flows from drain to source within the JFET model. This holds for both n-channel and p-channel JFETs.

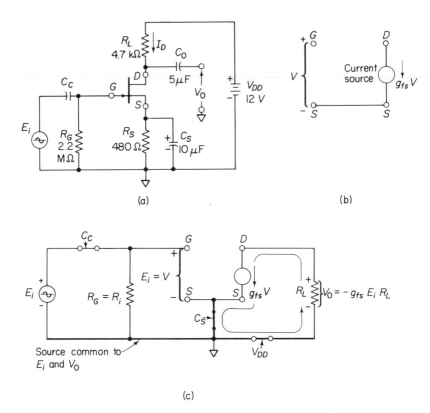

(a)   (b)

(c)

Figure 3-10

The JFET in (a) is modeled in (b) to show ac signal operation in (c).

Input Resistance

To analyze ac performance of a JFET in a circuit, redraw the circuit schematic by replacing the JFET with its ac model and both circuit capacitors and power supplies with their (hopefully) zero ac impedance. The procedure is illustrated in Fig. 3-10c.

By inspection of Fig. 3-10c, the ac input resistance presented to input signal voltage is equal to bias resistor R_G and symbolized by R_i.

$$R_i = R_G \qquad (3\text{-}12)$$

Voltage Gain

To find voltage gain, observe that $E_i = V$. Output voltage V_0 goes negative with respect to the source terminal when E_i goes positive with respect to the source terminal. Thus V_0 is 180° out of phase with E_i and the source terminal is common to both input and output. For this reason the circuit is classified as a common-source amplifier.

Ac output voltage V_0, in Fig. 3-10c, equals the product of ac drain current $g_{fs}V$ and R_L. But, since $V \approx E_i$, we may express V_0 directly in terms of E_i as

$$V_0 = -g_{fs}V R_L = -g_{fs}E_iR_L \qquad (3\text{-}13a)$$

and

$$A_V = \frac{V_0}{E_i} = -g_{fs}R_L \qquad (3\text{-}13b)$$

EXAMPLE 3-8:

Given: $I_{DSS} = 2.5$mA, $V_p = 2.0V$ for the JFET in Fig. 3-10a. Find voltage gain when (a) $R_S = 0$ and (b) R_S is adjusted for $I_D = I_{DSS}/2$.

SOLUTION:

(a) With $R_S = 0$, $V_{GS} = 0$ and transconductance g_{fs} is maximum at g_{fo}. From Eq. (3-4b)

$$g_{fo} = \frac{2I_{DSS}}{V_p} = \frac{(2)(2.5\text{mA})}{2V} = 2500\mu\mho.$$

From Eq. (3-13b)

$$A_V = -g_{fs} R_L = -(2500\mu\mho)(4.7\text{k}\Omega) = -11.7$$

(b) From Eq. (3-5b)

$$V_{GS} = V_p\left(1 - \sqrt{\frac{I_D}{I_{DSS}}}\right) = 2(1 - \sqrt{0.5}) = 0.6V$$

From Eq. (3-6)

$$g_{fs} = 2500\mu\mho\left(1 - \frac{0.6}{2}\right) = 1750\mu\mho$$

Finally, from Eq. (3-13b)

$$A_V = -(1750\mu\mho)(4.7\text{k}\Omega) = -8.2$$

Incidentally, $R_S = V_{GS}/I_D = 0.6V/1.25\text{mA} \cong 480\Omega$.

We conclude from Example 3-8 that voltage gains of the JFET are lower than those of the BJT for the same size load resistor. Also, JFET voltage gain goes down as operating point drain current, and consequently g_{fs}, is decreased. Zero gain occurs when $I_D = 0$ at pinch-off where $V_{GS} = V_p = 2.0V$. Any control device whose resistance varies with environmental changes can be inserted as a source resistor and vary gain. For example, thermistors, light-sensitive resistors (LSR), and strain gauges change resistance when subjected to variations in temperature, light level, and pressure, respectively. Resistance of an LSR decreases with increasing light level. An LSR source resistor would then lower V_{GS}, and raise I_D, g_{fs}, and gain, with increasing light level.

3-7 COMMON-DRAIN AMPLIFIER

A p-channel JFET is used in the circuit of Fig. 3-11a. Signal output voltage V_0 is developed across a load represented by source resistor R_S. For large values of R_S, V_{GS} would be close to pinch-off and g_{fs} would be low because of the resulting low I_D. Divider R_1 and R_2 provide a forward-bias voltage to offset some of the reverse bias due to $I_D R_S$. A variable R_2 allows adjustment of I_D to the desired dc operating point.

Input Resistance

From the ac circuit model in Fig. 3-11b, ac input resistance R_i is established primarily by R_G and is about 2.2MΩ. It is impractical to add the series resistance of $R_1 \| R_2$(50kΩmax) except for clarity.

Voltage Gain

Signal gate-source voltage V does *not* equal E_i. E_i furnishes V and output voltage V_0 where $V_0 = g_{fs}VR_S$, or

$$E_i = V + g_{fs}VR_S = V(1 + g_{fs} R_S) \tag{3-14a}$$

and

$$V = \frac{E_i}{1 + g_{fs}R_S} \tag{3-14b}$$

To find V_0 in terms of E_i, substitute for V from Eq. (3-14b) into

$$V_0 = g_{fs}VR_S = \frac{g_{fs}R_S E_i}{1 + g_{fs}R_S} \tag{3-14c}$$

If a load resistor R_L were inserted in the drain lead and V_0 taken from drain to ground, V_0 would equal $-g_{fs}VR_L$. Voltage gain of this common-source amplifier, with unbypassed source resistance would be

$$\frac{V_0}{E_i} = -\frac{g_{fs}R_L}{1 + g_{fs}R_S} \tag{3-14d}$$

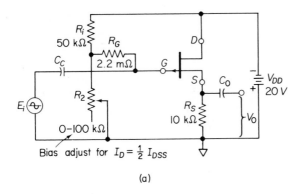

(a)

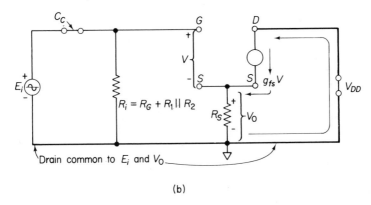

(b)

Figure 3-11

Circuit and ac model of common-drain amplifier.

EXAMPLE 3-9:

R_2 is adjusted for $I_D = I_{DSS}/2 = 1.25\text{mA}$ in Fig. 3-11a. Using the same transistor in Example 3-8, $g_{fs} = 1750\mu\text{℧}$ and $V_{GS} = 0.6V$. What is the (a) voltage gain, (b) dc voltage at the source terminal, and (c) dc voltage at the gate terminal?

SOLUTION:

(a) For Eq. (3-14c), $g_{fs}R_S = (1750\mu\text{℧})(10\text{k}\Omega) = 17.5$

$$A_V = \frac{V_0}{E_i} = \frac{g_{fs}R_S}{1 + g_{fs}R_S} = \frac{17.5}{1 + 17.5} \approx 1$$

(b) $V_S = I_D R_S = (1.25\text{mA})(10\text{k}\Omega) = 12.5V$ reverse bias.

(c) From Eq. (3-9), $V_{\text{Gate}} = V_{GG} = I_D R_S - V_{GS} = 12.5V - 0.6V = 11.9V$ forward bias. Therefore, dc gate to source voltage is $V_{GS} = 12.5V - 11.9V = 0.6V$.

Output Resistance

To find the resistance seen by looking into a source terminal, we make the test set up in Fig. 3-12. E is an ac test voltage. Signal generator E_i is turned down to zero volume and left in the circuit (source resistor R_S is removed for simplicity). With E_i replaced by its zero internal ac impedance, we see that E_i is developed between gate and source, causing current $g_{fs}E$ to flow as shown in Fig. 3-12. From Ohm's law, the ratio of total current I drawn from a generator E is the resistance seen by E, or

$$R_0 = \frac{E}{I} = \frac{E}{g_{fs}E} = \frac{1}{g_{fs}} \tag{3-15}$$

Since the source terminal is the output terminal, we can assign the output resistance symbol R_0 to resistance seen looking into the source terminal.

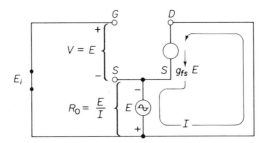

Figure 3-12

Common-drain output resistance.

EXAMPLE 3-10:

Find the resistance seen looking between source terminal and ground in the circuit of Ex. 3-9.

SOLUTION:

From Eq. (3-15)

$$R_0 = \frac{1}{g_{fs}} = \frac{1}{1750\mu\mho} = 560\Omega$$

Note that even including R_S in parallel with R_0 would not affect the result, since $10k\Omega \| 560\Omega \approx 560\Omega$.

3-8 LOW-FREQUENCY RESPONSE

If R_S is by-passed, the low-frequency cutoff f_L is established by C_S from

$$f_{LS} = \frac{1}{2\pi C_S(R_0 \| R_S)} \tag{3-16}$$

If R_S is *not* by-passed, coupling capacitor C_C sets the low-frequency cutoff at

$$f_{LC} = \frac{1}{2\pi C_C R_i} \tag{3-17}$$

Where both C_S and C_C are present in a circuit, the desired low-frequency cutoff should be set by C_S from Eq. (3-16). Then calculate a value for C_C for the same frequency from Eq. (3-17) and increase the value of C_C by a factor of 10. (The same general procedure is used for BJTs in section 4-2.)

EXAMPLE 3-11:
In Fig. 3-10a and Example 3-8b, $R_0 = 560\Omega$. Find the value of (a) C_S for a low-frequency response down to 100 Hz, and (b) required value of C_C.

SOLUTION:
(a) From Eq. (3-16)

$$C_S = \frac{1}{2\pi f_{LS}(R_0 \| R_S)} = \frac{10^6}{2\pi(100)(480 \| 560)} \simeq 6.2\mu F$$

(b) From Eq. (3-17)

$$C_C = \frac{1}{2\pi f_{LC} R_i} = \frac{1}{2\pi(100)2.2 \times 10^6} = 0.0007\mu F$$

Choose $C_S = 10\mu F$, $C_C = 10 \times 0.0007\mu F = 0.007\mu F \simeq 0.005\mu F$.

3-9 CONSTANT-CURRENT APPLICATIONS OF THE JFET

Constant Current Generator

A two-terminal, constant-current regulating diode (CRD) evolved naturally from the JFET. With gate shorted internally to source, the resulting diode would pass a constant current of I_{DSS}. By installing a source resistor as in Fig. 3-13a, current would be regulated to some constant value between I_{DSS} and 0, provided voltage across terminals AA' is maintained above $2V_p$. In data sheets regulator current is specified by the symbol I_p and is available from 0.1 to 10mA. Limiting voltage $2V_p$ is specified by the symbol V_L and is typically 1 to 3 volts.

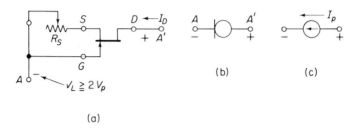

(a)

(b)   (c)

Figure 3-13

Construction (a), symbol (b), and circuit model (c) of a current-regulating diode CRD.

Low Reference Voltage

In Fig. 3-14a the CRD and R_L form a low reference-voltage that is below commonly available zener diodes. Load current through R_L in Fig. 3-14b remains constant as long as V_0 is below E by an amount necessary to hold the CRD in its saturation region Zener voltage will be extremely stable in Fig. 3-14c since zener current remains constant. One CRD and one capacitor form a linear ramp generator in Fig. 3-14d. When E is positive in Fig. 3-14e, during time interval O-A, D_A charges C at a constant rate to a peak voltage of

$$V_{0p} = I_{DA} \times \frac{T}{2C} = \frac{I_{DA}}{2fC} \qquad (3\text{-}18)$$

where $T =$ sine wave period in seconds, $f =$ frequency in Hertz, I_{DA} in μA, C in μF, and V in volts. During time interval A-B, D_B discharges C, and if $I_{DA} = I_{DB}$, V_{0p} will linearly decrease to zero. In practice make R_S (see Fig. 3-13a) of D_A a potentiometer equal to 3 times R_S of D_B. Then adjust R_S of D_A for a symmetrical wave.

3-10 FET SWITCH AND CHOPPER APPLICATIONS

A chopper device periodically breaks the connection between a signal source and load. One basic chopper application is to periodically connect and disconnect a low-level dc voltage to a load resistor. The resultant chopped, square wave can be amplified by a precise amount with a high gain ac amplifier, and rectified to obtain a higher dc voltage, precisely related to the original low-level dc signal. This technique circumvents the use of dc amplifiers with their inherent susceptibility to drift and gain changes.

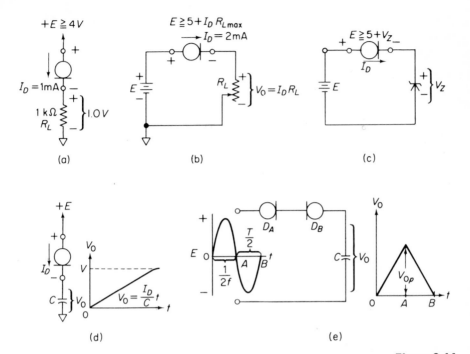

Figure 3-14

Applications for a CRD.

Series Chopper

The FET will perform as an excellent chopper if it is used in its ohmic region. This means V_{DS} should be restricted to less than $0.25V_p$. The most important FET electrical characteristic is drain-source resistance r_{ds}, which can be approximated in the ohmic region only from

$$r_{ds} \cong r_{ds(on)}\frac{V_p}{(V_p - V_{GS})}, \quad V_{DS} < V_p \tag{3-19}$$

Chopper voltage V_{GS} is actually a negative going square wave or rectified sine wave, but for clarity it is represented by a switch in Fig. 3-15(b). With $V_{GS} = 0$, or gate grounded and $E = 0.06V$, the operating point is located at point A on the 3kΩ load line drawn from $V_{DS} = E = 0.06V$. E divides between $r_{ds(on)}$ and $R_L + R$ to give an output voltage of

$$V_0 = E\frac{R_L}{R_L + R + r_{ds(on)}} = (0.06V)\frac{2700\Omega}{3250\Omega} \approx 0.05V \tag{3-20}$$

With $V_{GS} = 3V$, reverse-bias, the FET is pinched off, $V_{DS} = E$, and $V_0 = 0V$, as indicated by operating point B in Fig. 3-15a. When input E is reduced to

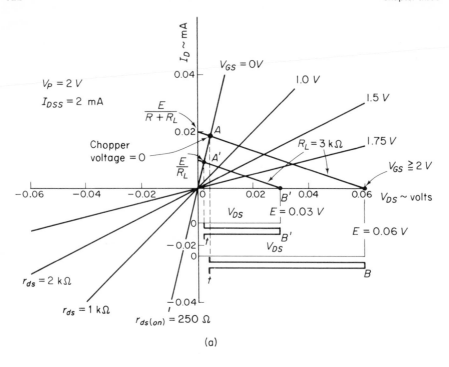

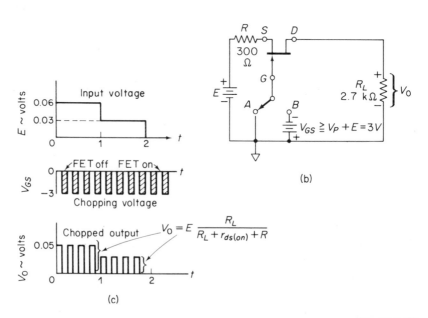

Figure 3-15

Ohmic characteristics in (a) are used in the chopper circuit of (b) to chop a
dc input voltage as in (c).

$0.03V$, operating points A' and B' locate on and off states of the FET respectively. As shown in Fig. 3-15c, output voltage V_0 is a chopped version of E and the peak value of V_0 is directly proportional to E. The gate should never be forward-biased. For improved chopper operation $r_{ds(on)}$ should be small and R_L should be large, at least ten times $r_{ds(on)}$. Since the JFET is in series with load and input voltage, Fig. 3-15 is classified as a series chopper.

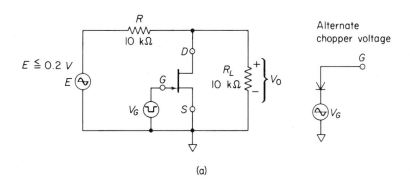

(a)

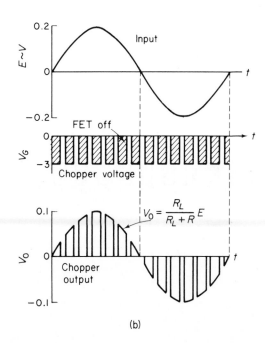

(b)

Figure 3-16

The shunt chopper in (a) generates voltage shapes in (b).

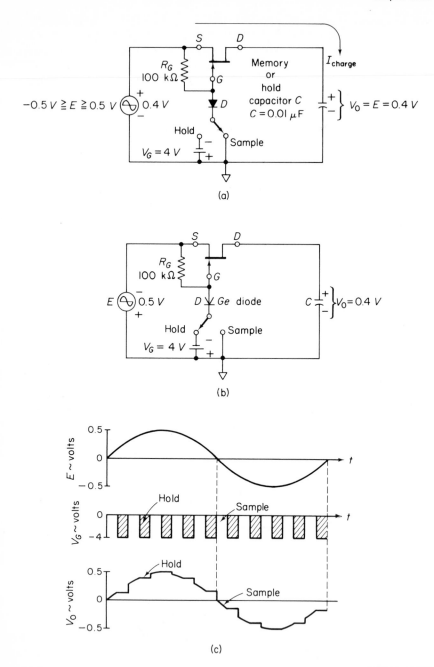

(a)

(b)

(c)

Figure 3-17

Sample-and-hold curcuit.

A shunt chopper circuit is given in Fig. 3-16a, where the FET is in parallel with load R_L. When $V_{GS} = 0$, $r_{ds(on)}$ parallels R_L and V_0 is negligible at

$$V_0 = \frac{r_{ds(on)}}{R} E = \frac{250\Omega}{10k\Omega} E = 0.02E, V_{GS} = 0 \qquad (3\text{-}21a)$$

For $V_{GS} \geq V_p$, $r_{ds} = \infty$, and E divides between R and R_L as

$$V_0 = \frac{R_L}{R + R_L} E = 0.5E, V_{GS} \geq V_p \qquad (3\text{-}21b)$$

Input voltage E may be a sine wave, or any other wave shape.

An analog voltage is one that varies with time and may be a transducer output that represents continuous measurement of pressure, temperature, volume, speed, or distance. Such information can be processed in a digital computer not only for record-keeping but also to initiate control signals to control, for example, fluid flow, material thickness, or a variety of manufacturing and maintenance processes.

The analog signal must be sampled piece-by-piece, and each piece must be held long enough for a conversion circuit to convert the analog signal to a digital signal for inputing to the digital computer. By modifying the chopper slightly, we have a sample-and-hold circuit. In Fig. 3-17a, $V_{GS} = 0$ and E is connected through r_{ds} of the JFET to *hold-capacitor* C. When E tries to make the source terminal negative with respect to the gate, diode D prevents the gate from becoming forward biased. V_0 charges to and follows E during this sample interval. In Fig. 3-17b, V_{GS} reverse-biases the gate and turns off the FET to disconnect E, and C holds its voltage during the hold interval. Waveshapes for input, output, and gate-control voltage V_{GS} are shown in Fig. 3-17.

3-11 MULTIPLEXING WITH JFETS

A simple but excellent circuit to demonstrate multiplexing can be built as in Fig. 3-18. Chopper voltage E_C turns off diodes D_2 during its negative half-cycle for approximately 5ms. (Frequency of E_C is 100Hz.) E_2 is a continuous 2kHz signal at $0.1V$ peak, so that about 10 cycles of E_2 are sent over the common communications link through Q_{2S} and Q_{2R} to R_{L2}, because $V_{GS} \cong 0$ for both

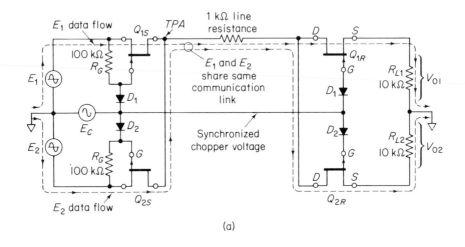

(a)

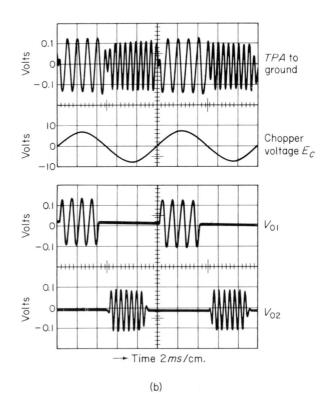

(b)

Figure 3-18

Multiplexer in (a) has waveshapes in (b).

JFETs. During this same interval $(-E_C)$, diodes D_1 are forward biased to pinch-off both Q_{1S} and Q_{1R}. Thus E_1 and R_{L1} are disconnected from the communications link.

During the positive half-cycle of E_C, Q_{2S} and Q_{2R} are pinched-off, disconnecting E_2 and R_{L2} from the communications link. $V_{GS} \approx 0$ for both Q_{1S} and Q_{1R}, so the 1,000Hz E_1 signal is connected for 5ms through the link to R_{L1}. In summary, Q_{1S} sends 5 cycles at 1,000Hz down the link, followed by 10 cycles at 2,000Hz from Q_{2S}. At the receiving end, Q_{2R} passes the 10 cycles of 2000Hz from E_2 to V_{02} and then blocks the following signals from E_1. The sequence of signals on the line are shown at test point A or TPA in Fig. 3-18b.

In Fig. 3-19a the chopper frequency is set at 1kHz. E_1 is a 200Hz signal and E_2 is a 100Hz signal. The voltage waveform from TPA to ground shows how Q_{1S} and Q_{2S} alternately connect E_1 and E_2 to the communication link for 1ms

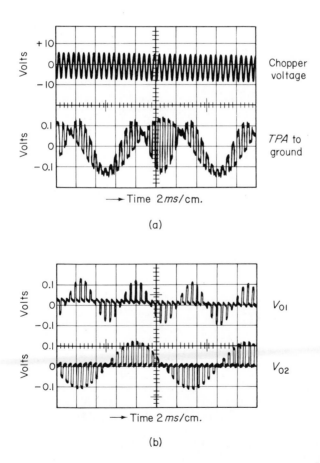

Figure 3-19

Multiplexer with chopper frequency larger than signaling frequencies.

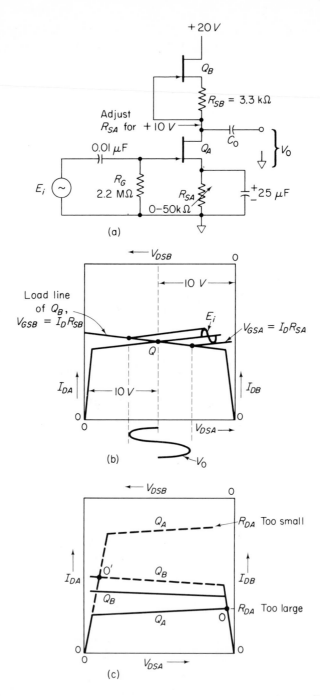

Figure 3-20

High-gain cascode amplifier.

intervals. In Fig. 3-19b we see how V_{01} and V_{02} have been separated, and are chopped versions of E_1 and E_2 respectively. We conclude that multiplexing is a time-sharing arrangement whereby time on a single communications link is shared by more than one transmitter-receiver.

3-12 SELECTED JFET APPLICATIONS

Cascode Amplifier

JFET Q_B is the ac load for Q_A in the unusual application of Fig. 3-20a. Bias resistor R_{S_A} is adjusted to divide the 20V dc supply voltage equally between each transistor-R_S combination. Operating point Q is located in Fig. 3-20b at the intersection of the Q_B load line and Q_A operating curve. Since the slope of the Q_B load line is practically flat, it represents a very high ac load, and voltage gain will be high. Expect gains between 30 and 500, depending on device characteristics. In Fig. 3-20c effects of maladjusting R_{S_A} are shown.

FET-BJT Amplifier

A *pnp* transistor provides current gain for the *n*-channel JFET in Fig. 3-21. Resistor R_1 is adjusted to set the collector terminal at $1/2 \, V_{DD}$. Most of signal (ac) drain current $g_{f_s}V$ enters the *pnp*'s base so that signal current through R_E

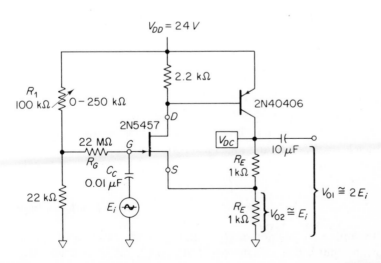

Figure 3-21

FET-BJT amplifier.

and R_F equals $\beta\, g_{f_s} V$ and $V_{01} = \beta\, g_{f_s} V (R_F + R_E)$. But E_i equals $\beta\, g_{f_s} R_E$, so voltage gains are

$$\frac{V_{01}}{E_i} \cong \frac{R_F + R_E}{R_E} = 2 \tag{3-22a}$$

and

$$\frac{V_{02}}{E_i}\frac{R_E}{R_E} = 1 \tag{3-22b}$$

By short-circuiting R_F and taking the output voltage across R_E, we have a source follower with (a) current gain approximately $\beta\, g_{f_s} V$, (b) output resistance $R_0 \cong 1/(\beta\, g_{f_s})$, and (c) voltage gain approximately 1. Input resistance depends on R_G and is very high; for Fig. 3-21, it is 22MΩ.

Multiple Input—Single Output

Isolating properties of the JFET are employed in the audio-mixer circuit of Fig. 3-22a. Bias resistors R_S of Q_1 and Q_2 are selected experimentally to set the drain terminals of Q_1 and Q_2 at $V_{DD}/2$ to insure both JFETs are in their saturated region. A typical voltage gain of about 3 is obtained between gate and drain of each FET. Individual volume controls R_1 and R_2 allow independent adjustment of signal levels from inputs E_1 and E_2. Q_3 is a common-drain amplifier whose stage gain may be adjusted by master-volume control R_3. Other input stages may be added as required. For example, a 4-input mixer could handle four microphones and allow you to: (1) boost the volume of a weak lead singer—MIC #1; (2) cut down the volume of a superloud guitar—MIC #2; (3) bring up the piano on his solo and fade him back into the background afterwards—MIC #3; (4) mix in a rhythm singing group or special effects—MIC #4; and (5) shut them all down with the master-volume control when the performance gets out of hand.

Signal Adding

If the gain from E_1 to V_0 is adjusted to 1, and gain from E_2 to V_0 is adjusted to 1, then $V_0 = E_1 + E_2$ and Fig. 3-22a performs as an adder. A CRO connected to display V_0 will show the instantaneous sum of any two input signals.

Single Input—Multiple Output

By shifting the position of Q_2, as in Fig. 3-22b, we obtain a circuit that will accept an input signal and deliver it to two (or more) different outputs. For example, we might want to tape a speech at out #2 while driving the auditorium's sound system with out #1.

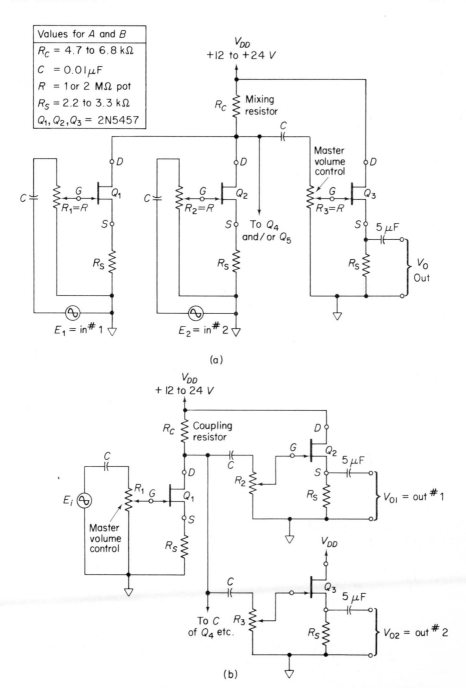

(a)

(b)

Figure 3-22

Audio-mixer or adder in (a) and single-input to multiple-outputs in (b).

Low Pass Filter

Another application of FETs is a low pass filter, shown in Fig. 3-23. Gate voltage V_G controls r_{ds} of the JFET. Input voltage E_i should be restricted to a low voltage to insure operation in the ohmic region. V_0 is expressed in terms of E_i and frequency

$$V_0 = \frac{E_i}{1 + f/f_c} \tag{3-23a}$$

where f is the frequency of E_i and

$$f_c = \frac{1}{2\pi r_{ds} C} \tag{3-23b}$$

For $V_{GS} = 0$, the JFET acts as a resistance equal to $r_{ds(on)}$ between drain and source. C by-passes high frequencies above f_c. Since V_{GS} controls r_{ds}, and r_{ds} determines f_c, we have a voltage-controlled filter. Its performance is analyzed in the following example.

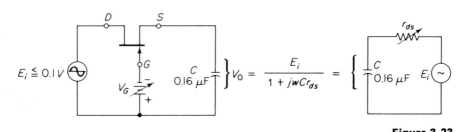

Figure 3-23

Voltage-controlled low-pass filter.

EXAMPLE 3-12:

If $r_{ds(on)} = 500\Omega$ and $V_p = 2.0V$, find the upper cutoff frequency in Fig. 3-23 for $V_{GS} = 0, 0.5, 1.0, 1.5$, and 1.75 volts.

SOLUTION:

From Eq. (3-19)

$$r_{ds} = (500\Omega)\frac{2V}{2V - V_{GS}}$$

Tabulating results:

V_{GS}	$0V$	$0.5V$	$1.0V$	$1.5V$	$1.75V$
r_{ds}	500Ω	660Ω	1000Ω	2000Ω	4000Ω

From Eq. (3-23b),

$$f_c = \frac{1}{2r_{ds}(0.16 \times 10^{-6})} = \frac{1 \times 10^6}{r_{ds}}$$

Tabulating results

V_{GS}	$0V$	$0.5V$	$1.0V$	$1.5V$	$1.75V$
f_c	2kHz	1.5kHz	1.0kHz	0.5kHz	0.25kHz

3-13 INTRODUCTION TO ENHANCEMENT-TYPE MOSFETS

The MOSFET (metal-oxide-silicon FET) or IGFET (insulated-gate FET) has an insulator (silicon dioxide) between the gate and other elements, rather than a *pn* junction like the JFET. The JFET is classified as type A (that is, it is operated in the depletion region), and may be subclassified as *n*- or *p*-channel. There are two basic types of MOSFET, type B and type C, according to whether or not a conducting channel is built between drain and source.

A type-C MOSFET is constructed as in Fig. 3-24a. With zero bias, no current flows between drain and source due to V_{DD} because of the *pn* junctions. Biasing the gate to attract electrons from the substrate *induces* an *n*-channel, as in Fig. 3-24b. Increasing the bias increases the number of available current carriers to *enhance* drain-current flow. Thus the enhancement-mode MOSFET is nonconducting with zero bias and can complement the depletion-mode JFET that has maximum conduction with zero bias.

Extremely high input resistance is one principal MOSFET advantage because of the insulating gate. The gate insulation can be considered to form a capacitor between substrate and gate metal. Diffusing *p*-type source and drain areas into an *n*-type substrate would form a type-*C*, *p*-channel, enhancement MOSFET. Note unlike the JFET the arrowhead is *not* placed on the circuit symbol's gate as illustrated in Fig. 3-24. The arrowhead is placed on the subbase and points toward the *n*-channel (induced or otherwise).

Threshold voltage $V_{GS(th)}$ is given on data sheets to indicate the minimum bias voltage required to induce a conducting channel. See Fig. 3-24. We deduce from characteristic curves in Fig. 3-24 that the same pinch-off mechanism limits drain current flow, with increasing V_{DS}, as in the JFET. Biasing an operating point for operation in the saturated region can be accomplished by adding divider resistors R_1 and R_2.

The gate should never be left floating with a MOSFET. Manufacturers install a brass ferrule to short-circuit leads. Otherwise even removing the MOSFET from its plastic package could build up a static gate voltage large enough to rupture gate insulation. Therefore, keep the leads shorted until installation and handle it by the case.

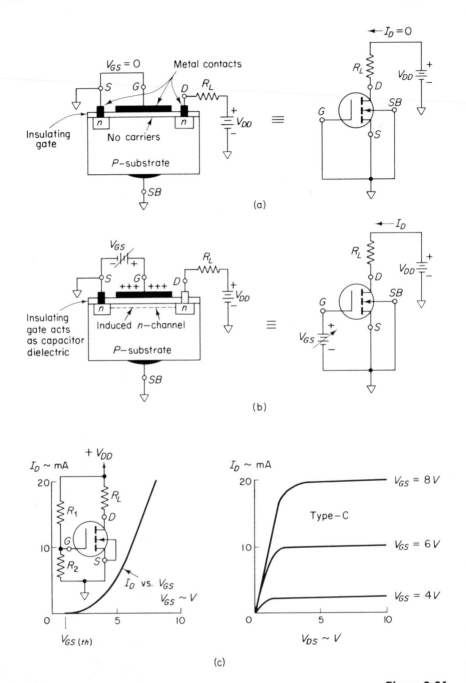

Figure 3-24

Enhancement type-C, n-channel MOSFET.

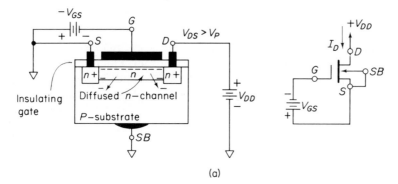

(a)

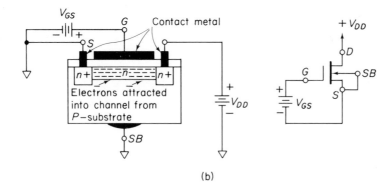

(b)

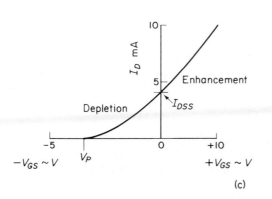

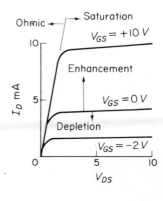

(c)

Figure 3-25

Depletion mode—type-B MOSFET.

3-14 INTRODUCTION TO DEPLETION-TYPE MOSFETS

A depletion-type B MOSFET in Fig. 3-25a is constructed with both an insulated gate and connecting channel between source and drain. Making gate voltage polarity (with respect to source) the same as the channel impurity type repels electrons out of the channel. This action depletes the channel of carriers and the MOSFET performs exactly like its n-channel JFET counterpart. When biased into the saturated region, as in Fig. 3-25c, and with negative gate voltage, the MOSFET is operated in its *depletion* mode.

As shown in Fig. 3-25, the insulating gate can also be biased (with a polarity opposite to that of the channel impurity type) to attract carriers into the diffused channel. This action enhances drain current conduction. Operation of the MOSFET is then similar to that of the type-C MOSFET.

Small-signal electrical characteristics of the MOSFET, g_{fs}, $r_{ds(on)}$, may be obtained graphically from characteristic curves or from data sheets. The small-signal model is identical with that of the JFET.

PROBLEMS

1. Typical values for the 3N128 n-channel JFET are given as $I_{DSS} = 15$mA. $V_p = 3.5V$. Calculate: (a) g_{fo}; (b) $r_{ds(on)}$.

2. Typical data sheet values for an n-channel JFET are given as $g_{fo} = 10,000$ μmhos, $I_{DSS} = 15$mA, $V_p = 3.0V$. Find: (a) I_D at $V_{GS} = 1V$; (b) g_{fs} at $V_{GS} = 1V$.

3. In Fig. 3-3b, $I_{DS} = 0.75$mA at $V_{GS} = 0.2V$, $I_{DS} = 0.440$mA at $V_{GS} = 0.4V$, $I_{DS} = 0.24$mA at $V_{GS} = 0.6V$, and $I_{DS} = 0.08$mA at $V_{GS} = 0.8V$. Calibrate the bias lines in terms of R_S. Refer to Example 3-2.

4. Given $R_S = 1.1$kΩ and $R_L = 3.9$kΩ Fig. 3-6. Locate the operating point.

5. Find ac output voltage swing V_0 (peak-to-peak) and gain for problem 4. Refer to Example 3-4.

6. Given $g_{fo} = 10,000\mu$mhos, $I_{DSS} = 15$mA, $V_p = 3.0V$. Use the simplified biasing procedure of Example 3-5 to: (a) choose $R_S \cong 1.2\, r_{ds(on)}$ and find (b) g_{fs} and (c) V_{GS} at $I_D = I_{DSS}/2$.

7. In Fig. 3-8 it is desired to restrict operating point current shift between 1.0 and 0.8mA. This corresponds to $V_{GSA} = 3.3V$ and $V_{GSB} = 0.2V$. Find required bias resistor R_S and bias voltage V_{GG}. Refer to Example 3-6.

8. Change R_1 in Fig. 3-9b to set $V_{GG} = 10.7V$.

9. If $V_p = 3.0V$, $I_{DSS} = 15$mA, and $g_{fo} = 10,000\mu$mhos for the JFET in Fig. 3-10a, find voltage gain: (a) when $R_S = 0$; (b) when R_S is adjusted for $I_D = I_{DSS}/2$.

10. In Fig. 3-11, R_2 is adjusted for $I_D = I_{DSS}/2 = 7.5\text{mA}$, but with $R_S = 1\text{k}\Omega$. g_{fs} is $7000\mu\text{mhos}$ and $V_{GS} = 0.9\text{V}$. (This is the JFET of problem 9). Find: (a) voltage gain; (b) dc voltage at the source terminal; (c) dc voltage at the gate terminal.

11. What resistance is seen between source terminal and ground in the circuit of problem 10?

12. In Fig. 3-11a, $R_i \cong 2.2\text{M}\Omega$. Pick C_C for a low-frequency cutoff of 30Hz.

13. What low-frequency cutoff is set by C_S in Fig. 3-10? Assume $R_0 \| R_s = 480 \| 560 = 250\Omega$.

14. The JFET with characteristics given in Fig. 3-6 is connected as a current regulating diode in Fig. 3-13. With $R_S = 1.1\text{k}\Omega$, what is the value of I_p?

15. Estimate the required R_S for the JFET of Fig. 3-6 to establish $I_D = 2\text{mA}$ in Fig. 3-14b.

16. Load resistor R_L is changed to 10kΩ in Fig. 3-15b. Find the peak value of V_0 for: (a) $E = 60\text{mV}$; (b) $E = 30\text{mV}$.

17. Load resistor R_L is changed to 40kΩ in Fig. 3-16a. For $E = 0.2\text{V}$, find V_0 for: (a) $V_{GS} = 0\text{V}$; (b) $V_{GS} \geq V_p$.

18. In Fig. 3-18, calculate V_{01} if $r_{ds(on)}$ of Q_{1s} and Q_{1R} equals 250Ω, and $E_i = 0.2\text{V}$.

19. $R_G = 10\text{M}\Omega$, $R_F = 5\text{k}\Omega$, and $R_E = 1\text{k}\Omega$ in Fig. 3-21. Find V_{01}/E_i and ac input resistance.

20. In Fig. 3-22, $R_1 = R_3 = 1\text{M}\Omega$, $R_C = 6.8\text{k}\Omega$, and $R_S = 3.3\text{k}\Omega$. Assume both wipers of R_1 and R_3 are adjusted to their upper limit so that all of E_1 is applied to the gate of Q_1 and all of the output voltage of Q_1 is applied to the gate of Q_3. Find voltage gain from: (a) E_1 to the gate of Q_3; (b) from gate of Q_3 to source of Q_3; (c) from E_1 to V_0. JFET characteristics are given in Fig. 3-6a, where $V_{GS} = 1.3\text{V}$, $I_D = 0.4\text{mA}$ at $R_s = 3.3\text{k}\Omega$, $I_{DSS} = 2.7\text{mA}$, and $V_p = 2\text{V}$.

chapter four
device limitations

4-0 INTRODUCTION

Measurement of device limitations is quite complicated and requires expensive, specialized equipment if we must push the device into performing at its outer limits. But if we are reasonable and have *no* desire to operate our devices on the edge of burnout, have a low-cutoff amplifier frequency of 1Hz driving a 12-inch cone (that will only reproduce down to 100Hz), and have a high-cutoff amplifier frequency of 25kHz when both tweeter and our ears ignore frequencies above 12kHz, then measurement of device limitations assumes less importance. We can more efficiently spend our time in learning how to use the devices, and accept manufacturers' typical data as applicable to our problem.

4-1 BJT LOW-FREQUENCY LIMITATIONS DUE TO
COUPLING CAPACITORS

In designing a BJT circuit to meet a low-frequency limitation, we have some design freedom because the circuit's external coupling and/or by-pass capacitor size control low-frequency cutoff. We employ the basic common-emitter amplifier of Fig. 4-1 to define low-frequency cutoff.

We begin measuring midfrequency voltage gain by setting the oscillator frequency at 1kHz. There are two methods of measuring gain—measure either the ratio of V_o to E_i or of V_o to E_g. A different low-cutoff frequency will be measured for each.

To measure E_g we must *remove* the transistor circuit and measure the signal generator's *open-circuit* signal output voltage with an ac VTVM for an rms measurement, or a CRO for a peak-to-peak measurement. When we reconnect the signal generator and transistor load, E_g divides between its internal resistance R_g and the circuit's input impedance consisting of C_c in series with R_{in}. (We can no longer measure E_g, only E_i.) At 1,000Hz assume C_c is an ac short circuit, so from Fig. 4-2(c)

$$V = E_g \frac{R_{in}}{R_g + R_{in}}; \; R_{in} = R_B \; // \; r_\pi \qquad (4\text{-}1a)$$

Gain from V to V_o is

$$\frac{V_o}{V} = - g_m R_{ac}; \; R_{ac} = R_c \; // \; R_L \qquad (4\text{-}1b)$$

so gain from E_g to V_o is the product of Eqs. (4-1a) and (4-1b), or

$$\frac{V_o}{E_g} = \frac{V}{E_g} \times \frac{V_o}{V} = - \frac{R_{in}}{R_g + R_{in}} g_m R_{ac} \qquad (4\text{-}1c)$$

As frequency of E_g is lowered, reactances of both C_c and C_o are increased according to $X_c = 1/(2\pi f C)$. To simplify the problem, temporarily assume C_o is infinitely large and remains an ac short circuit at frequencies where reactance of C_c increases. E_g must now divide in Fig. 4-1c among R_g, X_{C_c}, and R_{in}. We can now define lower cutoff frequency f_L as that frequency where signal voltage across R_{in} or V_o drops to 0.707 times its value at 1000Hz. At f_L, reactance X_{C_c} equals the net resistance in series with C_c. That is, we inspect the circuit to find what resistance R is "seen" by C_c, and express f_L as

$$f_L = \frac{1}{2\pi R C_c} \qquad (4\text{-}2)$$

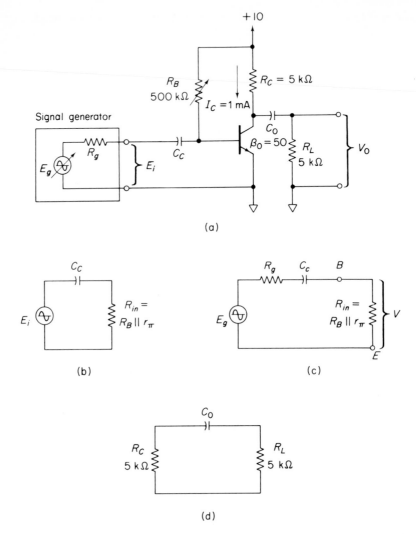

(a)

(b)

(c)

(d)

Figure 4-1

Low-frequency performance of the CE circuit in (a) depends on how the circuit is tested in (b) or (c) and on output coupling capacitor C_0 in (d).

where R is shown in Fig. 4-1(c) to be $R = R_g + R_B$. Furthermore, as V drops with decreasing frequency, V_o drops (from Eq. 4-1b). So the easiest way to measure f_L is to: (1) measure V_o at 1kHz with E_g set to 5mV; (2) decrease oscillator frequency until V_0 drops to 0.707 times its 1kHz value; and (3) read f_L from the oscillator dial at this point.

It is more common to hold E_i constant (by varying the volume control on E_g) while decreasing frequency. As f reduces, X_{C_c} increases, thus drawing less current through R_g and reducing signal voltage drop across R_g, thereby increasing E_i unless we make a volume adjustment. Thus a human operator provides corrective feedback to hold E_i constant no matter what current is drawn from E_i. Therefore E_i is made to be a perfect voltage source, and by definition has an internal impedance of zero. The low break-frequency f_L is established by C_c and the ac resistance, R_{in} as shown in Fig. 4-1b. Now when using Eq. (4-2) $R = R_{in}$.

EXAMPLE 4-1:

Choose C_C in Fig. 4-1 for a lower cutoff frequency of 125Hz if: (a) we hold E_i constant at 5mV; (b) we set E_g for 5mV and do not touch the signal generator's volume control. Given $R_B = 500k\Omega$, $r_\pi = 1250\Omega$, $R_g = 1250\Omega$.

SOLUTION:

(a) Find $R = R_{in} = 500k\Omega$ // $1250\Omega \cong 1250\Omega$ in Fig. 4-1b. Rewrite Eq. (4-2) as

$$C_c = \frac{1}{2\pi f_L R} = \frac{1}{2\pi(125)1250} \cong 1.0\mu F$$

(b) From Fig. 4-1(c), $R = R_g + R_{in} = 2500\Omega$, and from Eq. (4-2)

$$C_c = \frac{1}{2\pi(125)2500} = 0.5\mu F$$

EXAMPLE 4-2:

For part (a) in Example 4-1 find both V and V_o at (a) 1kHz and (b) f_L.

SOLUTION:

(a) At 1kHz, $E_i = V = 5mV$. From Eq. (4-1b)

$$V_o = -g_m R_{ac} E_i = -\left(\frac{1mA}{25mV}\right)(5k\Omega \text{ // } 5k\Omega)(5mV)$$
$$= -100 \times 5mV = -0.5V$$

(b) At f_L

$$V_o = 0.707 \times (-0.5)V = -0.35V \text{ and } V = 0.707E_i = 3.5mV$$

EXAMPLE 4-3:

For part (b) in Example 4-1 find both V and V_o at (a) 1kHz and (b) f_L.

SOLUTION:

(a) From Eq. (4-1a), $V = 5mV (1250/2500) = 2.5mV$. From Eq. (4-1c)

$$\frac{V_o}{E_g} = -\frac{R_{in}}{R_g + R_{in}} g_m R_{ac} = -(0.5)(100) = -50$$
$$\text{and } V_o = -50E_g = -0.25V$$

(b) At f_L

$$V_o = 0.707 \times (-0.25V) \cong -0.17V,$$

$$\text{and } V = 0.707 \times 2.5mV = 1.77mV$$

From Examples 4-1 to 4-3 we can conclude that doubling C_o or R reduces f_L by one-half, and that a reduction in gain is usually accompanied by a reduction in f_L.

We usually want only one capacitor to determine f_L. In the case of two coupling capacitors, design one to determine f_L and design the second (or any others) to have a lower break frequency of $0.1 \times f_L$. This procedure insures that the other capacitors will act as ac short circuits at f_L, and that voltage gain will be reduced by the first capacitor to a value at $0.1 f_L$ that will be low enough to avoid possible oscillations due to phase shifts accompanying increase of capacitor reactance at low frequencies.

EXAMPLE 4-4:

Choose a value for C_o in Fig. 4-1 so that C_c can set $f_L = 125Hz$.

SOLUTION:

Ac resistance of a BJT's collector is very large, so the resistance presented to C_o is shown in Fig. 4-1d to be $R = R_C + R_L = 10k\Omega$. Employ Eq. (4-2) to evaluate C_o for a 125Hz cutoff

$$C_o = \frac{1}{2\pi f_L R} = \frac{1}{2\pi(125)10^4} = 0.13\mu F$$

Now choose $C_o = 10 \times 0.13\mu F = 1.3\mu F$ or a commercially available $2\mu F$ to begin reducing gain at about $0.1 \times 125 = 12.5Hz$. Note the reactance of $C_o = 2\mu F$ is about 600Ω at 125Hz and is negligible with respect to R.

4-2 BJT LOW-FREQUENCY LIMITATIONS DUE TO BY-PASS CAPACITORS

In Fig. 4-2a, coupling capacitor C_C and by-pass capacitor C_E can interact at low frequency and complicate the problem of designing or analyzing the value of f_L. The most economical and simplest way to select C_E and C_C for a particular f_L is to proceed in two steps. First, find the equivalent ac resistance seen by C_E as in Fig. 4-2b. Express f_L as the reciprocal of their product and 2π as

$$f_L = \frac{1}{2\pi C_E R}, \text{ where } R = R_E \text{ // } \frac{r_\pi}{\beta_o + 1} \tag{4-3}$$

Remember from Chapter 2 that all base leg resistance seen from the emitter terminal is divided by $(\beta_o + 1)$. Usually R_E is large with respect to $r_\pi/(\beta_o + 1)$ and R equals $r_\pi/(\beta_o + 1)$. We assume that C_o is an ac short circuit during this calculation.

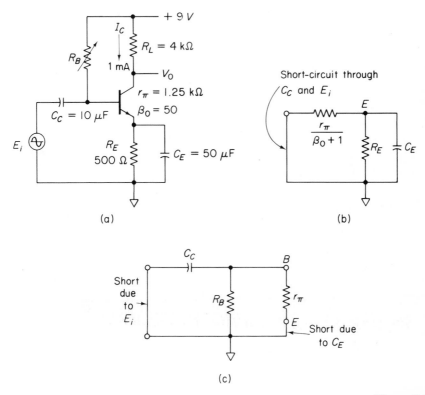

(a)

(b)

(c)

Figure 4-2

Emitter by-pass capitor C_E in (a) is selected from the ac model in (b), and C_C is selected from (c).

Second, find the ac resistance presented to C_c, assuming C_E is a short circuit. Then choose C_c to cut off at $0.1f_L$ from

$$0.1f_L = \frac{1}{2\pi C_c R} \tag{4-4}$$

R is found to approximately equal r_π in Fig. 4-2c. This procedure is essentially the same as the FET procedure in section 3-8.

EXAMPLE 4-5:

Choose values of C_E and C_c in Fig. 4-2 for $f_L = 125$Hz.

SOLUTION:

From Eq. (4-3), $R = 500 \text{ // } (1250/51) = 500 \text{ // } 25 = 25\Omega$.

$$C_E = \frac{1}{2\pi f_L R} = \frac{1}{2\pi(125)25} = 50\mu\text{F}$$

From Eq. (4.4)

$$C_c = \frac{1}{2\pi(0.1 f_L)R} = \frac{1}{2\pi(12.5)1250} = 10\mu\text{F}$$

An alternate solution to Example 4-5 is to pick $C_c = 1\mu\text{F}$ (for a 125Hz cutoff) and $C_E = 500\mu\text{F}$ (for a 12.5Hz cutoff). This $1\mu\text{F}$ and $500\mu\text{F}$ capacitor combination would be more expensive and bulky than the $10\mu\text{F}$ and $50\mu\text{F}$ combination. Thus economics dictates cutoff primarily by the by-pass capacitor, especially at values of f_L below 100Hz.

To analyze low-frequency behavior of Fig. 4-2, assume $E_i = 5$mV. At 1kHz, voltage gain would be $-g_m R_L = -4000/25 = -160$, with $V_o = -5$mV $\times 160 = -0.8V$. At f_L, C_c would be a short circuit and C_E would unby-pass R_E to the extent that V_o would be reduced to $0.707 \times 0.8 = 0.56V$ at f_L. Gain would also be reduced at f_L to $0.707 \times 160 = 113$. At $0.1 f_L$, R_E would essentially be unby-passed and gain would go to $-R_L/R_E = -8$ if C_c were not present. It is academic to think about any additional reduction in gain at $0.1 f_L$ due to C_c because the amplifier is no longer useful below f_L. More specifically, low-break frequency f_L is also called the 0.707, 3db, or *half power frequency*. This is because when any voltage across a load is reduced by 0.707, only half the original power remains for a power reduction of 3db. For frequencies below f_L, less than half power is available and the amplifier is not usually useful.

4-3 HIGH-FREQUENCY LIMITATIONS OF BJTs

Two *internal* transistor capacitances lower both input impedance and voltage gain at high frequencies. Base-emitter junction capacitance C_{be} and collector-base junction capacitance C_{cb} are added to the basic BJT model in Fig. 4-3a. Coupling and by-pass capacitors are modeled by short circuits at high frequencies where small transistor capacitance becomes important. Signal current, I_b, entering the base terminal flows through a small base bulk resistance, r_x, before encountering junction resistance r_π. Since r_x may only be significant because of its effect on high-frequency transistor performance, it was not included previously in the basic BJT model.

As frequency is increased, reactance of C_{be} decreases and shunts current away from r_π. Since only current through r_π is multiplied by β_o, $\beta_o I_b$ goes down and $V_o = -\beta_o I_b R_{ac}$ goes down. Since V_o goes negative and is larger than E_g, V_o actually pulls current I_f away from r_π via feedback capacitance C_{cb}. The

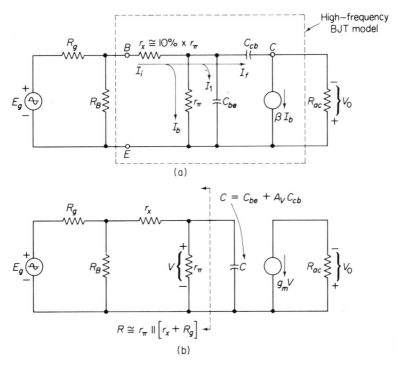

Figure 4-3

High-frequency BJT performance of the circuit in Figure 4-1(a) is analyzed from the high-frequency BJT model in (a), or simplified model in (b).

additional current I_f pulled through C_{cb} because of $-V_o$ makes C_{cb} act like a much larger capacitance (*Miller effect*) $A_v C_{cb}$ when viewed from the junction of r_π and r_x. The net effect may be simplified by an equivalent high-frequency capacitance C shown in Fig. 4-3b, where

$$A_v = g_m R_{ac} \tag{4-5}$$

and

$$C = C_{be} + A_v C_{cb} \tag{4-6}$$

Values for C_{cb} are given in data sheets and range from 2pf to 20pf. C_{cb} may also be represented by C_μ, C_{0B0}, or C'_{bc}. Values for C_{be} may be given directly or, more commonly, must be calculated from the data sheet characteristic f_T. f_T may be called the *high-frequency current gain, current gain-bandwidth product, alpha cutoff, f_α,* or f_{hfb}, and is defined as that frequency where β drops to a value of one. Once given f_T, calculate C_{be} from

$$C_{be} = \frac{g_m}{2\pi f_T} \tag{4-7}$$

Finally, upper cutoff frequency f_H is found from the reciprocal of the product of C and the resistance R presented to C. From Fig. 4-3b

$$f_H = \frac{1}{2\pi RC} \tag{4-8a}$$

where C is found from Eqs. (4-7) and (4-6) and

$$R = r_\pi \mathbin{/\!/} [r_X + R_B \mathbin{/\!/} R_g] \tag{4-8b}$$

Usually $R_B \gg R_g$ and Eq. (4-8b) reduces to

$$R \cong r_\pi \mathbin{/\!/} [r_X + R_g] \tag{4-8c}$$

EXAMPLE 4-6:

In Fig. 4-1, $R_g = 600\Omega$. The transistor is a 2N 5816 with f_T listed as 150 MHz, $\beta_o = 100$, and C_{cb} as 7pf. Find the expected high-frequency cutoff.

SOLUTION:

$$g_m = \frac{1\text{mA}}{25\text{mV}} = 0.040\mho \text{ and } C_{be} = \frac{0.040}{2\pi(150 \times 10^6)} = 42 \times 10^{-12}F$$

or $C_{be} = 42pF$. For Eq. (4-6), $A_v = g_m R_{ac} = 0.040 \times 2,500 = 100$ and $C = C_{be} + A_v C_{cb} = 42 + 100 \times 7 = 742pF$. Usually values for r_X are not given, although some data sheets may give values for r_X under the symbol r'_{bb}. Therefore, assume $r_X = 10\% r_\pi$. Evaluate r_π from $r_\pi = \beta_o/g_m$ $= 100/0.040 = 2500\Omega$, and $r_X = 0.1 \times 2,500 = 250\Omega$. From Eq. (4-8c), the resistance presented to C is

$$R = r_\pi \mathbin{/\!/} (r_X + R_g) = 2500\Omega \mathbin{/\!/} 850\Omega = 630\Omega$$

Finally, from Eq. (4-8a)

$$f_H = \frac{1}{2\pi RC} = \frac{10^{12}}{2\pi(630)(742)} = 340\text{kHz}$$

EXAMPLE 4-7:

If $E_g = 5\text{mV}$ in Example 4-6, what is: (a) V_o at 1kHz and (b) V_o at f_H?

SOLUTION:

(a) From Eq. (4-1), $R_{in} = r_\pi = 2500\Omega$ and

$$V_o = -E_g \frac{R_{in}}{R_g + R_{in}} g_m R_{ac} = -5\text{mV}\frac{2,500}{600 + 2,500}(100) = 0.4V$$

(b) At f_H, V_o is down to 0.707 times its 1kHz value, or $V_o = 0.707 \times 0.4 \cong 0.28V$.

EXAMPLE 4-8:

Using the same transistor of Example 4-6, find f_H for the circuit of Fig. 4-2, but assume $R_L = 2.5\text{k}\Omega$ and $I_C = 1\text{mA}$. Since $R_g = 0$, it shorts R_B in the equivalent high-frequency model of Fig. 4-4. Resistance presented to equivalent transistor input capacitance C is essentially equal to r_X. Modify Eq. (4-8a) from Fig. 4-4 to get

$$f_H = \frac{1}{2\pi RC} = \frac{10^{12}}{2\pi(250)(742)} = 860 \text{ kHz}$$

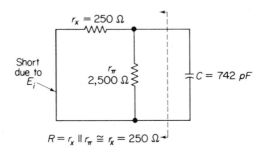

Figure 4-4

Circuit model for Example 4-8.

Compare Examples 4-6 and 4-8 to see that, for the same load and operating point current, f_H increases as R_g decreases, to a maximum at $R_g = 0$. Thus for high-frequency response we must use a low resistance source. In addition, as load resistance is increased to increase gain, C increases to reduce f_H, so we must use lower gain if higher-frequency response is desired.

4-4 THERMAL AND POWER LIMITATIONS

When current I flows between two terminals of a device and a voltage drop V exists across these terminals, power P_D is being absorbed by the device according to

$$P_D = IV \tag{4-9a}$$

Usually all of P_D appears as heat generated within the device, and causes a temperature rise. With no P_D, device temperature assumes that of its surroundings or ambient (usually air). The symbol for ambient temperature is T_A. P_D must be dissipated as heat from within the device by flowing from the region where it is generated into the ambient. In a BJT biased to perform as an amplifier, P_D is caused by collector current I_C flowing across the reverse-biased collector junction with voltage drop V_{CB}. Since V_{CE} is approximately equal to V_{CB}, we modify Eq. (4-9a) to apply to a BJT as

$$P_D = V_{CE}I_C \qquad (4\text{-}9b)$$

Heat generated at the collector junction flows through the collector body to the case and then to the ambient, as in Fig. 4-5a. Average junction temperature depends on the difference between average rate at which heat is generated and average rate at which heat is removed. The average rate at which heat is generated is P_D. Since power dissipated in the transistor is developed at the collector-base junction, then the operating limitation of a device is its *maximum*

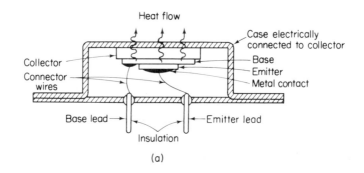

(a)

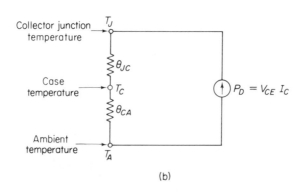

(b)

Figure 4-5

(a) Cross-sectional view illustrating path of heat flow and (b) heat model of a transistor.

junction temperature $T_{J(\text{max})}$. $T_{J(\text{max})}$ is given on manufacturers' data sheets, and typical values are 200°C for silicon transistors and 85-100°C for germanium transistors. Other maximum temperature limitations specified by manufacturers are for storage and soldering.

Thermal Resistance

Average power P_D will generate heat at the junction that will be removed at a rate directly proportional to the difference between *junction temperature* T_J and *ambient temperature* T_A, or

$$T_J - T_A = P_D \theta_{JA} \tag{4-10}$$

where θ_{JA} is the *thermal resistance* from junction to ambient, with units of °C/W. Equation (4-10) is valid for the small temperature differences and distances encounted in semiconductor devices. Since the heat generated has to flow from junction to case and then from case to ambient, θ_{JA} is actually the sum of two components. These components are (1) *thermal resistance* from *junction* to *case* θ_{JC}, and (2) *thermal resistance* from *case* to *ambient* θ_{CA}. Rewriting Eq. (4-10)

$$T_J - T_A = P_D(\theta_{JC} + \theta_{CA}) \tag{4-11}$$

Derating Curve

Transistor manufacturer's may give either θ_{JC} and θ_{JA} or the information contained in Eq. (4-11) as a *dissipation derating curve*. See Fig. 4-6.

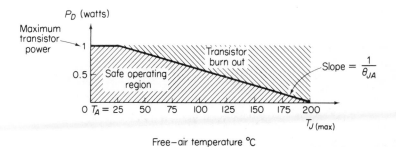

Figure 4-6

Typical dissipation derating curve for a TO-5 case.

The dissipation derating curve of Fig. 4-6 is for a silicon transistor encapsuled in a TO-5 case. Maximum transistor power dissipation is 1W up to 25°C ambient (free-air) temperature. Operating junction temperature range is −65°C to $+200°C$. If the transistor is dissipating 1W and ambient temperature T_A is 25°C, the junction temperature is 200°C ($T_{J(max)} = 200°C$). The transistor is right at the edge of its heat dissipating capability. If the ambient temperature is raised above 25°C, the power dissipation must be decreased so that the junction temperature does not rise above 200°C. From Fig. 4-6 we see that if the ambient temperature rises to 200°C, no heat can be dissipated by the transistor, or $T_{J(max)}$ will be exceeded.

Case Temperature

The horizontal axis of Fig. 4-6 is a plot of free-air or ambient temperature, and the slope is the reciprocal of θ_{JA} (thermal resistance junction to ambient). For some transistors (mostly power transistors) the horizontal axis is a plot of *case temperature*, T_C, and the slope is the reciprocal of θ_{JC} (thermal resistance junction to case). Values of θ_{JC} are much lower for power transistors then for small-signal transistors, because the collector junction of a power transistor is larger and is connected electrically to its larger case. For example, θ_{JC} for a 2N3055 power transistor is 1.5°C/W, while θ_{JC} for a 2N40408 small-signal transistor is 35°C/W. Note $\theta_{JC} = 1.5°C/W$ means that for every watt the transistor must dissipate, the junction temperature rises 1.5°C above the temperature of the case.

EXAMPLE 4-9:

(a) From Fig. 4-6, calculate θ_{JA}. (b) Use Eq. (4-10) and determine P_{Dmax} for an ambient of 125°C. Compare with Fig. 4-6.

SOLUTION:

(a)
$$\theta_{JA} = \frac{T_{Jmax} - T_A}{P_D} = \frac{(200 - 25)°C}{1W} = 175°C/W$$

(b) Rearranging Eq. (4-10)

$$P_{Dmax} = \frac{T_J - T_A}{\theta_{JA}} = \frac{(200 - 125)°C}{175°C/W} = 0.43W$$

From Fig. 4-6, $P_{Dmax} \cong 0.43W$, at $T_A = 125°C$.

EXAMPLE 4-10:

For the transistor in Example 4-9, calculate junction temperature T_J, if the transistor were required to dissipate 1.5W.

SOLUTION:

From Example 4-9, $\theta_{JA} = 175°C/W$, then

$$T_J = T_A + \theta_{JA}P_D = 25°C + (175°C/W)(1.5W) = 287°C$$

Since T_J exceeds $T_{J\text{max}}$ the transistor will be destroyed unless a heat sink (section 4-5) is used.

EXAMPLE 4-11:

A power derating curve is shown in Fig. 4-7 for a 2N3055 general purpose power transistor, with the following specifications: $\theta_{JC} = 1.5°C/W$, $P_{D(\text{max})} = 115W$ *at* $T_C = 25°C$, and $T_{J(\text{max})} = 200°C$. Calculate the maximum permissible dissipation in an ambient of 25°C.

SOLUTION:

Note θ_{JC} is given but could have been determined from Fig. 4-7. A 2N3055 power transistor is encapsuled in a TO-3 case whose top surface area is approximately 1.5 in². (If the transistor is mounted on a phenolic board, which is a poor conductor of heat, only the top surface can dissipate heat.) Assuming heat transfer from a metal semiconductor case is roughly 125°C/ W — in², then

$$\theta_{CA} = \frac{125°C/W - \text{in}^2}{1.5\text{in}^2} = 83°C/W$$

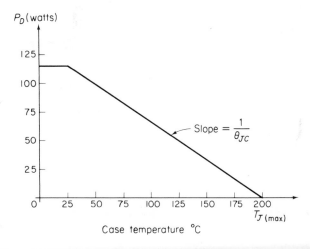

Figure 4-7

Typical dissipation curve for a power transistor encapsuled in a TO-3 case.

Rearranging Eq. (4-11)

$$P_D = \frac{T_{J(\text{max})} - T_A}{\theta_{JC} + \theta_{CA}} = \frac{(200 - 25)°C}{(1.5 + 83)°C/W} \simeq 2.0W$$

We conclude that power ratings of power transistors on manufacturer's data sheets are *not* the practical capability of the device alone. For this particular example, a transistor rated at 115W is capable of dissipating only 2.0W. A closer examination of the data sheet shows that the 115W rating is with a *case* temperature maintained at 25°C, *not* an ambient temperature of 25°C. It is impossible to keep the case at 25°C in practical applications unless it is immersed in a circulating fluid of 25°C. Therefore, to increase the power handling capabilities, a *heat sink* must be connected to the device.

4-5 HEAT SINKS

The function of a heat sink is to remove heat from the case of a device. The more efficient the heat sink, the more power a semiconductor can handle. Heat sink types range from a flat piece of metal to elaborate fins, from commercially available off-the-shelf to custom-made, from a simple clip-on to multiple mounting. The ability of any heat sink to transfer heat to the ambient depends on its material, volume, area, shape, contact between case and sink, and movement of air around the sink. Finned aluminum heat sinks yield the best heat transfer per unit cost, copper yields the best per unit volume, and magnesium the best per unit weight.

The path of heat generated at the collector-base junction is from junction to case, from case to sink, from sink to ambient. Each interface is a thermal resistance and represented by:

$$\text{junction to case} - \theta_{JC}$$
$$\text{case to sink} - \theta_{CS}$$
$$\text{sink to ambient} - \theta_{SA}$$

The thermal resistance model of a transistor in Fig. 4-5b is modified for a heat sink and shown in Fig. 4-8. Note that the purpose of a heat sink is to increase

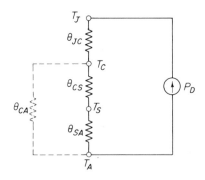

Figure 4-8

Heat model for a transistor using a heat sink.

the area of the case, but, to be effective, $\theta_{CS} + \theta_{SA} \ll \theta_{CA}$. Since this is true for all commercially available heat sinks, we may ignore θ_{CA} and rewrite Eq. (4-10) for heat sinks as:

$$T_J - T_A = P_D(\theta_{JC} + \theta_{CS} + \theta_{SA}) \tag{4-12}$$

Thermal resistance θ_{JA} is an inherent characteristic of the semiconductor design, and thus something over which we have no control except by choosing another transistor. The collector junctions of power transistors are usually connected electrically to the case. To insulate the transistor from heat sink and mounting screws, mica insulating separators and nylon shoulder washers or their equivalent must be used. See Fig. 4-9. Silicon grease (called thermal compound or heat sink compound) is used to fill air gaps between transistor, se-

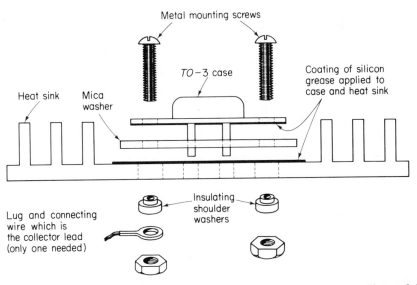

Figure 4-9

Mounting a TO-3 case to a heat sink.

parator, and heat sink, thereby, improving thermal conduction at this interface. Now the transistor must be securely mounted to the heat sink. Manufacturers of thermal compounds and heat sinks give a recommended torque to be used in mounting. Overtightening of mounting screws or studs may warp the surfaces and reduce contact area between transistor and heat sink. Typical values of θ_{CS} are $0.02 - 5°C/W$, depending on interface surfaces, mounting pressure, and whether a silicon grease is used.

Evaluating θ_{SA}

To determine thermal resistance, θ_{SA}, we refer to specification sheets sup-plied by manufacturers of heat sinks. Figure 4-10 is a typical set of heat sink characteristics—a plot of temperature rise above ambient versus the transistor's average power dissipation P_D. Curves of Fig. 4-10 are for natural convection, that is, a fan is *not* positioned near the heat sink to remove heat. Curves A, B, and C are for the same family of heat sinks, but of different sizes. Curve A is for the smallest heat sink, while curve C is for the largest.

To use Fig. 4-10, consider that the transistor is dissipating 5W. For heat sink A at 5W, the sink temperature, T_S, rise above ambient is 82°C. If ambient temperature T_A is 25°C, then $T_S = 82°C + 25°C = 107°C$. For heat sink B at 5W, sink temperature rise is 34°C, then $T_S = 34°C + 25°C = 59°C$. If heat sink C is mounted to the transistor, its sink temperature would be $T_S = 15°C + 25°C = 40°C$. Note that the sink temperature T_S is *not* the temperature measured with a bulb type thermometer placed on the heat sink. Manufacturers measure sink and case temperatures by drilling a small hole in the sink and inserting a tiny thermocouple between the case of the transistor and the heat sink. Obviously this measurement is impractical for average users of heat sinks who simply want to know if the heat sink at hand is adequate, or if they will have to buy or make a new one. To answer this question we must know more about thermal resis-tance *sink-to-ambient* θ_{SA}. From Fig. 4-10, $\theta_{SA} = \Delta T / P_D$. For heat sink A, $\theta_{SA} = 82°C/5W = 16.4°C/W$, for heat sink B, $\theta_{SA} = 34°C/5W = 6.8°C/W$, and for heat sink C, $\theta_{SA} = 15°C/5W = 3°C/W$. As would be expected, the largest heat sink C has the smallest thermal resistance because a larger size heat sink allows a faster transfer of heat to the surrounding air than a smaller one.

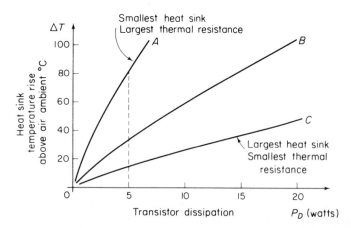

Figure 4-10

Heat sink characteristics for natural convection.

EXAMPLE 4-12:

A power transistor must dissipate 5W into the air. From manufacturer's data sheet $\theta_{JC} = 1.8°C/W$ and T_J is not to exceed 100°C. Assume a silicon-greased mica washer is used, and its thermal resistance is 0.2°C/W. From Fig. 4-10, which heat sink should be chosen? $T_A = 25°C$.

SOLUTION:

Rearranging Eq. (4-12)

$$\theta_{JC} + \theta_{CS} + \theta_{SA} = \frac{T_J - T_A}{P_D} = \frac{(100 - 25)°C}{5W} = 15°C/W$$

Since $\qquad \theta_{JC} + \theta_{CS} = 1.8°C/W + 0.2°C/W = 2°C/W$

then $\qquad \theta_{SA} = (15 - 2)°C/W = 13°C/W$

Choice of a proper heat sink must have its thermal resistance θ_{SA} equal to or less than (preferably less than) 13°C/W. Thus heat sink A is eliminated, leaving B and C. The final decision should be heat sink B because (1) at $P_D = 5W$ its $\theta_{SA} = 6.8°C/W$ is certainly adequate to the job, and (2) it is smaller and lighter than C, and therefore less expensive.

EXAMPLE 4-13:

If heat sink B is chosen for the solution of Example 4-12, calculate case, sink, and junction temperatures under operating conditions.

SOLUTION:

From the heat sink model of Fig. 4-8

$$T_C = T_A + \theta_{SA}P_D = 25°C + (6.8°C/W)(5W) = 59°C$$

$$T_S = T_A + (\theta_{CS} + \theta_{SA})P_D = 25°C + (0.2°C/W + 5.8°C/W)(5W)$$
$$= 60°C$$

or $\qquad T_S = T_C + \theta_{CS}P_D = 59°C + (0.2°C/W)(5W) = 60°C$

$\qquad T_J = T_A + (\theta_{JC} + \theta_{CS} + \theta_{SA})P_D$

$\qquad T_J = 25°C + (1.8 + 0.2 + 6.8)°C/W(5W) = 69°C$

or $\qquad T_J = T_S + \theta_{JC}P_D = 60°C + (1.8°C/W)(5W) = 69°C$

Heat sink B is certainly adequate under operating conditions because the junction of the transistor only reaches 69°C, well below the design specification $T_J = 100°C$. Note again units for thermal resistance are °C/W, which means for every watt the transistor must dissipate, the temperature rises 1°C. For this example, $\theta_{SA} = 6.8°C/W$ and $P_D = 5W$, then the sink temperature must be (6.8°C/W)(5W) = 34°C above ambient. Similarly, thermal resistance $\theta_{CS} = 0.2°C/W$ causes the case temperature to be (0.2°C/W)(5W) = 1°C above the temperature of the sink. Thermal resistance $\theta_{JC} = 1.8°C/W$ causes the junction temperature to be $(1.8°C/W)(5W) = 9°C$ above the case temperature. See Fig. 4-11

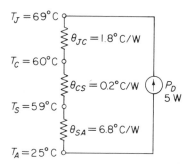

Figure 4-11

Heat model for solution of Example 4-13.

The heat sinks illustrated in Fig. 4-12 are not a complete set of all the heat sinks available, but are representative of the most common finned aluminum types. Generally, our type 1 would be used on small-signal transistor cases such as TO-5, TO-8, and TO-18 to extend their low P_{Dmax} rating of about 1W to about 3W. A type 2 sink would be used for plastic power transistors and SCRs. Type 3 sinks are for applications using transistors with TO-3 and TO-66 cases. Type 4 may be used for all power devices, the flat side allowing for narrow packaging. This type may be purchased in foot lengths and cut as needed. Type 5, unlike type 4, has serrated fins on both sides and is used in general purpose applications up to 50 watts. Type 6 heat sink is recommended for high power SCRs and rectifiers, with heat dissipation range in the vicinity of 100 watts.

The most inexpensive heat sinks are cut from sheet aluminum stock. For such cases, Fig. 4-13 should be helpful. All three plots are for 3/32″ bare aluminum mounted horizontally. Vertical positioning can increase the power dissipation rating by approximately 12 percent. Anything larger than 9″ × 9″ is certainly awkward to work with, and a commercially available heat sink should be purchased. For the transistor in Examples 4-12 and 4-13, T_C should be held to about 60°C or a rise of $60 - 25 = 35°C$ above ambient. From Fig. 4-13, 5 watts causes a temperature rise of 28°C in a 4″ × 4″ sink, and would be adequate.

EXAMPLE 4-14:

The total thermal resistance $(\theta_{JA} = \theta_{JC} + \theta_{CS} + \theta_{SA})$ of a power transistor and heat sink is 20°C/W. Calculate the maximum power dissipation for ambient temperatures of 25°C and 75°C. $T_{Jmax} = 200°C$.

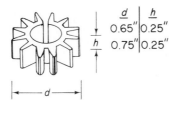

d	h
0.65"	0.25"
0.75"	0.25"

Type 1

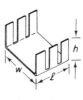

Type 2

$\ell = w = 1.8''$

h
0.5"
0.75"
1.0"
1.25"

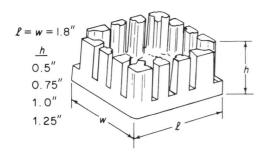

Type 3

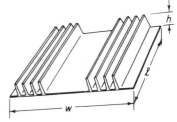

Type 4

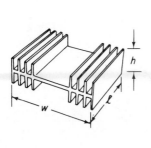

Type 5

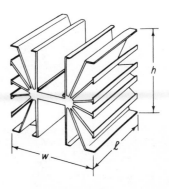

Type 6

Figure 4-12

Basic heat sink designs.

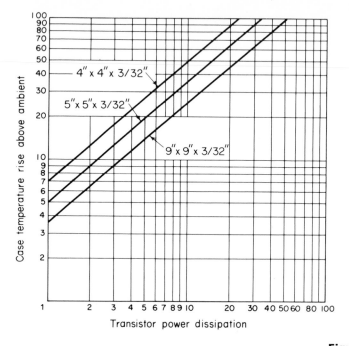

Figure 4-13

Heat sink data for 3/32" aluminum.

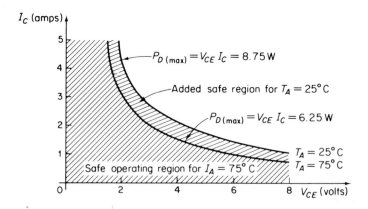

Figure 4-14

$P_{D(\text{max})} = V_{CE}I_C$ plotted on a power transistor's collector characteristics.

SOLUTION:

From Eq. (4-12)

$T_A = 25°C$

$$P_{D\max} = \frac{T_{J\max} - T_A}{\theta_{JA}} = \frac{(200 - 25)°C}{20°C/W} = 8.75W = V_{CE}I_C$$

$T_A = 125°C$

$$P_{D\max} = \frac{(200 - 75)°C}{20°C/W} = 6.25W = V_{CE}I_C$$

Assume values for I_C and calculate corresponding values of $V_{CE} = P_{D\max}/I_C$. The results are then plotted on the transistor's collector characteristics. This plot tells at a glance what operating points are allowed without exceeding thermal limitations. See Fig. 4-14.

4-6 Class A Power Amplifier

Figure 4-15a shows a class A direct-coupled amplifier along with its output characteristics and waveforms. As stated in Chapter 2, locating operating point Q midway on the load line permits maximum output voltage swing. From Fig. 4-15b, Q is located at $V_{CE} = V_{CC}/2$ and $I_C = V_{CC}/2R_L$. For linear operation, or to avoid distortion, the peak value of output voltage V_{op} must not be greater than $V_{CC}/2$. If V_{op} does not exceed $V_{CC}/2$, then the peak value of collector current I_{cp} will not exceed $V_{CC}/2R_L$. At this point we stress that when ac input signal E_i is applied, current and voltage waveforms are composed of a dc component plus an ac component.

Elements in the output loop—power supply V_{CC}, load resistor R_L, and collector-emitter terminals of the transistor—deliver or dissipate almost all the circuit's power. Let us investigate what average power is delivered or dissipated by each element due to the dc component and the ac component.

Supply Power

V_{CC} supplies power to the transistor of Fig. 4-15. With no input signal E_i, the total supply power P_S equals $V_{CC}(I_B + I_C)$. Since $I_B \ll I_C$, we neglect I_B and express P_S by

$$P_S = V_{CC}I_C \qquad (4-12)$$

where I_C is the dc operating point current.

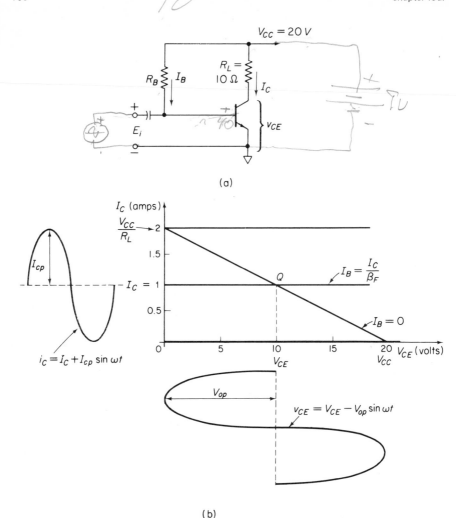

Figure 4-15

Class A power amplifier and output waveforms.

When the ac input signal E_i is applied, output current will vary sinusoidally around I_C, as shown in Fig. 4-15b as *total instantaneous* current $i_c = I_C + I_{cp}$ sin ωt. Total instantaneous power p_s is the product of V_{CC} and i_c, or

$$p_s = V_{CC}I_C + V_{CC}I_{cp} \sin \omega t \qquad (4\text{-}13)$$

The second term in Eq. (4-13) has an average value of zero, so that the *average* supply power is given by Eq. (4-12). Note *average* supply power is *independent* of an input signal. Although the average value of supply power is independent of I_{cp}, the supply must be capable of furnishing a maximum current value of $I_C + I_{cp}$. Therefore, a current-limiter on the power supply must be set at a value equal to or greater than $I_C + I_{cp}$, or the power supply will introduce distortion.

Load Resistor Power

When input signal E_i is applied, load resistor R_L conducts the dc current I_C plus the alternating current $I_{cp} \sin \omega t$. Both currents produce heat which R_L must dissipate. Therefore, load resistor power P_R consists of dc no-signal power P_{RD} (the dc current) plus ac output power P_o (due to the rms value of the alternating current). Expressing each component:

$$P_{RD} = I_c^2 R_L \tag{4-14}$$

$$P_o = I_{\mathrm{rms}}^2 R_L = \left(\frac{I_{cp}}{\sqrt{2}}\right)^2 R_L = \frac{I_{cp}^2}{2} R_L \tag{4-15a}$$

or in terms of voltage

$$P_o = \frac{V_{\mathrm{rms}}^2}{R_L} = \left(\frac{V_{op}}{\sqrt{2}}\right)^2 \frac{1}{R_L} = \frac{V_{op}^2}{2R_L} \tag{4-15b}$$

Total power delivered to the load is

$$P_R = P_{RD} + P_o \tag{4-16a}$$

$$P_R = I_c^2 R_L + \frac{V_{op}^2}{2R_L} \tag{4-16b}$$

Figure 4-15b shows the maximum, ideal, possible output voltage swing obtainable for the circuit of Fig. 4-15a, where $V_{op} = V_{CC}/2$. Although this expression may be substituted into Eq. (4-15b), we will leave the equation in general terms of V_{op}. This way V_{op} may represent any value from 0 to $V_{CC}/2$.

Transistor Dissipation

As with R_L, the transistor's dissipation depends on a dc component P_{DD} due to I_C and an ac component P_o due to the output voltage and current swing. Average power dissipated by a transistor is expressed by

$$P_D = P_{DD} - P_o \tag{4-17a}$$

where dc power at no signal is P_{DD}

$$P_{DD} = V_{CE}I_C \tag{4-17b}$$

and

$$P_o = \begin{pmatrix} \text{rms value of} \\ \text{output voltage} \end{pmatrix} \begin{pmatrix} \text{rms value of} \\ \text{output current} \end{pmatrix}$$

$$P_o = \left(\frac{V_{op}}{\sqrt{2}}\right)\left(\frac{I_{cp}}{\sqrt{2}}\right) \tag{4-17c}$$

The minus sign[1] in Eq. (4-17a) means that the transistor is transferring power P_o to the load. As a result, the transistor (under class A operation only) must dissipate less power when a signal is applied, and therefore runs cooler. For class A operation, the maximum power a transistor must dissipate is when: (1) it is biased for maximum output swing, and (2) no ac input signal is present.

We do not use the expressions $I^2 R$ or V^2/R for a transistor because there is no easily calculated value for R corresponding to the operating resistance between collector and emitter. Note whatever power P_o is dissipated by R_L must be delivered by the transistor. Therefore, Eq. (4-17c) must equal Eq. (4-15b).

Efficiency

Efficiency η is the ratio of average useful power output P_o (Eq. 4-15b) to average supply power input P_s (Eq. 4-12), or, in percent

$$\eta = \frac{V_{op}^2}{2V_{cc}I_cR_L} \times 100 = \frac{P_o}{P_s} \times 100 \tag{4-18a}$$

When biased for maximum output where $I_c = V_{cc}/(2R_L)$, efficiency is expressed by

$$\eta = \frac{V_{op}^2}{V_{cc}^2} \times 100, \text{ for } I_c = \frac{V_{cc}}{2R_L} \tag{4-18b}$$

When also driven to maximum output, with $V_{op} = V_{cc}/2$, circuit efficiency is maximum at 25 percent because $P_D = V_{cc}^2/(8R_L)$ from Eq. (4-15b), and $P_s = V_{cc}^2/(2R_L)$ from Eq. 4-12.

Maximum transistor power at the maximum output bias point ($V_{CE} = V_{cc}/2$, $I_c = V_{cc}/(2R_L)$ is found from Eq. (4-17b) to be (no signal)

$$P_{D\max} = \frac{V_{cc}^2}{4R_L}$$

[1]The minus sign can be shown by multiplying instantaneous collector current times instantaneous collector-emitter voltage

$$P_d = (I_C + I_{cp} \sin \omega t)(V_{CE} - V_{op} \sin \omega t)$$

$$P_d = V_{CE}I_C - I_CV_{op} \sin \omega t + V_{CE}I_{cp} \sin \omega t - V_{op}I_{cp} \sin^2 \omega t$$

The first term is Eq. (4-17b). The average value of the second and third terms is zero. The average value of the fourth term is the rms value of V_{op} times the rms value of I_{cp}, and is given by Eq. (4-17c).

So the ratio of maximum class A output power to maximum transistor power is

$$\frac{P_{o\max}}{P_{D\max}} = \frac{V_{CC}^2/8R_L}{V_{CC}^2/4R_L} = \frac{1}{2}$$

This limitation on class A operation is severe enough to restrict its applications to outputs of 1/2 watt or less.

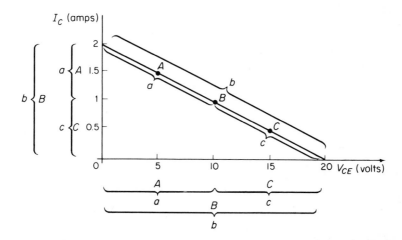

Figure 4-16

Maximum peak to peak undistorted output swing (*a, b, c*) for the three operating points *A, B,* and *C*, respectively.

EXAMPLE 4-15:

For each of the three operating points *A, B, C* on Fig. 4-16, calculate the average powers for the load resistor, transistor, and power supply for the circuit of Fig. 4-15a with (a) no signal and (b) maximum undistorted output swing.

SOLUTION:

Operating point A:

Load

no signal $\quad P_{RD} = I_C^2 R_L = (1.5A)^2(10\Omega) = 22.5W$

signal—in Fig. 4-16, interval *a* shows the maximum undistorted peak-to-peak output swing possible for operating point *A*. From Eq. (4-15a)

$$P_o = \frac{I_{cp}^2}{2}R_L = \frac{(0.5)^2}{2}(10\Omega) = 1.25W$$

$$P_R = P_{RD} + P_o = 22.5W + 1.25W = 23.75W$$

Transistor

no signal $P_{DD} = V_{CE}I_C = (5V)(1.5A) = 7.5W$

signal $P_o = \left(\dfrac{V_{op}}{\sqrt{2}}\right)\left(\dfrac{I_{cp}}{\sqrt{2}}\right) = \left(\dfrac{5V}{\sqrt{2}}\right)\left(\dfrac{0.5A}{\sqrt{2}}\right) = 1.25W$

$P_D = P_{DD} - P_o = 7.5W - 1.25W = 6.25W$

Supply   With or without a signal, Eq. (4-12) is applicable

$$P_S = V_{CC}I_C = (20V)(1.5A) = 30W$$

Also note that supply power = load power + transistor power.

no signal $P_S = P_{RD} + P_{DD} = 22.5W + 7.5W = 30W$

signal $P_S = P_{RD} + P_o + P_{DD} - P_o$

$P_S = 22.5W + 1.25W + 7.5W - 1.25W = 30W$

Operating point B:
Load

no signal $P_{RD} = I_C^2 R_L = (1A)^2(10) = 10W$

signal—maximum undistorted peak-to-peak swing possible for operating
point B is illustrated by b.

$$P_o = \frac{I_{cp}^2}{2}R_L = \frac{(1A)^2}{2}(10) = 5W$$

$$P_R = P_{RD} + P_o = 10W + 5W = 15W$$

Transistor

no signal $P_{DD} = V_{CE}I_C = (10V)(1A) = 10W$

signal $P_o = \left(\dfrac{V_{op}}{\sqrt{2}}\right)\left(\dfrac{I_{cp}}{\sqrt{2}}\right) = \left(\dfrac{10V}{\sqrt{2}}\right)\left(\dfrac{1A}{\sqrt{2}}\right) = 5W$

$P_D = P_{DD} = P_o = 10W - 5W = 5W$

Supply

no signal $P_S = P_{RD} + P_{DD} = 10W + 10W = 20W$

signal $P_S = P_{RD} + P_o + P_{DD} - P_o$

$P_S = 10W + 5W + 10W - 5W = 20W$

Using Eq. (4-12) as a check, $P_S = V_{CC}I_C = (10W)(2A) = 20W$.

Operating point C:
Load

no signal $\qquad P_{RD} = I_C^2 R_L = (0.5A)^2(10) = 2.5W$

signal—maximum possible peak-to-peak undistorted swing is shown by *c*.

$$P_o = \frac{I_{cp}^2}{2} R_L = \frac{(0.5A)^2}{2}(10\Omega) = 1.25W$$

$$P_R = P_{RD} + P_o = 2.5W + 1.25W = 3.75W$$

Transistor

no signal $\quad P_{DD} = V_{CE}I_C = (15V)(0.5A) = 7.5W$

signal $\qquad P_o = \left(\frac{V_{op}}{\sqrt{2}}\right)\left(\frac{I_{cp}}{\sqrt{2}}\right) = \left(\frac{5V}{\sqrt{2}}\right)\left(\frac{0.5A}{\sqrt{2}}\right) = 1.25W$

$$P_D = P_{DD} - P_o = 7.5W - 1.25W = 6.25W$$

Supply

no signal $\quad P_S = P_{RD} + P_{DD} = 2.5W + 7.5W = 10W$

signal $\qquad P_S = P_{RD} + P_o + P_{DD} - P_o$

$$P_S = 2.5W + 1.25W + 7.5W - 1.25W = 10W$$

Using Eq. (4-12) as a check, $P_S = V_{CC}I_C = (20V)(0.5A)$
$\qquad\qquad\qquad\qquad\qquad\qquad\qquad = 10W.$

Table 4.1 summarizes the results of this example.

Table 4.1

Comparison of Maximum Output Powers for Three Operating Points

Operating Point	P_R Load	+	P_D Transistor	+	P_o Output	=	P_S Supply
A no signal	22.5	+	7.5	+	0	=	30
A signal	23.75	+	6.25	+	1.25	=	30
B no signal	10	+	10	+	0	=	20
B signal	15	+	5	+	5	=	20
C no signal	2.5	+	7.5	+	0	=	10
C signal	3.75	+	6.25	+	1.25	=	10

EXAMPLE 4-16:

Calculate junction temperature T_J for the three operating points in Example 4-15. Assume ambient temperature is 25°C and total thermal resistance is $\theta_{JA} = 10°C/W$.

SOLUTION:

Using Eq. (4-10), $T_J = T_A + \theta_{JA}P_D$; since $T_A = 25°C$ and $\theta_{JA} = 10°C/W$, then for:

Operating point A:

no signal $\quad$ $T_J = 25°C + (10°C/W)(7.5W) = 100°C$

signal $\qquad$ $T_J = 25°C + (10°C/W)(6.25) = 87.5°C$

Operating point B:

no signal $\qquad$ $T_J = 25°C + (10°C/W)(10W) = 125°C$

signal $\qquad$ $T_J = 25°C + (10°C/W)(5W) = 75°C$

Operating point C:

no signal $\quad$ $T_J = 25°C + (10°C/W)(7.5W) = 100°C$

signal $\qquad$ $T_J = 25°C + (10°C/W)(6.25W) = 87.5°C$

This example demonstrates a previous statement that, for class A operation, a transistor "runs" cooler when an ac input signal is applied. Note for operating point B the transistor's junction is 50°C cooler under signal conditions than under stand-by conditions (transistor is biased, but no ac input signal).

EXAMPLE 4-19:

Calculate the efficiency for each operating point in Example 4-16.

SOLUTION:

Using Eq. (4-18a), $\eta = P_o \times 100/P_s$, then for:

Operating point A:

$$\eta = \frac{1.25W}{30W} \times 100 = 4.16\%$$

Operating point B:

$$\eta = \frac{5W}{20W} \times 100 = 25\%$$

Operating point C:

$$\eta = \frac{1.25W}{10W} \times 100 = 12.5\%$$

Note that the efficiency at operating point B is twice that at C and six times that at A. Therefore, for class A operation, biasing a transistor so that its operating point is located midpoint on the load line is the ideal position. The advantages are: (1) maximum output voltage swing; (2) maximum efficiency; (3) coolest junction temperature under signal conditions. To overcome disadvantages and practical limitations for using class A operation (such as stand-by dissipation, efficiency and maximum output swing) in the output stage of a power amplifier, we turn to the complementary amplifier.

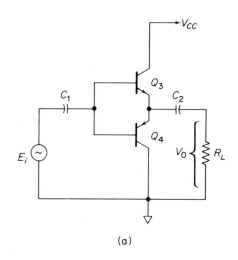

(a)

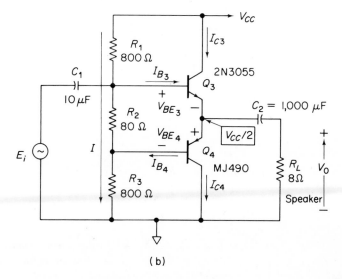

(b)

Figure 4-17

Basic complementary amplifier in (a) is biased to eliminate crossover distortion in (b).

4-7 COMPLEMENTARY AMPLIFIER

To avoid load power limitations of the class A amplifier, we add a transistor to form the *complementary* power amplifier of Fig. 4-17a. No collector current flows, when E_i equals zero, to eliminate standby current drain. When E_i goes positive, Q_3 is forward-biased and Q_4 is reverse-biased. When E_i goes negative, Q_3 is reverse-biased and Q_4 is forward-biased. Thus Q_4 conducts on negative half-cycles of E_i to complement the conduction of Q_3 on positive half-cycles. Each acts as an emitter follower, with voltage gain of 1 and current gain determined by β_F of Q_3 and Q_4.

Since Q_3 does not conduct until E_i rises above $0.6V$, and Q_4 does not conduct until E_i drops below $-0.6V$, serious *crossover distortion* results. That is, we lose output signal when E_i crosses over 0 volts in either positive or negative direction. To eliminate crossover distortion we must install a biasing network to insure that Q_3 and Q_4 are just barely forward-biased under a no signal condition. Then, any change in E_i will be coupled by either Q_3 or Q_4 through C_2 to load R_L.

Biasing

Voltage divider network R_1, R_2, and R_3 are installed in Fig. 4-18 to establish a voltage drop across R_2 that will forward-bias Q_3 and Q_4 just enough to insure that small idle (no signal) collector current of 5 to 50mA flows in Q_3 and Q_4. Junction of emitters of Q_3 and Q_4 should be biased to a dc voltage of $V_{CC}/2$ (at no signal). Voltage drop across R_2 should be: (a) $1.2V$ if both Q_3 and Q_4 are silicon, or (b) $0.8V$ if Q_3 is silicon and Q_4 is germanium. Since (a) voltage drop across R_2 is small with respect to drops across R_1 and R_3, (b) R_1 must equal R_3 to set the emitter junctions at $V_{CC}/2$, and (c) R_3 should be less than the average β_F (or h_{FE}) of Q_3 and Q_4 times R_L, to insure enough current drive. These conditions are summarized by

$$R_1 = R_3 = \frac{V_{CC}}{2I} \leq \beta_F R_L \qquad (4\text{-}19)$$

Assuming Q_3 and Q_4 are silicon, pick R_2 to develop $1.2V$ from

$$R_2 = \frac{1.2}{I} V \qquad (4\text{-}20)$$

Note that dc idle current $I_{C3} = I_{C4}$ in Fig. 4-17 will approximately equal dc bleeder current I. Therefore, idle base current $I_{B3} = I_{B4}$ will be small with respect to I, and allow us to use simplified Eqs. (4-19) and (4-20).

EXAMPLE 4-18:
If $R_L = 8\Omega$, $\beta_{F3} = \beta_{F4} = 100$, $V_{CC} = 24V$, pick values for R_1, R_2, and R_3.

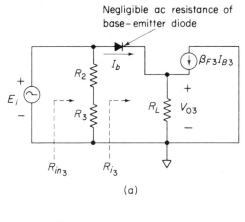

Negligible ac resistance of base–emitter diode

(a)

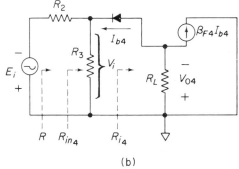

(b)

Figure 4-18

Ac complementary amplifier model for positive inputs in (a) and negative inputs in (b).

SOLUTION:

(a) From Eq. (4-19), $R_1 = R_3 = 100 \times 8 = 800\Omega$.

$$I = \frac{V_{CC}}{2R_3} = \frac{24}{2 \times 800} = 15\text{mA}$$

From Eq. (4-20), $R_2 = 1.2/0.015 = 80\Omega$.

In practice, R_2 is adjusted to eliminate crossover distortion at V_o with low levels of E_i ($\leq 1.0V$).

AC Analysis

Voltage gain and input resistance are analyzed with (1) Q_3-on and Q_4-off for positive input signals, and (2) Q_3-off and Q_4-on for negative input signals.

Figures 4-18a and b show large signal models used for power transistors. Ac resistance of the base-emitter diode is negligible and develops negligible ac voltage drop across it. From Fig. 4-18a, input resistance:

$$R_{i3} = (\beta_{F3} + 1)R_L \tag{4-20a}$$

and
$$R_{in3} = (R_2 + R_3) \,/\!/ \, R_{i3} \tag{4-20b}$$

Note any resistance in the emitter lead of a power transistor is reflected into the base lead by multiplying it by $(\beta_F + 1)$. (This is similar to small signal transistors studied in Chapter 2, except β_o instead of β_F was used.) Since ac voltage drop across the base-emitter diode is negligible, output voltage V_{o3} is approximately E_i and thus

$$Av_3 = \frac{V_{o3}}{E_i} \simeq 1 \tag{4-21}$$

Resistance R_1 of Fig. 4-17a has been neglected in the models and analyses of Fig. 4-18 because R_1 is usually replaced by a driver transistor. When E_i goes negative, Q_3 is off and Q_4 is on and we employ the large signal model of Fig. 4-18b, where input resistance is given by

$$R_{i4} = (\beta_{F4} + 1)R_L \tag{4-22a}$$

If $\beta_{F4} = \beta_{F3}$, then $R_{i4} = R_{i3}$, and

$$R_{in4} = R_3 \,/\!/ \, R_{i4} \tag{4-22b}$$

and the resistance "seen" by signal generator E_i is

$$R = R_2 + R_{in4} \tag{4-22c}$$

Output voltage V_{o4} is approximately V_i *not* E_i. The ratio

$$\frac{V_{o4}}{V_i} \simeq 1$$

E_i divides between R_2 and R_{in4}, so

$$\frac{V_i}{E_i} = \frac{R_{in4}}{R_2 + R_{in4}} \tag{4-23a}$$

and the voltage gain of Fig. 4-19b is

$$A_{v4} = \frac{V_{o4}}{E_i} = \frac{V_i}{E_i} \times \frac{V_{o4}}{V_i} = \frac{R_{in4}}{R_2 + R_{in4}}(1) \tag{4-23b}$$

If R_2 is made excessively large (several hundred ohms), the negative input signal reaching the base of Q_4 will be much smaller than the positive input signal reaching the base of Q_3. Replacing R_2 with a diode wherever possible is desirable because the incremental (ac) resistance of the diode minimizes differences between V_{o3} and V_{o4}.

Load Resistor Power

The output voltage waveform across R_L is a full sine wave. Thus power delivered to load R_L is $P_o = V_{rms}^2/R_L$, where $V_{rms} = V_{op}/\sqrt{2}$ and

$$P_o = \left(\frac{V_{op}^2}{\sqrt{2}}\right)\frac{1}{R_L} = \frac{V_{op}^2}{2R_L} \tag{4-24}$$

Transistor Dissipation

Instantaneous power is the product of instantaneous current and voltage.

$$P_D = (I_{cp}\sin \omega t)\left(\frac{V_{CC}}{2} - V_{op}\sin \omega t\right) \tag{4-25}$$

Equation (4-25) is the expression for instantaneous power for one transistor, and the average value is

$$P_D = \frac{V_{op}V_{CC}}{2\pi R_L} - \frac{V_{op}^2}{4R_L} \quad \text{(average power dissipated by one transistor)} \tag{4-26}$$

Multiplying Eq. (4-26) by 2 yields

$$P_D = \frac{V_{op}V_{CC}}{\pi R_L} - \frac{V_{op}^2}{2R_L} \quad \text{(average power dissipated by both transistors)} \tag{4-27}$$

Supply Power

Supply power P_S must equal transistor dissipation plus load resistor power.

$$P_S = P_D + P_o \tag{4-28a}$$

Substituting Eqs. (4-24) and (4-27) into Eq. (4-28a) gives

$$P_S = \frac{V_{CC}V_{op}}{\pi R_L} \tag{4-28b}$$

The maximum value of V_{op} for a complementary amplifier is $V_{CC}/2$.

EXAMPLE 4-19:

For the circuit of Example 4-18, what is the (a) load resistor power, (b) dissipation in each transistor, and (c) supply power at maximum output?

SOLUTION:

(a) At full output, $V_{op} = V_{CC}/2 = 12V$. From Eq. (4-24)

$$P_o = \left(\frac{V_{CC}}{2}\right)^2 \times \frac{1}{2R_L} = (12)^2 \times \frac{1}{2 \times 8} = 9W$$

(b) From Eq. (4-26)

$$P_D = \frac{V_{CC}^2}{4\pi R_L} - \frac{V_{CC}^2}{16R_L} = \frac{24^2}{4\pi \times 8} - \frac{24^2}{16 \times 8} = 5.7 - 4.5 = 1.2W$$

Total dissipation in both transistors is $2 \times 1.2 = 2.4W$.

(c) From Eq. (4-28b)

$$P_S = \frac{24 \times 12}{\pi 8} = 11.4W$$

Checking from Eq. (4-28a)

$$P_S = 11.4 = 2.4 + 9 = 11.4$$

4-8 INTEGRATED CIRCUIT AMPLIFIERS

Figure 4-19 is a simplified schematic of a typical low-power (1 to 2W) integrated circuit (IC) amplifier. Bias resistors R set the base of voltage amplifier Q_1 at $V_{CC}/2$. Driver transistor Q_2 replaces R_1 of Fig. 4-18b and Q_3-Q_4 form the complementary output stage. Feedback resistor R_F provides 100 percent dc feedback from the dc output terminal to the $(-)$ input terminal. Only a small dc bias current flows through R_F, so the dc output terminal is biased slightly below the $+$ input terminal (due to V_{BE1}). This biasing arrangement locks the dc output terminal at $V_{CC}/2$ for maximum swing, so bias points do not depend critically on the value of V_{CC}.

As will be shown in Chapter 6, voltage gain depends only on external precision resistors R_F and R_E according to

$$\frac{V_o}{E_i} = \frac{R_F + R_E}{R_E} = \frac{10,000 + 200}{200} \cong 50 \tag{4-29}$$

R_3 is returned through R_L instead of directly to ground in an arrangement known as *bootstrapping* so as not to waste signal current through R_3, and to raise the input resistance of the complementary pair to approximately $\beta_F R_L$. Bootstrapping almost eliminates signal shunting by R_3.

All of the resistors and transistors in Fig. 4-19 may be fabricated into one IC to give power amplifiers of up to 2W. Such ICs are usually intended for use with printed circuit boards, and may use much of the copper foil as a heat sink.

Feedback resistors R_E and R_F may be included in some IC package for a fixed gain amplifier, or the $(-)$ input and dc output leads may be brought out in other designs for the user to set his own gain according to Eq. (4-29).

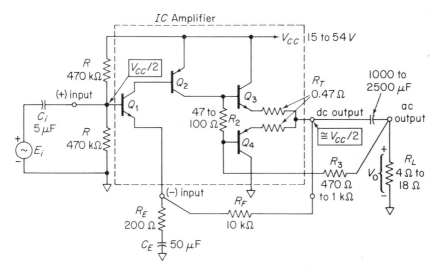

Figure 4-19

Simplified integrated circuit amplifier.

The large electrolytic capacitors are added by the user, since they are not contained in the tiny IC package.

In more exotic types of IC amplifiers, Q_1 and Q_2 are replaced on the IC chip by more complex circuitry to maximize gain and signal swing. For more power, Q_3 and Q_4 (or a more complex arrangement) may be installed on separate heat sinks by the user. Manufacturers also wire an IC amplifier together with a separate complementary amplifier unit in one *hybrid* package to yield a hybrid IC amplifier with power capabilities of up to 100 watts or more.

PROBLEMS

1. Repeat Example 4-1 for a lower-cutoff frequency of 30Hz.
2. If C_c in Fig. 4-1 is 10μF, what is the lower-cutoff frequency? $R_g = 1.25\ k\Omega$. Neglect effects of output capacitor C_o.

3. Repeat Example 4-4 for a lower-cutoff frequency of 30Hz.

4. Calculate new values for C_c and C_E in Fig. 4-2 so that the lower-cutoff frequency is 30Hz. Reference, Example 4-5.

5. In Fig. 4-2, if I_C is halved to 0.5mA and C_c and C_E remain unchanged, calculate f_L.

6. In the circuit of Fig. 4-1, the transistor is a 2N40408 with $f_T = 100\ MHz$, $C_{cb} = 15\ pF$, $\beta_o = 50$ and $R_g = 600\ \Omega$. Calculate the high-cutoff frequency, f_H.

7. Repeat Problem 4-6, but with C_o and R_L in Fig. 4-1 removed, thus the ac load is $R_c = 5\ k\Omega$.

8. A silicon transistor with a maximum junction temperature of 200°C is to operate in ambient of 30°C. With no heat sink $\theta_{JA} = 120°C/W$, calculate P_{Dmax}. If a heat sink is connected and $\theta_{JA} = 20°C/W$, recalculate P_{Dmax}.

9. Thermal power ratings of a silicon transistor are $P_{Dmax} = 10W$, $T_{Jmax} = 200°C$, $\theta_{JC} = 2°C/W$, and $\theta_{CS} = 0.5°C/W$. If $T_A = 40°C$ find the maximum θ_{SA} requirement for the heat sink.

10. $T_{Jmax} = 100°C$ and $\theta_{JA} = 5°C/W$. Will a transistor with these thermal characteristics be able to dissipate 10W in an ambient of 25°C?

11. A power transistor with a $T_{Jmax} = 200°C$ and $\theta_{JC} = 1.5°C/W$ is mounted on a heat sink with thermal characteristics of $\theta_{CS} = 0.5°C/W$ and $\theta_{SA} = 10°C/W$. If the transistor must dissipate 10W, calculate the maximum T_A for safe operation. From Fig. 4-13 determine an adequate size for a heat sink, $T_A = 25°C$.

12. In Fig. 4-15 V_{CC} is halved to 10V. Calculate the supply, load, and transistor power with no ac signal applied and with maximum output.

13. The maximum power a transistor can dissipate is 5W. What is the maximum signal power output obtainable and supply power required for class A operation?

14. In Fig. 4-15a R_L is halved to 5Ω. Calculate average powers for the load resistor, transistor and power supply. Let $V_{op} = V_{CE}$. Compare answers with operating point B Example 4-15.

15. In Fig. 4-17b must the biasing resistors be changed if (a) $R_L = 4\Omega$, (b) $R_L = 16\Omega$. (c) $\beta_{F3} = \beta_{FA} = 100$?

16. Using the values of Fig. 4-17 determine (a) R_{i3} and R_{in3} in Fig. 4-18a and (b) R_{i4}, R_{in4} and R in Fig. 4-18b.

17. For Fig. 4-17b what is the (a) load resistor power, (b) dissipation of each transistor and (c) supply power at maximum output if (a) $R_L = 4\Omega$ and (b) $R_L = 16\Omega$? Compare answers with Example 4-19.

chapter five
basic dc power supplies

5-0 INTRODUCTION TO POWER SUPPLIES

Most electronic devices and circuits require a direct (dc) voltage to operate. Since the most convenient and economical source of power is the ac wall outlet, it is advantageous to be able to convert the alternating voltage (usually 110V rms) to dc voltage (usually smaller in value). This conversion is accomplished by a rectifier power supply.

Unregulated Power Supply

In contrast to a regulated power supply, an unregulated power supply's dc terminal voltage is affected significantly by the loading network. For example, a multistage amplifier with no input signal applied may draw a steady current of 50mA from a power supply whose terminal voltage is 50V dc. The amplifier could be modeled by a resistor equal to 50V/50mA = 1kΩ. If an input signal is applied to the amplifier, average current drawn from the dc supply could increase to 500mA. The increased current flowing through the supply's internal resistance would perhaps reduce terminal voltage to 40V dc. Under this condition (input signal) the amplifier appears as an equivalent load resistor of $40V/500mA = 80\Omega$.

Unfortunately, our load model is not very precise, because when an input sine wave is applied to the amplifier, the current drawn from the dc power supply also varies sinusoidally. Thus, the amplifier acts like a resistor that is varying sinusoidally. In this chapter we will learn how terminal voltage of a dc supply varies with this type of load.

Regulated Power Supply

A regulated power supply is capable of delivering the necessary current to an amplifier or any other load without a significant reduction in terminal voltage. We will add a voltage regulating section to the unregulated supply to make a regulated supply. A zener diode and resistor is the cheapest type of regulator. The zener regulates terminal voltage at a fairly constant value regardless of what current is drawn from the now regulated supply. Although terminal voltage variation is minimized by the zener, the zener introduces limitations on maximum load current because of its low heat-dissipating ability. To overcome the heating problem caused by this zener limitation, we add a power transistor. The reason we add the transistor (rather than change the type of zener) is that a low-power zener and high-power transistor cost less than a single high-power zener.

Objectives

Restating and expanding this introduction, we will use a half-wave rectifier to learn a simple technique of studying the performance of any power supply. That is, we will analyze the dc performance or voltage regulation first, and then the ac performance or ripple. Next, we will note how the full-wave rectifier improves load current capacity over the half-wave, but does not significantly improve regulation. Finally, we will see how basic regulators improve performance of an unregulated full-wave rectifier. Then we will find performance limits of these basic regulators.

5-1 INTRODUCTION TO THE HALF-WAVE RECTIFIER

Transformer T is usually required in the half-wave rectifier of Fig. 5-1b to step 60Hz line voltage down to give required values of dc load voltage. Figure 5-1a shows the input voltage sine wave measured between terminals 1 and 2

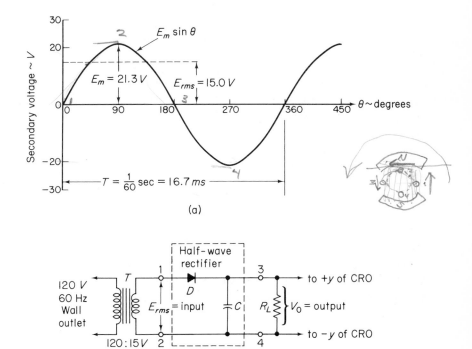

(a)

(b)

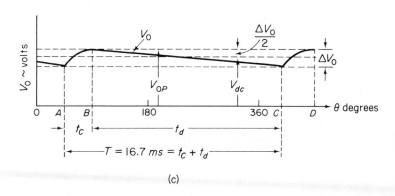

(c)

Figure 5-1

Half-wave rectifier with input and output voltages.

with a cathode ray oscilloscope. We will use the symbol V_o to mean rectifier output voltage or load voltage. The only instrument that will measure V_o is a CRO, since it displays voltage magnitude at each instant of time. V_o is more specifically defined as total instantaneous voltage, measured across the load.

We shall start with the simplest load—resistor R_L. V_o (shown in Fig. 5-1c) has a peak value of V_{op} and a peak-to-peak change in voltage of ΔV_o. (Δ means "change in.")

As discussed briefly in section 1-7, capacitor C charges in the time interval t_c and discharges during time t_d in Fig. 5-1. Both time intervals make up the period T. Period T is the time interval for one complete cycle and is found by dividing frequency f of the voltage wave into 1. These statements are summarized by

$$\frac{1}{f} = T = t_c + t_d \qquad (5\text{-}1)$$

For 60Hz line frequencies, $T = 1/60 = 16.7$ milliseconds.

Unregulated Supply Performance Measurements

To test dc rectifier operation we connect a dc voltmeter across R_L (terminals 3 and 4) and measure dc load voltage V_{dc}. This dc voltmeter finds the average value between V_{op} and $V_{op} - \Delta V_o$, and reads

$$V_{dc} = V_{op} - \frac{\Delta V_o}{2} \qquad (5\text{-}2)$$

Dc load current I_L is measured by inserting a dc milliammeter in series with R_L. However, if we know R_L, we can calculate I_L with the dc voltmeter reading from

$$I_L = \frac{V_{dc}}{R_L} \qquad (5\text{-}3)$$

Dc meters tell us nothing about ac ripple voltage ΔV_o. If a CRO is not available we must connect an ac voltmeter across R_L. The usual ac voltmeter does not measure peak-to-peak readings of an ac voltage, but measures rms voltage. Since ΔV_o is approximately triangular, we can measure rms ripple voltage V_r with an ac voltmeter and convert, if necessary, to ΔV_o from

$$\Delta V_o = 2\sqrt{3}\, V_r = 3.5 V_r \qquad (5\text{-}4)$$

(Some multimeters have an ac voltmeter section that measures the sum of dc component V_{dc} and V_r. If abnormally high rms ripple is read, connect a 1μF-600V capacitor in series with the ac voltmeter to remove the dc component.)

Study of rectifier performance is divided into two parts. First, we see how dc load voltage varies with dc load current. This is strictly a dc analysis, and we temporarily neglect ac ripple. Second, we use only ac meters to analyze how ac ripple voltage varies with dc load current, and temporarily neglect the dc problem. Finally we will superimpose both ac and dc problems to see how they are interrelated.

5-2 DC VOLTAGE REGULATION OF THE HALF-WAVE RECTIFIER

The half-wave rectifier is redrawn in Fig. 5-2 to show how dc load current I_L and dc load voltage V_{dc} are measured. When R_L is an open circuit ($R_L = \infty$), capacitor C will charge to the peak positive value of secondary ac voltage. Under this special load condition, $I_L = 0$, and C never discharges. Thus V_{dc} will equal V_{op} and also E_m at no load, because ac ripple voltage is zero, or

$$V_{\text{dc no-load}} = V_{op} = E_m = (1.41)E_{\text{rms}} \tag{5-5}$$

As R_L is reduced to draw more dc load current, dc load voltage decreases. Corresponding values of I_L and V_{dc} are plotted and points are connected to draw the *dc voltage regulation curve*. Regulation curves are records of how load voltage varies with load current. Note that V_{dc} has dropped to 18.3V at full-load current of 120mA from 21V at no load. This change in dc load voltage, ΔV_{dc}, occurs because, as dc current is drawn from the transformer-diode-capacitor combination, it flows through an equivalent *internal* or *output resistance* symbolized by R_o. As load current increases, internal voltage drop across R_o increases, to leave less dc terminal voltage. R_o is complex and depends on such factors as wire

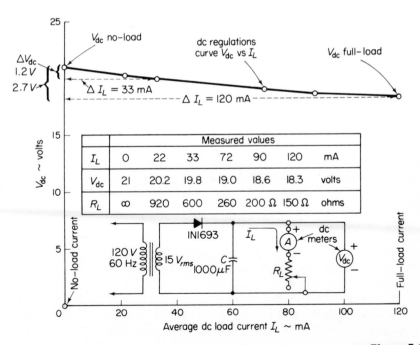

Figure 5-2

Regulation curve for a half-wave rectifier.

resistance, diode IV characteristic, capacitor size, transformer construction, and even the length and type of power run to the wall outlet. It is usually easier to build the rectifier and measure R_o than to measure each of the above factors (if possible) and then try to calculate R_o from an unwieldy formula.

Measurement of R_o

From the regulation curve of Fig. 5-2 we use R_o to tell how much V_{dc} will drop between no-load and full-load current. Expressed mathematically

$$V_{dc\ full\text{-}load} = V_{dc\ no\text{-}load} - (I_{L\ full\text{-}load})(R_o) \tag{5-6a}$$

Since $V_{dc\ no\text{-}load}$ divides between R_o and R_L, then

$$V_{dc\ full\text{-}load} = V_{dc\ no\text{-}load} \frac{R_L}{R_o + R_L} \tag{5-6b}$$

Equation (5-6a) can also be used to measure R_o at any intermediate value of load current, as in the following example.

EXAMPLE 5-1:

From the data of Fig. 5-2, calculate R_o for maximum dc load currents of: (a) 33mA, and (b) 120mA.

SOLUTION:

From Eq. (5-6) and data of Fig. 5-2

(a) $V_{dc\ at\ 33mA} = V_{dc\ no\text{-}load} - (I_{L\ at\ 33mA})(R_o)$

$$19.8V = 21V - (33mA)R_o$$

$$R_o = \frac{\Delta V_{dc}}{\Delta I_L} = \frac{(21 - 19.8)V}{(0 - 33)mA} = \frac{1.2V}{-33mA} = -40\Omega$$

The minus sign signifies that V_{dc} goes up $(+)$ as I_L goes down $(-)$. Observe that R_o is also the ratio of a *change* in V_{dc} or ΔV_{dc} to a corresponding *change* in I_L or ΔI_L.

(b) $V_{dc\ full\text{-}load} = V_{dc\ no\text{-}load} - I_{L\ full\text{-}load}R_o$, or

$$R_o = \frac{V_{dc\ no\text{-}load} - V_{dc\ full\text{-}load}}{I_{L\ no\text{-}load} - I_{L\ full\text{-}load}} = \frac{(21 - 18.3)V}{(0 - 120)mA} = -22.5\Omega$$

We can conclude from Example 5-1 that R_o apparently decreases with increasing load. This observation is in agreement with the fact that the regulation curve is not a straight line. As a matter of fact, for load currents exceeding 1 ampere, expect values of R_o to be in the order of a few ohms.

EXAMPLE 5-2:

A 120 to 24V transformer is connected to a half-wave rectifier. Based on experience, we estimate R_o will be about 20Ω when load current is 100mA. What is the value of V_{dc} (a) at no load; (b) at 100mA?

SOLUTION:

(a) From Eq. (5-5)

$$V_{dc\ no\text{-}load} = (1.41)(24V) = 34V$$

(b) From Eq. (5-6)

$$V_{dc\ full\text{-}load} = 34V - (100mA)(20Ω) = 32V$$

EXAMPLE 5-3:

What transformer would we need to build a half-wave rectifier that would supply 24V at full-load current of 200mA? Assume $R_o = 20Ω$.

SOLUTION:

From Eq. (5-6)

$$V_{dc\ no\text{-}load} = V_{dc\ full\text{-}load} + I_L R_o = 24V + (200mA)(20Ω) = 28V$$

From Eq. (5-5), $E_m = V_{dc\ no\text{-}load} = 28V$, and secondary rms voltage must be

$$E_{rms} = \frac{E_m}{1.41} = \frac{28V}{1.41} \approx 20V$$

We need a 120/20V transformer.

Transformer Selection

When we try to build the rectifier of Example 5-3, we will probably find that a 120/20V transformer is not a stock item. We could get a 120/24V stock transformer at substantial savings over a special 20V order. This would give a no-load V_{dc} of 34V, so we need to reduce this 34V by a drop of $34 - 24 = 10V = 200$ mA $\times R_o$, or $R_o = 50Ω$. Thus we must add a 30Ω in series with R_L to increase the existing 20Ω output resistance to 50Ω. If load current now goes down to, for example, 10mA, load voltage will rise to about 34V. Suppose the load is an amplifier designed to work at 24V, and suddenly its supply voltage is 34V. Bias voltages will change and some components may even be damaged because of excessive voltage. These last remarks illustrate that getting the right dc voltage, at the right dc load current, at a cost we can afford, is often a matter of luck with a typical half-wave rectifier. When we add a resistor to get the desired voltage, we get poorer regulation. *Both* problems of correct dc voltage and improved regulation will be solved by adding a voltage regulating zener (as shown in section 5-7).

5-3 AC RIPPLE VOLTAGE OF THE HALF-WAVE RECTIFIER

For each value of dc load current from a half-wave rectifier, there is a particular value of dc load voltage. Riding on or superimposed on this dc load voltage is an ac ripple voltage ΔV_o. We approximate actual ripple voltage in Fig. 5-1c by a triangular wave in Fig. 5-3a. One cycle of ripple voltage occurs over the same period T as line voltage frequency. For convenience, time A is chosen as refer-

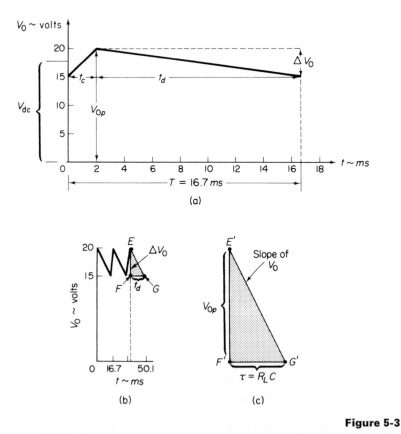

(a)

(b)   (c)

Figure 5-3

Ac ripple voltage in (a) has a discharge slope of $\Delta V_o/t_d$ in (b), and also $V_{op}/(R_L C)$ in (c).

ence. Usually we need to know what size to make filter capacitor C in order to reduce ΔV_o to a minimum value. We must find a relationship among load R_L, rectifier voltage, C, and ΔV_o. In Fig. 5-3b, triangle EFG, the slope, is determined by ΔV_o and discharge time t_d. In Fig. 5-3c the slope of V_o, during t_d, is extended to the x-axis along line $E'G'$ to form a similar triangle, $E'F'G'$. During the initial discharge of C, its voltage V_o has a slope of V_{op} divided by the circuit time constant. Since the diode is reverse-biased during t_d, C can discharge only through R_L, so that circuit time constant τ equals R_LC. Equating slopes:

$$\frac{\Delta V_o}{t_d} = \frac{V_{op}}{R_LC} \tag{5-7}$$

where t_d is in seconds, R_L in ohms, and C in farads.

Selection of Filter Capacitor C

Each factor in Eq. (5-7) can be measured—ΔV_o, V_{op}, and t_d with a CRO, R_L and C with a bridge. We may not have a CRO available, or may want to predict ΔV_o before building the circuit. There are difficulties because V_{op} and t_d depend on load current, and somewhat on R_o. That is, as load current increases, V_{op} decreases, since a large capacitor charging current during t_C reduces E_m because of internal transformer resistance. Also, as load current increases, C must charge and discharge more to furnish added load current and ripple increases.

We can make a good approximation of t_d for typical load currents as $t_d \approx 0.8T$. We can also approximate V_{op} for typical load currents as $V_{op} \approx 0.8E_m$. Making these substitutions in Eq. (5-7) and also substituting f for $1/T$ gives

$$\Delta V_o \approx \frac{0.6E_m}{fR_LC} \tag{5-8a}$$

and for a 60Hz line voltage substitute $f = 60$Hz:

$$\Delta V_o \approx \frac{E_m}{100R_LC} \tag{5-8b}$$

Equations (5-8a) and (5-8b) are only for a half-wave rectifier. Do not expect exact results if you try to calculate ΔV_o from Eq. (5-8b) (or from any other similar equation in the literature, for that matter). Measured capacitance of electrolytic capacitors may be 20 percent or more than the labeled value. Electrolytic capacitors also have an equivalent series loss resistance of 2 to 30 ohms, and for closer results this value should be added to R_L. Finally, there can be up to 20 percent difference between measured and labeled value of R_L.

Ripple voltage is also specified as the ratio of rms ripple voltage V_r to dc full-load voltage V_{dc}. Multiplying the ratio by 100 gives % ripple where

$$\% \text{ ripple} = 100 \frac{V_r}{V_{dc}} \tag{5-9}$$

EXAMPLE 5-4:

Predict p/p ripple voltage for the circuit of Fig. 5-2 at (a) full load, (b) $I_L = 33\text{mA}$, and (c) no load.

SOLUTION:

$$E_m = V_{dc\ no\text{-}load} = 21\text{V}$$

(a) From Eq. (5-8b)

$$\Delta V_o = \frac{21\text{V}}{100(150\Omega)(1000\mu f)} = 1.4\text{V}$$

(b) $$\Delta V_o = \frac{21\text{V}}{100(600\Omega)(1000\mu f)} = 0.35\text{V}$$

(c) $$\Delta V_o = 0\text{V, at no load}$$

We learn from Example 5-4 that ripple voltage increases directly with a decrease in R_L and with an increase in load current. That is, if I_L doubles because R_L is halved, then ripple voltage doubles. Note also that if C is doubled, ripple voltage will be reduced by one-half. Next we superimpose dc and ac solutions onto a half-wave rectifier problem.

EXAMPLE 5-5:

A half-wave rectifier is built from a 120/12.6V transformer. Assume R_o will equal 20 ohms at full-load current of 200mA. (a) Find $V_{dc\ full\text{-}load}$. (b) Choose C to give 2 percent ripple at full load. (c) What value of V_{op} would be seen on a CRO at full load?

SOLUTION:

(a) From Eq. (5-5)

$$E_m = 1.41 \times 12.6\text{V} = 17.8\text{V} = V_{dc\ no\text{-}load}$$

From Eq. (5-6)

$$V_{dc\ full\text{-}load} = 17.8\text{V} - (200\text{mA})(20\Omega) = 13.8\text{V}$$

(b) From Eq. (5-9)

$$V_r = \frac{\% \text{ ripple} \times V_{dc}}{100} = \frac{2 \times 13.8V}{100} = 0.27V$$

From Eq. (5-4)

$$\Delta V_o = 3.5 \times 0.27V \cong 1V$$

Rearranging Eq. (5-3)

$$R_L = \frac{V_{dc}}{I_L} = \frac{13.8V}{200mA} = 69\Omega$$

Because we do not know V_{op} or t_d, we must use approximate Eq. (5-8b) rather than Eq. (5-7)

$$C \approx \frac{E_m}{100R_L\Delta V_o} = \frac{17.8V}{(100 \ c/s)(69\Omega)1V} \cong 2600 \times 10^{-6}\mu F$$

Choose $C \approx 3,000\mu F$.
(c) From Eq. (5-2)

$$V_{op} = V_{dc} + \frac{\Delta V_o}{2} = 13.8V + \frac{1V}{2} = 14.3V$$

EXAMPLE 5-6:
Refer to the CRO measurement of V_o in Fig. 5-3a. What are the values of: (a) V_{dc}? (b) ΔV_o? (c) V_r? (d) % ripple? (e) If $R_L = 200\Omega$, what is the value of C?

SOLUTION:
From Fig. 5-3a, (a) $V_{dc} = 17.5V$, (b) $\Delta V_o = 5V$. (c) From Eq. (5-4), $V_r = \Delta V_o/3.5 = 5V/3.5 = 1.43V$. (d) From Eq. (5-9), % ripple = $1.43V \times 100/17.5V = 8\%$. (e) For Eq. (5-7) find $V_{op} = 20V$ from Fig. 5-3a, and $t_d = 14.7$ ms.

$$C = \frac{V_{op} \times t_d}{\Delta V_o \times R_L} = \frac{(20V)(14.7ms)}{(5V)(200\Omega)} = 294 \times 10^{-6}F \approx 300\mu F$$

Maximum Diode Current Ratings

We have seen that, as load current increases, ripple increases. During discharge time t_d, C must furnish an average load current I_L with a total charge equal to $I_L t_d$. The same amount of charge must be replaced in C during charge time t_C. Since t_C is much shorter than t_d, average charging current I_{CH} is much larger than discharge current I_L. We can predict the maximum current capacity required of the diode by equating charge

$$I_{CH}\, t_C = I_L\, t_d \tag{5-10a}$$

Maximum charging current occurs at full load of $I_{L\,max}$, so the diode must be able to withstand a *peak repetitive forward current* $I_{FM(rep)}$ of

$$I_{FM(rep)} = I_{CH} = I_{L\,max}\frac{t_d}{t_c} \tag{5-10b}$$

$I_{FM(rep)}$ is usually 5 to 10 times *maximum average forward current rating* I_o. For safety pick a diode with a maximum I_o rating equal to twice $I_{L\,max}$.

In section 5-13 we will learn how to make a simple measuring circuit that will display the regulation curve and ripple voltage simultaneously on a CRO.

5-4 FULL-WAVE BRIDGE-TYPE RECTIFIER

The same transformer and secondary sine-wave voltage of Fig. 5-1a are used with the full-wave bridge rectifier of Fig. 5-4a. If load and filter capacitor are removed, a CRO would show V_o to be the dashed rectified sine wave in Fig. 5-4b. Adding C would give $V_{dc\,no\text{-}load} = 21V$ with no ripple, because R_L is not present to discharge C. When R_L is then connected to draw $I_L = 175mA$, V_{dc} is reduced to 17.8V because of output resistance R_o, as predicted by Eqs. (5-3) and (5-6). Ripple voltage ΔV_o is shown in Fig. 5-4c. During time interval AB, the positive input half-cycle exceeds V_o enough to forward-bias D_1 and D_2 to charge C. Likewise, during time interval CD, the negative input half-cycle forward-biases D_3 and D_4 to recharge C. At all other times the diodes are reverse-biased. Maximum *peak-inverse voltage* (PIV) or reverse-bias applied to each diode pair is $E_m + V_{op} \approx 2E_m$. But since two diodes are always in series, they divide the peak-inverse voltage almost equally. Therefore, the diodes used in a full-wave bridge only need a PIV rating greater than E_m.

Selection of Filter Capacitor C

One full cycle of ripple voltage occurs in only one-half the period of the line frequency. Equation (5-1) is modified for full-wave operation

$$t_c + t_d = \frac{T}{2} = \frac{1}{2F} \tag{5-11}$$

The ratio expressed in Eq. (5-7) still applies to the full-wave rectifier. But now we

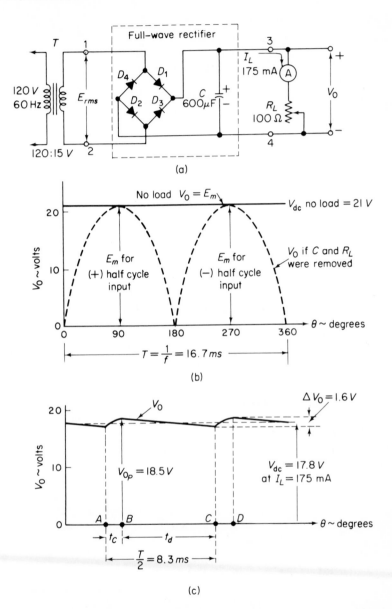

Figure 5-4

A full-wave rectifier in (a) has output voltage V_O shown at no-load in (b) and at full-load in (c).

substitute the approximation $t_d \approx 0.8T/2$ and $T = 1/f$ into Eq. (5-7) to obtain

$$\text{full-wave } \Delta V_o \approx \frac{0.3E_m}{fR_LC} \tag{5-12a}$$

and for a 60Hz line voltage substitute, $f = 60$Hz yields

$$\Delta V_o \approx \frac{E_m}{200 R_L C} \tag{5-12b}$$

where ΔV_o is the ripple voltage and E_m is the no-load voltage.

It is clear from a comparison of Eqs. (5-8) with Eqs. (5-12) that, for the same load resistance and capacitance, the full-wave rectifier will give one-half as much ripple as a half-wave rectifier. For the same ripple at the same load, we need only half as much capacitance for a full-wave rectifier as we do for a half-wave rectifier.

EXAMPLE 5-7:

(a) From Fig. 5-4 find R_o for a load current change of 0 to 175mA. (b) Assume we have measured $V_{dc} = 17.8$V at $I_L = 175$mA, and $C = 600\mu$F. Predict the value of ΔV_o.

SOLUTION:

(a) From Fig. 5-4b read $V_{dc\ no\text{-}load} = 21$V, and from Fig. 5-4c read $V_{dc} = 17.8$V at $I_L = 175$mA. From Eq. (5-6)

$$17.8\text{V} = 21\text{V} - (175\text{mA})R_o$$

and

$$R_o = \frac{3.2\text{V}}{175\text{mA}} = 18.3\Omega$$

(b) From Eq. (5-3)

$$R_L = \frac{V_{dc}}{I_L} = \frac{17.8\text{V}}{175\text{mA}} = 102\Omega$$

From Eq. (5-12b)

$$\Delta V_o = \frac{21}{(200)(102)(600 \times 10^{-6})} = 1.7\text{V}$$

As will often be the case, calculated ΔV_o is larger than measured ΔV_o. (The reason is that the capacitor actually has about 20 ohms equivalent series resistance, and would change the calculated ΔV_o to ≈ 1.4V.)

Measurement of Full-wave Rectifier Performance

A dc regulation curve for the full-wave rectifier of Example 5-7 was measured and plotted as a solid line in Fig. 5-5. Capacitor C was then doubled to 1200μF and dc voltage regulation was essentially the same as with $C = 600\mu$F. Ripple voltage ΔV_o was measured at each load current and plotted by centering ΔV_o on V_{dc} in accordance with Eq. (5-2). Short-dash lines in Fig. 5-5 trace out the

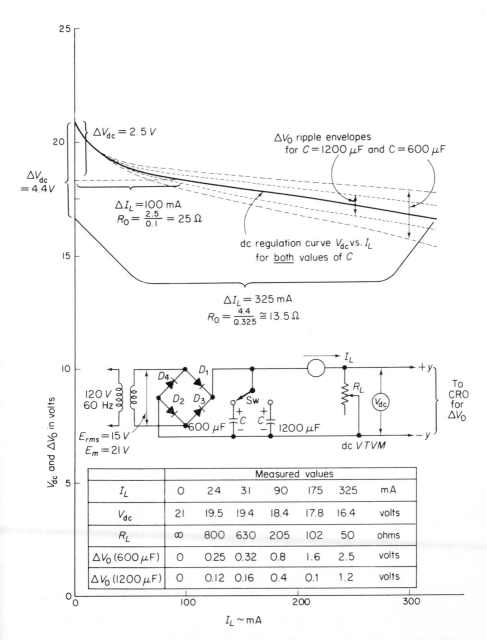

Figure 5-5

Typical full-wave dc voltage regulation curve with ac ripple superimposed.

envelope of ΔV_o with $C = 1200\mu\text{F}$. Long-dash lines trace out the envelope of ΔV_o with $C = 600\mu\text{F}$. We get a picture of ripple increasing with increasing load current at the same time that V_{dc} is decreasing. As expected, ΔV_o is halved when C is doubled. For example, at $I_L = 90\text{mA}$, $\Delta V_o = 0.8\text{V}$ for $C = 600\mu\text{F}$ and $\Delta V_o = 0.4\text{V}$ for $C = 1200\mu\text{F}$.

Because C is charged twice during each power cycle in a full-wave rectifier, about twice as much load current can be furnished compared to the half-wave for the same diode $I_{FM(\text{rep})}$ rating. Thus the full-wave rectifier is superior, with respect to ripple and load current capability, for the same size filter capacitor.

Negative and Positive Voltage Supplies

Transformer T in Figs. 5-1 and 5-4 isolates load R_L from the ac line voltage. One side of the transformer primary will always be electrically grounded, depending on how the power plug is inserted into the wall outlet. Since there is no direct connection from either primary wire to the secondary, we can choose either dc output terminal as reference and call it ground. For example, if we wire the $(-)$ output terminal to the chassis and make all voltage measurements with respect to the chassis, we can call the chassis ground. The dc rectifier is then a positive supply because loads are connected to the "hot" $(+)$ output terminal and returned to the $(-)$ ground terminal. Alternately, we can wire the $(+)$ terminal to the chassis and create a negative supply. In either case the dc supply is classified as "floating" in the sense that it is not directly attached to ac or water pipe ground.

5-5 FULL-WAVE CENTER-TAPPED BRIDGE RECTIFIER

As will be explained from Fig. 5-6a, a center-tapped transformer converts a bridge rectifier from a single-pole dc supply to a bipolar (two-pole) dc supply. Diodes D_1 and D_3, together with $C+$, form a positive voltage power supply in Fig. 5-6b. Charging current direction is shown at the positive input half-cycle peak. While the transformer secondary is specified as 24V rms, only the top half of the secondary voltage charges C to $12\text{V} \times 1.41 = 17\text{V}$. Diode D_3 is reverse-biased by $V_o \approx 17\text{V}$ in series with 17V from the bottom secondary transformer half. Hence D_3 must have a PIV rating greater than $17\text{V} + 17\text{V} = 34\text{V}$, or twice the ac peak voltage between center tap and either transformer terminal 1 or 2. During the negative input half-cycle, D_3 conducts I_{CH} to charge C back up to $\approx 17\text{V}$.

Diodes D_2 and D_4, together with $C-$, form a full-wave negative power supply in Fig. 5-6c. For the same positive input half-cycle, D_2 conducts current (at the same time that D_1 conducts in Fig. 5-6b) from the bottom secondary winding to charge $C-$. During the negative input half-cycle peak, D_4 connects the top secondary half to recharge $C-$.

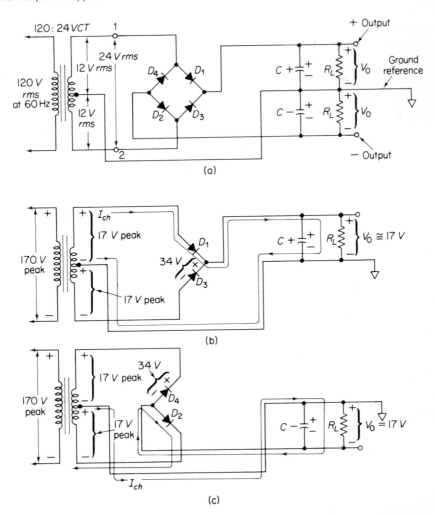

Figure 5-6

The full-wave center-tapped bridge rectifier in (a) is made up of the positive output part in (b) and negative output part in (c).

Comparison of Full-wave Center-tapped and Bridge-type Supplies

The single-pole full-wave rectifiers of Figs. 5-6b and 5-6c can be used in place of the full-wave bridge rectifier of Fig. 5-4. However, for the same dc output voltage, for example $V_{\text{dc no-load}} = 21V$, we would need a 120:15 volt transformer for the full-wave bridge. The center-tapped (CT) full-wave transformer required would be 120:30 volt CT, with more wire (and bulk) in the secondary. Usually it is more economical to use a center-tapped bridge rectifier for a bipolar dc supply and a regular bridge rectifier for a single-pole dc supply.

   Dc regulation formulas and ac ripple formulas are the same for either
bridge or center-tapped full-wave rectifiers. The only difference is in the value
of E_m. For the center-tapped full-wave rectifier, E_m is the peak value of one-half
of the entire secondary voltage. For example, a 120:12.6 volt CT transformer
has an E_m of $(12.6 \text{ V}/2) \times 1.4 = (6.3\text{V})(1.4) = 9\text{V}$. More specifically, we would
measure 12.6V rms or 18V peak between terminals 1 and 2 in Fig. 5-6a, and
6.3V rms $= 9\text{V}$ peak between 1 and the center tap.

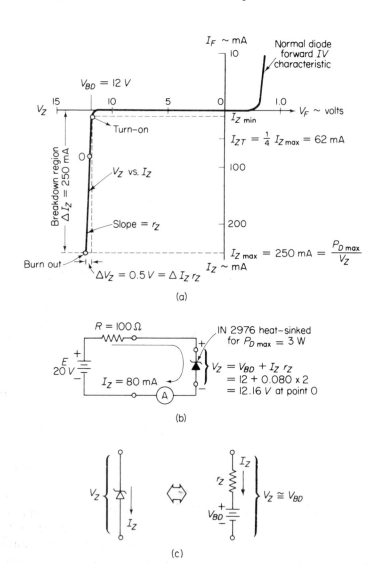

Figure 5-7

IV zener characteristic in (a) is measured from the test circuit in (b) to give
the circuit model in (c).

5-6 ZENER DIODE MEASUREMENTS

The zener diode compensates for major deficiencies in the half- or full-wave rectifier because the zener has low output resistance to improve dc regulation, and we can buy a zener that will give us almost any required value of V_{dc} between 1.5V and 200V. As will be shown in section 5-7, voltage across the zener V_Z is the actual output voltage V_o. To predict how V_o will vary with changes in input voltage or load current, we must know precisely how zener voltage varies with these same changes. For this reason we review the zener introduction in section 1-9 and go deeper into measurements of zener performance.

Zener Voltage

When the zener is used as a voltage regulator, it must always be reverse-biased and operate along its breakdown characteristic. In Fig. 5-7b, test voltage E reverse-biases the zener. As E is increased from 0V, negligible leakage current flows until E equals breakdown voltage V_{BD}. In order to insure that the zener has entered breakdown, E must be increased slightly until a minimum zener current I_{Zmin} (typically 5 to 10mA) flows. Then, as E is increased, V_Z increases slightly as more zener current flows because of voltage drop across an internal zener resistance, r_Z. Figure 5-7b shows current and voltage values for zener operating point 0 in Fig. 5-7a. Technically, the voltage drop across the zener is expressed as

$$V_Z = V_{BD} + I_Z r_Z \qquad (5\text{-}13a)$$

Realistically, for most applications, dc zener voltage is expressed by the excellent and simpler approximation

$$V_Z \approx V_{BD} \qquad (5\text{-}13b)$$

A circuit model for the zener and Eq. (5-13a) is given in Fig. 5-7c. This model is valid only when the zener operates in the breakdown region.

Zener Power

Maximum zener power dissipation P_{Dmax} is given in data sheets and sets the maximum limit of I_Z. Since zener power dissipation is determined by zener current and zener voltage, maximum dissipation is expressed as

$$P_{Dmax} = I_{Zmax} V_Z \qquad (5\text{-}14)$$

Zeners are available with power ratings between 250mW and 50W.

EXAMPLE 5-7:
(a) What is maximum allowable zener current in the circuit of Fig. 5-7?
(b) What value of E must not be exceeded to avoid zener burnout?

SOLUTION:

(a) From Eqs. (5-14) and (5-13b)

$$I_{Z\max} = \frac{P_{D\max}}{V_Z} = \frac{3W}{12V} = 250mA$$

(b) 250mA flowing through $R = 100\Omega$ develops a 25V drop across R. E must not exceed $V_R + V_Z = 25V + 12V = 37V$.

Zener Regulation Action

It is clear from Fig. 5-7 and Example 5-7 that the zener must always operate in the breakdown region, with a current between 5mA and 250mA, to insure turn-on and avoid burnout. Now we get to the key point on how the zener regulates. If zener current must always range between $I_{Z\min}$ and $I_{Z\max}$, then voltage across the zener can only change by an amount ΔV_Z. Since $I_{Z\min}$ is usually small in respect to $I_{Z\max}$, the *maximum change* in zener current will be approximately $I_{Z\max}$. ΔV_Z is related to $I_{Z\max}$ by zener resistance r_Z as

$$\Delta V_Z = (I_{Z\max})(r_Z) \tag{5-15}$$

(As a point of caution, actual change in zener current, ΔI_Z, in our circuit applications may be smaller than the power rating derived, $I_{Z\max}$.) Finally, *change* in output voltage ΔV_o will equal ΔV_Z, since the zener terminals will be the dc output terminals. Consequently, *change* in output voltage due to any ΔI_Z for whatever reason (regulation or ripple) will always equal ΔV_Z, or

$$\Delta V_o = \Delta V_Z = \Delta I_Z r_Z \tag{5-16}$$

Measurement of r_Z

It is most important that we be able to obtain an accurate measurement of r_Z, because r_Z determines how much dc output voltage will change [Eq. (5-16)]. For purposes of comparison, manufacturers measure r_Z from the zener's slope around a dc *test-zener current* I_{ZT}. I_{ZT} is defined as the current necessary to develop one-quarter rated power ($1/4P_{D\max}$), and therefore

$$I_{ZT} = \frac{I_{Z\max}}{4} \tag{5-17}$$

Since slope of the zener current will vary, becoming steeper away from the "knee," we must measure the slope over the same distance along the characteristic for comparison. Standard procedure is to vary I_Z over an rms range of 10 percent I_{ZT}, or peak-to-peak range of $10 \times \sqrt{2} \times 2 = 28\% \ I_{ZT}$, at 60Hz. This procedure is shown in Fig. 5-8. To measure r_Z for the zener of Fig. 5-7:

(1) Increase E until the dc milliammeter reads $I_{ZT} = 250\text{mA}/4 \approx 62\text{mA}$. (2) Ac voltmeter V_1 reads rms zener current in terms of ac voltage drop across current-sensing resistor R. Rms current should be $0.10 \times 62\text{mA} = 6.2\text{mA}$. So increase the variac until V_1 reads $6.2\text{mA} \times 100\Omega = 620\text{mV}$. The rms variation in I_Z or

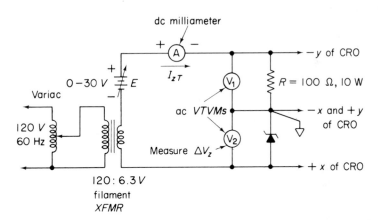

Figure 5-8

Measurement of r_Z at I_{ZT}.

$I_{Z\text{rms}}$ can be calculated from V_1 by $I_{Z\text{rms}} = V_1/100\Omega$. (3) Read the corresponding rms variation in V_Z or ΔV_Z from voltmeter V_2. (The ac VTVM gives much more sensitive and accurate measurement of the small variations in ΔV_Z than is available from the CRO.) Calculate r_Z from meter readings.

$$r_Z = \frac{\Delta V_Z}{\Delta I_Z} = \frac{V_2}{V_1/100\Omega} = 100\frac{V_2}{V_1} \qquad (5\text{-}18)$$

EXAMPLE 5-9:

Measurements are taken from the circuit of Fig. 5-8 as $V_1 = 620\text{mV}$, $V_2 = 21\text{mV}$. Find r_Z.

SOLUTION:

From Eq. (5-18)

$$r_Z = 100\frac{0.021}{0.620} = \frac{100}{30} \approx 3.3\Omega$$

A CRO may connected, as in Fig. 5-8, to display (1) dc operating point in the third quadrant with 0 variac voltage, or (2) slope of r_Z at I_{ZT} when E and the variac are adjusted for test conditions, or (3) the reverse breakdown I-V characteristic if variac voltage is increased.

5-7 DESIGN OF A ZENER DIODE REGULATOR

Resistor R_A and a zener diode form a regulator section inserted between an unregulated supply and load in Fig. 5-9a. To analyze or design the regulator we must know output requirements of the regulator, usually given as dc load voltage and maximum load current, and unregulated supply regulation curve.

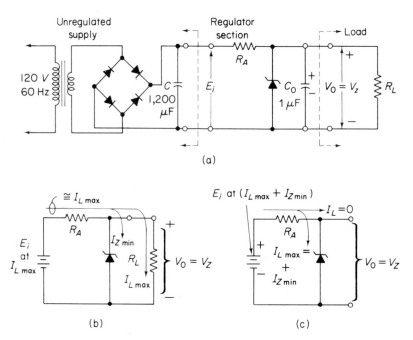

(a)

(b)   (c)

Figure 5-9

The zener regulated power supply in (a) has full-load and no-load
conditions shown in (b) and (c), respectively.

To understand zener regulator performance, consider the two dc load extremes of no-load and full-load. Neglect ac ripple for the present. When full-load current is drawn from the regulator in Fig. 5-9b, the zener must draw at least $I_{Z\text{min}}$ to stay in the breakdown region and maintain regulation. As will be shown, R_A will be designed to insure that the zener is not starved for current. When the load is removed, all of the previous maximum load current is absorbed by the zener. So, in a sense, the zener acts as a current switch to absorb all current not demanded by the load. Under both extremes of no-load and full-load current resistor R_A carries the *same* current (approximately $I_{L\text{max}}$). We can make this statement because $I_{Z\text{min}}$ is small with respect to $I_{L\text{max}}$. This means that current drawn from the unregulated supply will be constant, and we gain a distinct advantage. E_i will be stabilized and equal V_{dc} at I_{Lmax} of the unregulated supply. To restate this point, the output dc voltage of the *unregulated* supply does *not* ride up and down its regulation curve as load changes, but stays constant at a V_{dc} set by I_{Lmax}. Thus for any design or analysis problem we need to know V_{dc} of the unregulated supply at I_{Lmax} to find E_i.

Design of R_A

R_A can be increased to a maximum value in Fig. 5-9b, where it will just allow minimum zener current to flow. Since V_z is on one side of R_A, E_i at I_{Lmax} is on the other, and R_A carries $I_{Lmax} + I_{Zmin}$.

$$R_{Amax} = \frac{(E_i \text{ at } I_{Lmax}) - V_z}{I_{Lmax} + I_{Zmin}} \qquad (5\text{-}19)$$

There are three points to be considered in using Eq. (5-19). First, if you want to buy a 12V zener, data sheets and price data list zener voltage as *nominal zener voltage* $V_z = 12V$. However, there are three common grades of V_z tolerance at 20 percent, 10 percent, and 5 percent, plus special premium zeners guaranteed for 1 percent. Closer tolerance costs more money, so if we bought the more

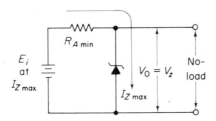

Figure 5-10

Circuit condition at maximum allowable zener power.

expensive 5 percent zener we would be guaranteed a V_z between $12V - 0.05 \times 12V = 11.4V$ and $12V + 0.05 \times 12V = 12.6V$. Sanity dictates that we can use 12V for V_z in Eq. (5-19). Second, I_{Zmin} is usually small with respect to I_{Lmax} and can be neglected in Eq. (5-19). Third, there will be a ripple voltage superimposed on E_i at I_{Lmax} corresponding to ΔV_o of the *unregulated* supply. Our dc value of E_i will have an instantaneous minimum of $E_i - \Delta V_o/2$, and if we selected R_A right at R_{Amax} the zener could come out of regulation for a few milliseconds during each half-cycle. However, for simplicity we will always pick R_A less than R_{Amax}. A good design procedure is to connect a $0.1\mu F$ to $1\mu F$ capacitor across the zener to hold the zener on during minimums of E_i and to short-circuit zener-generated noise.

R_A can be reduced to a minimum value in Fig. 5-10, where it will limit zener current to a maximum value of $I_{Z\,max}$. E_i is now specified at $I_{Z\,max}$ rather than at $I_{L\,max}$ because we are duplicating the actual circuit condition at the edge of zener burnout. By inspection

$$R_{A\,min} = \frac{(E_i \text{ at } I_{Z\,max}) - V_Z}{I_{Z\,max}} \qquad (5\text{-}20)$$

In the next examples we will show that a zener should be selected with an $I_{Z\,max}$ at least 2 times $I_{L\,max}$ to give us some design freedom in the choice of R_A.

EXAMPLE 5-10:

Design a regulator to supply 12V at a maximum load current of 100mA. Use the unregulated supply of Fig. 5-5 and the zener of Fig. 5-7b ($I_{Z\,max}$ = 3W/12V = 250mA).

SOLUTION:

Allow 5mA for $I_{Z\,min}$. From Fig. 5-5, $V_{dc} = 18.2$V at $I_L = I_{L\,max} + I_{Z\,min}$ = (100 + 5)mA = 105mA. From Fig. 5-11a and Eq. (5-19)

$$R_{A\,max} = \frac{(18.2 - 12)\text{V}}{(100 + 5)\text{mA}} = \frac{6.2\text{V}}{105\text{mA}} = 59\Omega$$

From Fig. 5-5, $V_{dc} = 17$V at $I_L = I_{Z\,max} = 250$mA. From Eq. (5-20) and Fig. (5-11b)

$$R_{A\,min} = \frac{(17 - 12)\text{V}}{250\text{mA}} = \frac{5\text{V}}{250\text{mA}} = 20\Omega$$

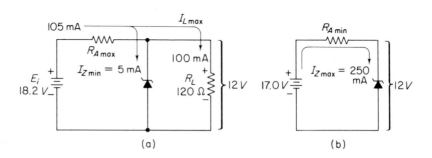

(a)   (b)

Figure 5-11

Calculation of $R_{A\,max}$ in (a) and $R_{A\,min}$ in (b) for Example 5-10.

As will be shown in section 5-8, R_A should be as large as possible for best ripple reduction, so choose $R_A = 47\Omega$ as a close standard 10 percent resistor to $R_{A\,max}$. Connect a 1μF capacitor across the zener.

EXAMPLE 5-11:

Assume we buy the lowest cost 12V zener rated at 1.0W. Try to use it in the regulator of Ex. 5-10.

SOLUTION:

From Eq. (5-14), $I_{Zmax} = P_{omax}/V_Z = 1W/12V \approx 83mA$. R_{Amax} remains at 59Ω. However, when the unregulated supply delivers 83mA, $V_{dc} = 18.5V$ $= E_i$ at I_{Zmax}. Substituting into Eq. (5-20)

$$R_{Amin} = \frac{(18.5 - 12)V}{83mA} = \frac{6.5V}{83mA} = 78\Omega$$

Now R_A cannot be lower than 78Ω or the zener will burn out when load is removed. But if $R_A = 78\Omega$, it will be larger than R_{Amax}, the zener will come out of regulation, and V_o will drop below 12V. Thus we cannot use the 1.0W Zener.

We conclude that zener power dissipation rating is the fundamental limitation on how much current a zener regulator can supply. Therefore, maximum available load current should be roughly one-third to one-half of I_{Zmax}.

5-8 ANALYSIS OF A ZENER REGULATOR

As has been stated, changes in load voltage ΔV_o of the zener regulator are determined by the zener's r_Z. The relationship was given by Eq. (5-16) and is repeated here for convenience.

$$\Delta V_o = \Delta I_Z r_Z$$

Based on our study in section 5-7, we concluded that the zener carries $I_{Zmin} = 5mA$ at full-load and $I_Z = I_{Lmax}$ at no-load. Thus maximum *change* in zener current is $\Delta I_Z = I_{Lmax} - I_{Zmin}$. Since I_{Zmin} is usually small with respect to I_{Lmax}, we can state maximum ΔV_o in more useful terms of I_{Lmax} by substituting into Eq. (5-16) to get a measure of voltage regulation from

$$\Delta V_{omax} = I_{Lmax} r_Z \qquad (5\text{-}21a)$$

Since the zener absorbs any change in load current ΔI_L, Eq. (5-21a) can also be written in general terms to give change of output voltage for any change in load current.

$$\Delta V_o = \Delta I_L r_Z \qquad (5\text{-}21b)$$

But since the ratio of a change in output voltage to a change in output current is by definition *output resistance*, let us rearrange Eq. (5-21b) as

$$\frac{\Delta V_o}{\Delta I_L} = r_z = R_{oz} \qquad (5\text{-}21\text{c})$$

where R_{oz} = output resistance of the zener regulator.

We conclude that any signal currents generated in the load (for example, if the load is an audio amplifier) will see a low power supply impedance of r_z.

The Cost of Regulation

We pay for regulation with loss of power through heat. For example, if V_{dc} of the unregulated supply equals $2V_Z$ or $2V_o$, then the voltage drop across R_A will equal $V_o \doteq V_Z$ at any load condition. See Figs. 5-9b and c, where $I_{R_A} \approx I_{Lmax}$ at full-load and no-load. Thus, full-load power of $V_o I_{Lmax} = V_Z I_{Lmax}$ equals heat power dissipated in R_A of $V_Z I_{Lmax}$. Note carefully that, at no-load, zener power is $V_Z I_{Lmax}$ and R_A still dissipates $V_Z I_{Lmax}$.

Only half of the unregulated supply's power ($2V_Z I_{Lmax}$) is used at full-load, and none of the *same* unregulated supply power is used at no-load. We conclude that the unregulated supply should be chosen for an E_i of about $1.5V_Z$ at $I_L = I_{Lmax}$, to minimize heat dissipation, yet allow a reasonable value for R_A.

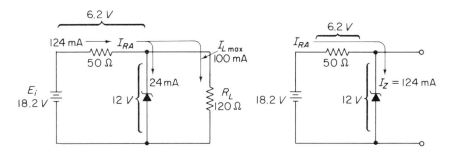

Figure 5-12

Solution to Example 5-12.

EXAMPLE 5-12:

Choosing $R_A = 50\Omega$ for the solution of Example 5-10, what currents flow in the zener at no-load and at full-load of 100mA?

SOLUTION:

Full-load conditions are shown in Fig. 5-12a, where E_i is estimated *initially* at $I_{Lmax} = 105$mA, as in Fig. 5-11a. By inspection or modifying Eq. (5-19), current through R_A is

$$I_{R_A} = \frac{E_i - V_Z}{R_A} = \frac{(18.2 - 12)V}{50\Omega} = \frac{6.2V}{50\Omega} = 124\text{mA}$$

and

$$I_Z = I_{R_A} - I_{L\text{max}} = 124\text{mA} - 100\text{mA} = 24\text{mA}$$

Now recheck our estimate of E_i by finding $V_{dc} = 18.1$V at $I_L = 124$mA from Fig. 5-5. If we changed E_i to 18.1V in Fig. 5-12, I_{R_A} would be 122mA and the correction is not worth the effort (in this particular case). Under no-load conditions in Fig. 5-12b, all of I_{R_A} flows through the zener.

EXAMPLE 5-13:

Assuming $r_Z = 3\Omega$ in the regulator of Example 5-12, what is the maximum possible change in output voltage for load current change between (a) no-load and full-load, (b) 25mA and 75mA?

SOLUTION:

(a) From Eq. (5-21a), $\Delta V_{o\text{max}} = (100\text{mA})(3\Omega) = 0.3$V. (b) $\Delta I_L = 75$mA -25mA $= 50$mA. From Eq. (5-21b), $\Delta V_o = (50\text{mA})(3\Omega) = 0.15$V.

Zener Ripple Voltage Reduction

It is useful to examine how the zener regulator affects ripple *both* at the unregulated supply terminals and at load terminals. We use the unregulated supply of Fig. 5-5 (with $C = 1200\mu$F) and the regulator of Example 5-12 to explore the problem. Since the regulator draws 124mA at 18.2V from the unregulated supply, it acts like a *dc load* of 18.2V/124mA $= 147\Omega$. (Note that this dc load does *not* equal $[R_A + R_L]$ because the zener draws some current.) From Fig. 5-5 we would expect a ripple from the unregulated supply of approximately ΔV_o (unreg.) $= 0.6$V. However, the ac load resistance seen by C of the unregulated supply is actually $R_A + r_Z = 53\Omega$. Employ Eq. (5-12a) to see that ripple increased from

$$\text{unreg. } \Delta V_o \approx \frac{0.3 \times 21}{60(147)(1,200)} = 0.6\text{V, assuming dc load} = 147\Omega$$

to

$$\text{unreg. } \Delta V_o = \frac{0.3 \times 21}{60(53)1,200} = 1.6\text{V, with actual ac load} \cong 53\Omega$$

In other words, adding the zener actually increased ripple in the unregulated supply roughly by the ratio $(R_A + R_L)/(R_A + r_Z)$, where R_L is the actual load on the regulator.

However, as shown in the ac model of Fig. 5-13, ripple voltage coming into the regulator, ΔE_i = unreg. ΔV_o, divides between R_A and r_Z. Only that part of the ripple developed across r_z appears in the load as Regulator ΔV_o. By inspection of Fig. 5-13, and noting that R_L and R_A are normally large with respect to r_Z

$$\text{Regulator } \Delta V_o = \frac{r_Z}{R_A + r_Z}\, \Delta E_i \approx \frac{r_Z}{R_A}\, \Delta E_i \tag{5-22}$$

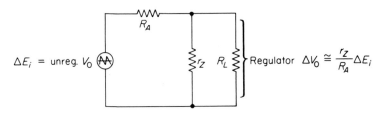

Figure 5-13

Ripple from the unregulated supply ΔE_i is reduced by division between R_A and r_Z.

EXAMPLE 5-14:

By what fraction will ripple voltage be reduced by the regulator of Fig. 5-12 if $r_Z = 3\Omega$ and $R_A = 50\Omega$?

SOLUTION:

From Eq. (5-22)

$$\frac{\Delta V_o}{\Delta E_i} = \frac{3}{50}$$

EXAMPLE 5-15:

What will be the net ripple voltage developed across the load in the regulator of Ex. 5-12?

SOLUTION:

It has been shown that ripple voltage coming into the regulator is 1.6V. This will be reduced, from Example 5-14, by 3/50 to

$$\text{Regulator } \Delta V_o = 1.6\text{V} \times \frac{3}{50} = 0.096\text{V}$$

Refer to Eq. (5-22) to conclude that R_A should be made as large as possible to improve ripple reduction. Furthermore, we reason that if a zener can reduce ripple by 3/50 after increasing ripple by a factor of 3, its net ripple reduction is $3 \times 3/50 \approx 1/5$. To reduce ripple by 1/5 with a larger capacitor (instead of a zener) we would need 5 times as much capacitance. In the examples thus far, we would need to increase C from $1200\mu\text{F}$ to $6000\mu\text{F}$. So as far as performance is concerned, zener cost is offset by savings in capacitor size and cost, and we obtain a dramatic improvement in regulation.

High-power zeners are more expensive than high-power transistors. For this reason we will add a transistor in the next section to remove economically the power limitation of the zener. The transistor will absorb power formerly absorbed by the zener.

5-9 SERIES REGULATOR ACTION

Addition of a high-power transistor to the zener regulator eliminates the zener's heat limitation. The zener-transistor pair acts as a power zener, although the zener acts more like a voltage reference and the transistor does the regulating. The resulting regulator, shown in Fig. 5-14, is known as an *emitter-follower* regulator. Since transistor Q is in series with load and unregulated supply, this regulator arrangement is classified as a series regulator.

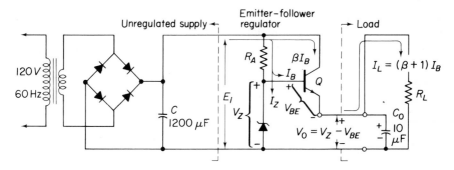

Figure 5-14

Series emitter-follower regulated power supply.

In Fig. 5-14, resistor R_A conducts current to turn the zener on. Of course E_i must always be larger than V_Z. Base-to-ground voltage of Q is clamped at V_Z by the zener. Q's base-emitter junction is forward-biased, so its emitter voltage will always be lower than (or follow) its base voltage by V_{BE}. Since Q's emitter is the regulator output terminal, output voltage V_o is

$$V_o = V_Z - V_{BE} \qquad (5\text{-}23)$$

Circuit Currents

Load current I_L is set by R_L, and V_o from Eq. (5-23). β of the transistor has absolutely nothing to do with I_L if the regulator is working properly. All β does is determine how I_L will divide between base and collector terminals of Q. At full-load, I_{Lmax}, the zener carries minimum current and base current I_B is maximum at I_{Lmax}/β. V_{BE} will be at its maximum value of around 0.8 to 1.0V for large load currents. At very small or zero load current, V_{BE} is small at 0.4 to 0.5V and the zener absorbs I_{Lmax}/β. From these statements we compare improvements given by the emitter-follower regulator over the straight zener regulator.

Comparison of Zener and and Emitter-Follower Regulator Action

First, zener current change from full- to no-load is no longer ΔI_{Lmax}, but has been reduced by β of the transistor to $\Delta I_{Lmax}/\beta$. This means that when load is removed the zener absorbs I_{Lmax}/β, not I_{Lmax}, so power dissipation in the zener is reduced almost by a factor of β and we can use a low-cost zener.

Second, zener voltage variation in the emitter-follower regulator, because of load current changes, is reduced from $(I_{Lmax})(r_Z)$ to $(I_{Lmax}/\beta)r_Z$. At first glance this reduction in zener voltage variation would appear to improve output voltage regulation. However, this is not the case. Assume that load current variations in V_Z are negligible. At no-load, V_{BE} is about 0.5V, and at full-load V_{BE} is about 0.9V. Emitter-to-ground voltage or V_o must fluctuate by $(0.9 - 0.5)V = 0.4V$. Since the emitter current is load current, changes in load current will cause changes in V_{BE}, depending on the $I_E - V_{BE}$ characteristic of the transistor. Thus, changes in V_{BE} determine $\Delta V_o/I_L = \Delta V_{BE}/I_L$, or voltage regulation. If we need better regulation, we have reached the end of the line with this circuit and must change to one of the feedback regulators in the next chapter.

5-10 DESIGN OF AN EMITTER-FOLLOWER REGULATOR

Design procedure will be illustrated by an example similar to the zener regulator so that we may learn from comparison. Refer to Examples 5-9 to 5-12. Given $r_Z = 3\Omega$ and P_{max} of the zener $= 500\text{mW}$ (we change to a low-cost, low-power zener), design an emitter-follower to supply 12V at a maximum load current of 100mA. Use the unregulated supply of Fig. 5-5.

Design of R_{Amax}

Full-load conditions are shown in Fig. 5-15a. Base current I_B is found from

$$I_B = \frac{I_{Lmax}}{\beta + 1} = \frac{100\text{mA}}{50} = 2\text{mA} \qquad (5\text{-}24)$$

Allowing 5mA to hold the zener in breakdown, unregulated dc supply voltage is found from Fig. 5-5 to be $E_i = V_{dc} = 18.2\text{V}$ at $I_L = 105\text{mA}$. Remember, as long as the zener is on, $I_B + I_C$ will not change. They are set by R_L and $(V_Z - V_{BE})$. So if we increase R_A, only zener current will be reduced. R_A cannot be

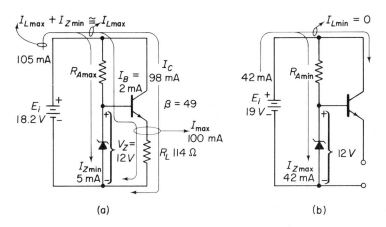

Figure 5-15

Design aids for calculation of $R_{A\max}$ and $R_{A\min}$ in (a) and (b) at full-load and at zener burnout, respectively.

increased beyond a value where I_Z drops below $I_{Z\min}$. This maximum value of R_A is expressed from Fig. 5-15a.

$$R_{A\max} = \frac{E_{i\text{ at full-load}} - V_Z}{I_{Z\min} + I_{L\max}/(\beta + 1)} = \frac{(18.2 - 12)\text{V}}{[5 + (100/50)]\text{mA}} = 890\Omega \qquad (5\text{-}25)$$

Design of $R_{A\min}$

From a 500mW zener, $I_{Z\max}$ is found from Eq. (5-14) as $I_{Z\max} = 500\text{mW}/ 12\text{V} \approx 42\text{mA}$. No load conditions at burnout are shown in Fig. 5-15b. R_A cannot be reduced below a value called $R_{A\min}$, that is necessary to hold zener current less than $I_{Z\max}$. Note that $E_i = 19\text{V}$ at 42mA from the unregulated supply. Contrast this situation with those in Figs. 5-11b and 5-12b. By inspection of Fig. 5-15b

$$R_{A\min} = \frac{E_{i\text{ at }I_{Z\max}} - V_Z}{I_{Z\max}} = \frac{(19 - 12)\text{V}}{42\text{mA}} = 167\Omega \qquad (5\text{-}26)$$

(Note that we could have used a cheaper 250mW zener, with $I_{Z\max} = 21\text{mA}$, $R_{A\min} \cong 330\Omega$.) Select a standard size resistor below $R_{A\max}$, such as $R_A = 680\Omega$.

5-11 ANALYSIS OF AN EMITTER-FOLLOWER REGULATOR

To analyze performance of the regulator designed in section 5-10, begin with the full-load situation in Fig. 5-16a. We know $I_L = 100\text{mA}$ and I_Z will be somewhere between 5 and 42mA. *Estimate* current from the unregulated supply at $100 + (5 + 42)/2 \cong 125\text{mA}$ to get an estimate of $E_i = 18.2\text{V}$ from Fig. 5-5. Now calculate current through R_A from

$$I_{R_A} = \frac{E_{i\text{ at full-load}} - V_Z}{R_A} = \frac{(18.2 - 12)\text{V}}{680\Omega} = 9.1\text{mA} \qquad (5\text{-}27)$$

Checking our assumption, unregulated supply current will be $I_{R_A} + I_C \cong 9 + 98 = 107\text{mA}$, and $V_{dc} = E_i$ is still approximately 18.2V. Since $I_B = 2\text{mA}$, zener current will be

$$I_Z = I_{R_A} - I_B = 9.1\text{mA} - 2\text{mA} = 7.1\text{mA} \qquad (5\text{-}28)$$

At no-load, zener current should increase but remain below $I_{Z\text{max}} = 42\text{mA}$. Assume initially that I_Z equals $I_{Z\text{max}}/2 = 21\text{mA}$, and obtain $E_i = 19.5\text{V}$ at 21 mA from Fig. 5-5. From Fig. 5-16b

$$I_Z = \frac{E_{i\text{ at no-load}} - V_Z}{R_A} = \frac{(19.5 - 12)\text{V}}{680\Omega} = 11\text{mA} \qquad (5\text{-}29)$$

(Now that we have a better estimate of I_Z, we could correct E_i to 20V at 11mA and employ Eq. (5-28) to get I_Z actually equal to 11.8mA.)

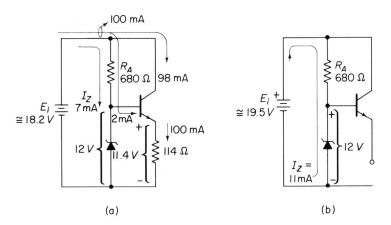

(a)   (b)

Figure 5-16

Evaluation of full-load and no-load currents, in (a) and (b) respectively, for an emitter-follower regulator.

Comparison of Zener and Emitter-follower Regulator Performance

Compare Fig. 5-16b with Fig. 5-12b. With the emitter-follower we have lost one advantage of the zener regulator, with which unregulated supply voltage did not vary. But we have gained a significant advantage, in that we do not waste a lot of power in the zener at no-load.

Another advantage of the emitter-follower is that R_A can be made larger (680Ω versus 47Ω). This means ripple voltage in the unregulated supply is *not* increased significantly by the addition of the regulator. Furthermore, the reduction in ripple is still expressed by Eq. (5-22) and is better because R_A is larger.

EXAMPLE 5-16:

If $\Delta V_o = 1.0\text{V}$ for the unregulated supply, at full-load, in the emitter-follower of our design example, what is ΔV_o of the regulated supply? $R_A = 680\Omega$ and $r_Z = 3\Omega$.

SOLUTION:

From Eq. (5-22)

$$\text{Regulator } \Delta V_o = \frac{r_Z}{R_A}\,\Delta E_i = \frac{3\Omega}{680\Omega}(1\text{V}) \cong 4\text{mV}$$

Output resistance R_{oE} of the emitter-follower to the flow of signal currents is essentially equal to the low ac emitter-base resistance of Q in series with r_Z. Both resistances are in the same order of magnitude, and R_{oE} can be *approximated* by

$$R_{oE} \cong 2r_Z \tag{5-30}$$

An electrolytic capacitor of 1 to $10\mu\text{F}$ is usually added across the load. In addition, a $0.1\mu\text{F}$ capacitor may be added across the electrolytic. Both capacitors reduce ac output resistance over an extended frequency range. Adding a small capacitor ($0.1\mu\text{F}$) across the zener also helps to hold the zener in regulation at larger load currents, and reduces effects of zener noise.

5-12 TRANSISTOR DISSIPATION IN THE EMITTER-FOLLOWER REGULATOR

Maximum power dissipation of the transistor, $P_{D\text{max}}$, occurs when the transistor conducts full-load current. For our design and analysis example in sections 5-10 and 11 we evaluate $P_{D\text{max}}$ by reference to Fig. 5-16.

$$P_{D\text{max}} = (E_{i\,\text{at full-load}} - V_o)\,I_{L\text{max}} \tag{5-31}$$

$$P_{D\text{max}} = (18.2 - 11.4)(0.1) = 0.68\text{W}$$

where V_o is found from Eq. (5-23).

If we redesigned R_A for a full-load current of 325mA, the full-wave supply of Fig. 5-5 gives an E_i of 16.4V. Poor regulation of the unregulated supply actually helps protect the transistor, because E_i drops to reduce voltage across the transistor. At a regulator output of 325mA, $P_{D\text{max}}$ would be $(16.4 - 11.4)$

$(0.325) = 1.6\text{W}$. If E_i had remained at 18.2V, $P_{D\text{max}}$ would have been 2.2W, or about a third more power. This observation will show us how to add a measure of short-circuit protection.

Short-circuit Protection

When the load is short-circuited, as in Fig. 5-17, the transistor's base is clamped to 0.6V. This turns off the zener, disconnecting the zener from further effect on short-circuit operation, and also protecting it. Since we have no idea what short-circuit current will flow, we can't estimate terminal voltage of the

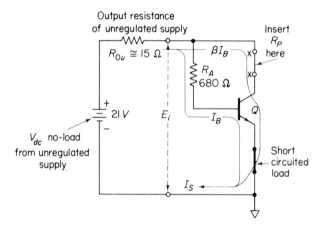

Figure 5-17

Short-circuit load conditions in an emitter-follower regulator. $\beta = 49$ for transistor Q.

unregulated supply But we can accurately predict no-load voltage and estimate output resistance R_{ou} of the unregulated supply, as in Fig. 5-17. (Refer to section 5-1 and the dc regulation curve in Fig. 5-5.) Figure 5-17 resembles a common-emitter bias circuit with collector-base feedback, and its emitter current or short-circuit load current I_{SC} is expressed almost exactly by

$$I_S = \frac{E_m - V_{BE}}{R_{ou} + (R_A/\beta)} = \frac{(21 - 0.6)\text{V}}{[15 + (680/49)]\Omega} = \frac{20.4\text{V}}{(15 + 13.9)\Omega} = 0.7\text{A} \quad (5\text{-}32)$$

Terminal voltage E_i of the unregulated supply is evaluated by modifying Eq. (5-6)

$$E_{i\ \text{short-circuit}} = V_{dc\ \text{no-load}} - I_S R_{ou} \quad\quad\quad\quad (5\text{-}33)$$

$$E_{i\ \text{short-circuit}} = 21V - (0.7\text{A})(15\Omega) = 10.5\text{V}$$

Power Dissipation in the Transistor

Short-circuit transistor power dissipation $P_{D_{SC}}$ is expressed by modifying Eq. (5-31) for short-circuit conditions in Fig. 5-17 as

$$P_{D_{SC}} = (E_{i\,\text{short-circuit}} - 0)\,I_S = (10.5\text{V})(0.7\text{A}) = 7.4\text{W} \qquad (5\text{-}34)$$

For short-circuit protection, Q must be capable of dissipating 7.4W. To relieve the heating load on Q we can transfer some heat to a power dissipation resistor, R_p, inserted between points X-X in Fig. 5-17. For example, try adding $R_p = 10\Omega$ to take up $I_S R_p = (0.7\text{A})(10\Omega) = 7\text{V}$ of $E_{i\,\text{short-circuit}}$. ($R_p$ does not affect the value of I_S predicted by Eq. [5-32]). This leaves 3.5 volts for Q. Q now absorbs $(3.5\text{V})(0.7\text{A}) = 2.5\text{W}$ and R_p absorbs $7\text{V} \times 0.7\text{A} \cong 5\text{W}$. Of course, at full-load current, under *normal* operation R_p should not drop more than half the difference between $E_{i\,\text{full-load}}$ and V_o. In our 100mA regulator R_p would drop only $10\Omega \times 0.1\text{A} = 1\text{V}$, and would not affect normal operation.

5-13 MEASURING VOLTAGE REGULATION WITH A CRO

To display the total instantaneous voltage regulation curve for the $600\mu\text{F}$ power supply of Fig. 5-5, we add 20 ohm current sensing resistor R_S in Fig. 5-18a. Connect a CRO with its amplifiers direct-coupled and on external sweep. The CRO's spot is zeroed at bottom left and x-sensitivity set at 1V/cm to calibrate the x-axis directly in *total instantaneous load current* at $1\text{V}/20\Omega = 50\text{mA}/$ cm. This gives an average horizontal right deflection of 6.5cm at full-load dc current = 325mA. Since $V_{\text{dc no-load}}$ is 21V, we select y-sensitivity of 5V/cm to give an average vertical up deflection of $21\text{V}/(5\text{V/cm}) = 4.2\text{cm}$ at no-load. At $I_{L\text{max}} = 325\text{mA}$, the average vertical deflection would be $16.4\text{V}/(5\text{V/cm}) = 3.28\text{cm}$. Centered around point C, at $I_L = 325\text{mA}$, $V_{\text{dc}} = 16.4\text{V}$, the spot would deflect vertically, due to ripple voltage ΔV_o, a distance of $2.5\text{V}/(5\text{V/cm}) = 0.5$ cm. Of course line current would also ripple horizontally around the 6.5cm deflection, because ΔV_o is developed across R_L. Horizontal ripple deflection ΔI_L equals $\Delta V_o/R_L = 2.5\text{V}/50\Omega = 50\text{mA}$ and deflects a horizontal distance of 1cm. The resulting CRO display is line AB in Fig. 5-18b.

Measuring Load Resistance with a CRO

There is a simpler method of predicting trace AB by: (1) locating the dc full-load operating point C; (2) connecting C to the origin with a construction line to get the slope of R_L; and (3) extending the slope symmetrically around point C for a *vertical* distance of ΔV_o. It follows that we can measure R_L from the slope of the CRO trace and gain an added benefit from the CRO measure-

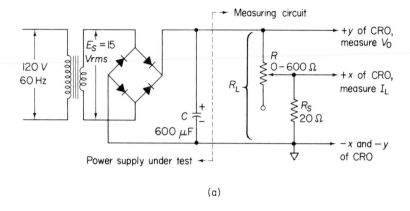

(a)

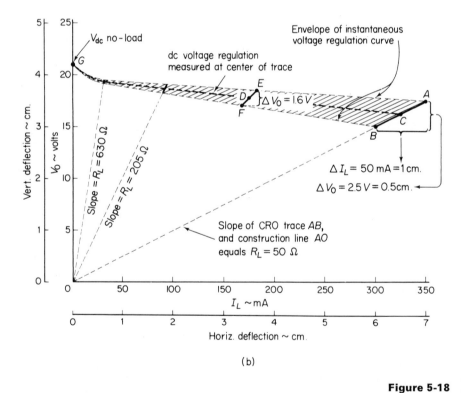

(b)

Figure 5-18

Total instantaneous voltage regulation curve in (b) is displayed on a CRO
by the test circuit in (a).

ment. For example, from data in Fig. 5-18b, point D is located in the CRO trace's center at $I_L = 175\text{mA}$, $V_{dc} = 17.8\text{V}$. Draw a line from D to origin and find its slope of rise over run is $17.8\text{V}/175\text{mA} = 102\Omega$. Read $\Delta V_o = 1.6\text{V}$ from the vertical distance between E and F. The cross-hatched area is an envelope containing all possible CRO traces as I_L is varied from 0 to 325mA. We illustrate the enormous amount of information contained in the CRO measurements with an example.

EXAMPLE 5-17:

A full-wave bridge rectifier is plugged into a wall outlet (120V, 60Hz). You are given no-load condition as point G and trace EF in Fig. 5-18b as full-load condition. What is: (a) output resistance of the power supply, (b) load resistance, (c) size of filter capacitor, (d) turns ratio of power transformer, (e) percent ripple at full load?

SOLUTION:

Information given from point G is $V_{dc\ no\text{-}load} = E_m = 21\text{V}$. With trace EF, find $\Delta V_o = 1.6\text{V}$ from the vertical displacement between tips of the trace. Estimate the trace's center to read $I_{L\max} = 175\text{mA}$, $V_{dc\ full\text{-}load} = 17.8\text{V}$.
(a) From Eq. (5-6)

$$R_o = \frac{V_{dc\ no\text{-}load} - V_{dc\ full\text{-}load}}{I_{L\ full\text{-}load}} = \frac{(21 - 17.8)\text{V}}{175\text{mA}} = 18.3\Omega$$

(b) From Eq. (5-3), $R_L = V_{dc}/I_L = 17.8\text{V}/175\text{mA} = 102\Omega$.
(c) From Eq. (5-12b)

$$C \simeq \frac{E_m}{200R_L V_o} = \frac{21}{200 \times 102 \times 1.6} \simeq 640\mu\text{F}$$

(d) From Eq. (5-5), $E_{rms} = E_m/1.41 = 21\text{V}/1.41 = 15\text{V}$. Transformer turns ratio is primary rms voltage/secondary rms voltage, or 120V/15V = 8 to 1.
(e) From Eq. (5-4), $V_r = \Delta V_o/3.5 = 1.6\text{V}/3.5 = 0.46\text{V}$. From Eq. (5-9)

$$\% \text{ ripple} = 100\frac{V_r}{V_{dc}} = 100\frac{0.46}{17.8} = 2.5\%$$

Thus from one point and one trace we can reconstruct the circuit, including component values.

The display circuit in Fig. 5-18 can demonstrate, quickly and graphically, dc regulation, output resistance, and ripple increase with load. However, R is usually an expensive wire wound variable resistor or power decade box. R_L is

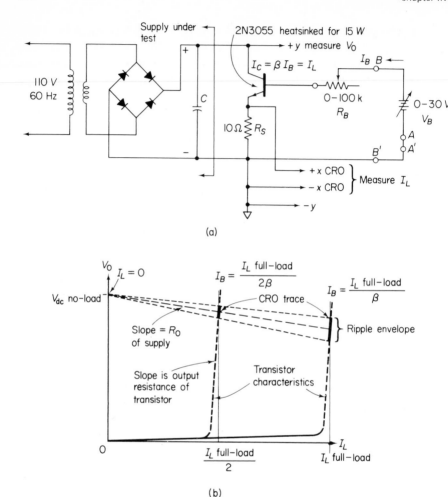

(a)

(b)

Figure 5-19

R_B controls I_B in (a) to vary load current via the transistor and obtain the CRO display in (b). When capacitor C is removed, the CRO displays $I_C - V_{CE}$ characteristic of the transistor.

not varied smoothly and continuously. We therefore proceed to the next section to learn how to obtain a continous display of voltage regulation, and to study how the supply behaves when it drives a load that varies at high frequency.

5-14 DYNAMIC MEASUREMENT OF POWER SUPPLY PERFORMANCE

For the same power rating, a power transistor and rms heat sink cost about the same as a variable power resistor. Initially we replace R_L with a 2N3055 transistor in Fig. 5-19a. Base current I_B can be controlled smoothly by low-power potentiometer R_B together with bias voltage V_B. As I_B is varied from 0 to I_{Lmax}/β, the voltage regulation curve is drawn smoothly across the CRO screen from $I_L = 0$ to I_{Lmax}. As shown in Fig. 5-19b, the CRO trace locates the intersection of the power supply I-V characteristic with the load's I-V characteristic. But in this case the load characteristic is essentially the $I_C - V_{CE}$ transistor characteristic. (Actually, emitter current is measured, but $I_E \cong I_C = I_L$. Voltage drop across R_S is small with respect to V_o, so $V_o \cong V_{CE}$.)

One distinct advantage gained by using a transistor as a load is that we can make dynamic load tests of the power supply. For example, assume I_B is set to draw $I_{Lmax}/2$ from the power supply. Now insert a signal generator or audio signal between points AA' to vary load current at an audio rate. The CRO will display current-voltage behavior of the power supply when it is driving one type of amplifier. If a square wave generator is connected between points AA' and a high-frequency power transistor is substituted for the 2N3055, we can test power supply response to square waves.

An additional feature can be incorporated into the test circuit of Fig. 5-19a. Set $I_B = I_{Lmax}/2$ and remove capacitor C. The power supply now functions as a sweep circuit to display the transistor $I_C - V_{CE}$ characteristic on the CRO. The test circuit thus performs as a curve tracer.

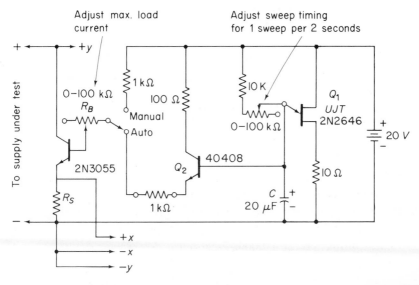

Figure 5-20

Automatic measurement of power supply performance.

5-15 AUTOMATIC MEASUREMENT OF POWER SUPPLY PERFORMANCE

With only a slight modification to the circuit of Fig. 5-19a we can incorporate the feature of automatic measurement. In Fig. 5-20 we add UJT Q_1 as a low-frequency relaxation oscillator that develops a saw tooth voltage across

C. Pnp transistor Q_2 couples the sawtooth through a switch and base current adjusting resistor R_B to drive the base of load transistor 2N3055. Base current is increased almost linearly and then drops to zero every one to two seconds. This base current sweep circuit replaces a human control on R_B of Fig. 5-19a. Load current of the power supply will be increased from no-load to full-load, depending on the adjustment of R_B. The CRO displays corresponding values of power supply output voltage V_o for each load current, and consequently displays the voltage regulation curve. Since the voltage regulation curve is displayed over and over again, the measurement is automatic.

The switch has a manual position, so we can stop the CRO trace and examine any desired load current-voltage combination of the supply. Once the test circuit is set up, any power supply can be tested quickly throughout its range.

PROBLEMS

1. What type of measuring instrument (s) do you need to make the following measurements on a power supply? (a) Total instantaneous load voltage. (b) Peak-to-peak ripple voltage. (c) Effective or rms ripple voltage. (d) Average load voltage. (e) Average load current. (f) Dc voltage regulation curve.

2. Load voltage measured with an ac voltmeter is $V_r = 1$V, and with a dc voltmeter is $V_{dc} = 24$V. What are the maximum and minimum instantaneous values of V_o?

3. From Fig. 5-2 calculate R_o for a full-load dc current of 60mA. Compare answer with Example 5-1.

4. With a line voltage of 120V, choose a transformer for a half-wave rectifier to supply 12V dc at full-load current of 100mA. Assume $R_o = 25\Omega$.

5. A half-wave rectifier has a 120/20V transformer and $R_o = 20\Omega$. Calculate dc output voltage at a full-load current of 100mA.

6. Replace the capacitor in Fig. 5-2 with a 500μF cap. Calculate ripple voltage ΔV_o at (a) full-load, (b) no-load, and (c) $I_L = 33$mA.

7. Repeat problem 6 for a 2,000μF capacitor. From your results, what effect does capacitor size have on ripple?

8. For each load current in Example 5-4 calculate % ripple.

9. A half-wave rectifier delivers 100mA full-load current at 20V at 3% ripple. $R_o = 30\Omega$. Find: (a) rms ripple voltage; (b) peak-to-peak ripple voltage; (c) no-load voltage; and (d) filter capacitance.

10. A half-wave rectifier with $C = 1,000\mu$F, $R_o = 20\Omega$, and a no-load voltage of 40V is connected to a load resistor of 200Ω. What is the value of (a) V_{dc} and (b) % ripple?

11. In the measurement of Fig. 5-3a, $V_{dc} = 17.5$V, $V_{op} = 20$V, and $I_L = 100$ mA. Find: (a) R_L; (b) C; (c) % ripple; and (d) I_{CH}. Hint: use Eq. (5-7).

12. A full-wave rectifier must deliver 500mA at 24V dc, with 2% ripple. Assume $R_o = 10\Omega$. Find: (a) transformer size; (b) ripple voltage ΔV_o; and (c) filter capacitor C.

13. (a) Evaluate % ripple in Fig. 5-4c. (b) If $t_c = 1_{ms}$, estimate I_{CH}.

14. If the half-wave rectifier of Fig. 5-2 is replaced with a full-wave rectifier, calculate peak-to-peak ripple voltage at (a) full-load, (b) no-load, and (c) $I_L = 33$mA. Compare answers with Example 5-4 and problem 7.

15. At $I_L = 200$mA in Fig. 5-5, calculate: (a) R_L; (b) R_o; and (c) % ripple.

16. An IN 4774 zener has a nominal zener voltage V_Z of 15V at I_{ZT}. $P_{Dmax} = 1$W and tolerance is 10%. Find: (a) I_{Zmax}; (b) I_{ZT}; (c) I_{Zrms}. (d) What range of zener voltage would you expect to measure for a large number of IN 4774? (e) If $r_Z = 5\Omega$ and zener current varies by $\Delta I_Z = 10$mA, what is the variation in V_Z?

17. A 10V zener, heat-sinked for 3W, is used in a zener regulator to deliver $I_{LFL} = 150$mA. Use the supply of Fig. 5-5 and, assuming $I_{Zmin} = 5$mA, calculate the range for R_A.

18. If $r_Z = 2\Omega$, what is the maximum change in output voltage from no-load to full-load for the zener of problem 17?

19. For the zener regulator of problem 17, calculate peak-to-peak output ripple voltage across (a) unregulated supply, (b) the load. Let $R_A = 47\Omega$ and $r_Z = 2\Omega$; C in the unregulated supply is 600μF.

20. If the zener of problem 17 is used in the circuit of Fig. 5-14, find R_{Amin} and R_{Amax}. $\beta_F = 99$.

21. Using the unregulated supply of Fig. 5-5, $C = 1,200\mu$F, select R_A to make a 15V emitter-follower regulator to supply $I_{Lmax} = 250$mA. Assume $\beta_F = 49$, $V_Z = 15.6$V, $r_Z = 3\Omega$, and zener $P_{Dmax} = 500$mW.

22. Assume you selected $R_A = 180\Omega$ in the regulator designed for problem 21. (a) Find no-load and full-load zener current (assume $I_{LNL} = 10$mA at $V_{dc} = 20$V). (b) If ripple from the unregulated supply is $\Delta V_o = 1.0$V, what is ripple voltage from the regulator?

23. For the regulator of problems 21 and 22, what power is developed in the: (a) transistor at full load; (b) transistor at short-circuit load; (c) unregulated supply at short-circuit load? Assume $R_{ou} = 20\Omega$ and $E_m = V_{dcNL} = 21$V for unregulated supply.

24. For the same transformer, R_L, and C, a full-wave rectifier has with respect to a half-wave rectifier (a) _____ ripple, (b) _____ dc load voltage, (c) _____ charging current, (d) _____ no-load voltage, (e) _____ capacitor discharge time, (f) _____ reverse-bias on nonconducting diodes, (g) _____ cost. (Code: A = twice as much, B = half as much, C = more, D = less and/or E = the same.)

chapter six
feedback voltage regulators
using integrated circuits

6-0 INTRODUCTION

Four of the most misunderstood, maligned, or difficult subject areas (according to students) are negative feedback, feedback voltage regulators, operational amplifiers, and systems, in that order. Basic negative feedback theory does not usually agree with laboratory results because of loading by the feedback network. Discrete component feedback voltage regulators have long and involved design and performance equations, even when they are simplified by approximations. Even worse, the resultant regulator is not much of an improvement over the emitter-follower regulator unless complex circuitry is added. Operational amplifiers often oscillate when they are supposed to amplify and amplify or latch when they are supposed to oscillate. There is always concern with learning the systems concept of circuit analysis whereby groups of components can be treated as building blocks in a system.

It is perhaps ironic that all of these concerns can be combined to make more interesting the study of that heretofore unpopular subject, feedback regulators. An integrated circuit operational amplifier (op-amp) is the element that simplifies and allows us to coordinate the practical difficulties inherent in a study of each subject area.

In the rest of this chapter, basic feedback theory exactly fits the feedback regulator. Proper choice of an op-amp that won't oscillate makes the feedback theory come true. We will study a basic 100mA feedback regulator that performs incredibly well, considering the cost—about 50 cents for a zener, a dollar for the op-amp, and a dollar for the transistor plus three resistors. Treating the 100mA regulator as a system, we will add on options, either individually or collectively, as desired. The options are current boost, short-circuit protection or current-limiting, remote load voltage sensing, and programming. Best of all, the op-amp reduces design analysis equations of the basic regulator to three simple expressions involving the selection of three resistors.

We will conclude with a brief examination of the amazing advances in semiconductor technology when all of the above options are built into one low-cost integrated circuit package. With these new system-packages the user must go to the manufacturer for application notes that specify how to use them. In many applications the installation procedure is as simple as connecting two wires from the unregulated supply to the regulator and two wires from the regulator to the load.

6-1 INTRODUCTION TO THE BASIC FEEDBACK REGULATOR

The basic feedback regulator in Fig. 6-1 consists of three main elements—pass transistor Q_1, differential amplifier A, and a zener-established reference voltage.

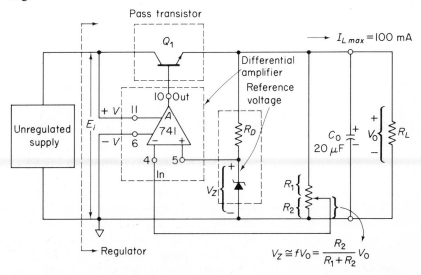

$$V_Z \cong fV_0 = \frac{R_2}{R_1 + R_2} V_0$$

Figure 6-1

Basic feedback regulator. Pin connections shown are for dual-inline package (DIP).

Resistor R_D furnishes a few mA to hold the zener in its breakdown region. The differential amplifier's positive input draws negligible current ($< 1\mu A$) and is clamped at V_Z.

Differential Amplifier

The differential amplifier (also called *error* or *comparison* amplifier) has a positive input terminal pin 5 and negative input terminal pin 4. Due to feedback action (explained in sections 6-2 and 6-3), the negative input will be held within a few microvolts of reference voltage V_Z. Pin 4 is wired to the junction of R_1 and R_2 and fixes current through R_2 at V_Z/R_2. Since the negative amplifier input also draws negligible current, the same current of V_Z/R_2 flows through *both* R_1 and R_2 to establish regulator output voltage V_o at

$$V_o = \frac{V_Z}{R_2}(R_1 + R_2) = \frac{V_Z}{f} \tag{6-1}$$

where $f = R_2/(R_1 + R_2)$.

Pass Transistor

Q_1 is a series regulating or pass transistor. Assuming its emitter is clamped at V_o, its collector-emitter voltage will absorb variations in E_i. E_i must always exceed V_o by 2 to 3 volts to insure that Q_1 does not saturate. This observation defines one basic specification for proper performance of any feedback amplifier—the *minimum input-output differential voltage*. We defer the explanation of how V_o is established and held constant until after a study of the differential amplifier in section 6-2.

The 741 Operational Amplifier

Successful regulator operation depends mainly on the differential amplifier in Fig. 6-1. To avoid most of the problems that accompany one's first experience with a high-gain differential amplifier, select a type 741, integrated circuit (IC), operational amplifier (op-amp). Complete specifications on the 741 op-amp are included in Appendix 1. The 741 is internally compensated and won't oscillate if lead lengths are short. It will withstand an indefinite short circuit from output pin 10 to ground or to either of supply terminals 6 or 11. Output current is internally limited to a maximum of 25mA. The maximum differential input voltage, between pins 4 and 5, can be up to ± 15V or $\pm$ supply voltage, whichever is less. Maximum power dissipation is 500mW. Cost of the 741 is below \$2.00 (even less from discount houses) and will certainly become cheaper.

Not only is the performance of an op-amp regulator incredibly good, but this circuit presents a unique opportunity to learn negative feedback action under almost classical, ideal conditions. We begin with a brief introduction to the op-amp.

6-2 CHARACTERISTICS OF THE OP-AMP

As shown in Fig. 6-2, *open loop gain A* is defined as the ratio of single-ended output voltage V_o (pin 10 to ground) to *differential input voltage E_d* (voltage between pin 4 and pin 5). Open loop gain is very high at low frequencies and is typically greater than 200,000. When *inverting* or $(-)$, input terminal 4 is driven

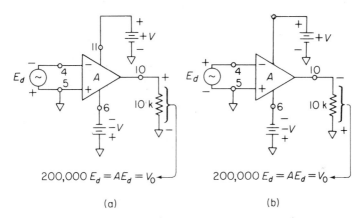

$$200,000\,E_d = AE_d = V_0$$

(a)

$$200,000\,E_d = AE_d = V_0$$

(b)

Figure 6-2

Basic characteristics of an op-amp.

negative with respect to *noninverting* or $(+)$ input terminal pin 5 in Fig. 6-2a, output V_o goes positive with respect to ground. That is, signal voltage V_o always goes in the same polarity direction as the $(+)$ input terminal, and in the opposite or inverted direction from the $(-)$ input terminal. For example in Fig. 6-2b, V_o goes negative when the $(+)$ input goes negative or when the $(-)$ input goes positive. Either pin 4 or pin 5 can be grounded, or both can be above, below, or on either side of ground. Regardless of the input configuration, V_o is always related by A to the *difference* in voltage between pins 4 and 5 by

$$V_o = AE_d \qquad (6\text{-}2)$$

The maximum value that will ever occur for V_o will be roughly 30V, and so the maximum value of E_d will be found from Eq. (6-1).

$$E_d = \frac{V_o}{A} = \frac{30}{200,000} = 150\mu\text{V} = 0.15\text{mV}$$

We really can't read $150\mu\text{V}$ with the usual inexpensive laboratory or shop test equipment, but we can state that for all practical purposes the *potential at the* $(-)$ *input terminal will approximately equal the potential at the* $(+)$ *input terminal.* (Caution—do *not* test any high-gain op-amp as in Fig. 6-2. Noise and tiny unbalances within the op-amp will force V_o to positive or negative saturation when V_o latches about 2 volts away from $+V$ or $-V$.)

The three significant characteristics of an op-amp can be summarized:

1. High gain between differential input and single-ended output.
2. Differential input voltage $E_d = V_o/A$ is always almost equal to zero.
3. Input impedance between $(-)$ and $(+)$ terminal or between either input terminal and ground is so high that currents of only a fraction of a micro-ampere will flow *into* pins 4 or 5.

6-3 OPERATION OF A BASIC FEEDBACK REGULATOR

In Fig. 6-3 bleeder current I_{BL} is chosen large with respect to amplifier input currents. Select $R_1 + R_2$ from a knowledge of the required V_o and a typical value of $I_{BL} = 1\text{mA}$.

$$R_1 + R_2 = \frac{V_o}{I_{BL}} \tag{6-3}$$

Bias currents entering pins 4 or 5 are found from the specification sheet to be a maximum of $500\eta\text{A}$ or $0.5\mu\text{A}$. Therefore, V_o divides proportionally between sensing resistors R_1 and R_2 and feeds back a fraction f of V_o to the inverting input terminal. Feedback fraction f is found simply from the voltage divider

$$f = \frac{R_2}{R_1 + R_2} \tag{6-4}$$

so that R_1 and R_2 sense the regulator's output voltage V_o and feed back a fraction of V_o to the $(-)$ input as

$$fV_o = \frac{R_2}{R_1 + R_2}V_o \tag{6-5}$$

Biasing the Zener Diode

Zener diode current I_Z is set by V_o and R_D to keep the zener in breakdown. We pick a value of I_Z to be well around the zener knee, typically 5 to 10mA.

Since V_o will be very constant, and since input bias current to the amplifier is very small, all of I_Z flows through the zener. Furthermore, I_Z will not vary, so V_Z will be exceeding stable. For all practical purposes the zener is supplied by an ideal constant current source with current determined by

$$I_Z = \frac{V_o - V_Z}{R_D} \qquad (6\text{-}6)$$

Deriving V_o in terms of V_Z

To face squarely the issue of what controls V_o, write the amplifier's input loop equation of Fig. 6-3

$$V_Z = E_d + fV_o \qquad (6\text{-}7)$$

Op-amp output is applied to the base of Q_1 and regulator output V_o is taken from the emitter of Q_1. Thus Q_1 is an emitter-follower with a voltage gain of 1. Consequently, our basic gain relationship of the op-amp is unchanged by the addition of Q_1 and substituting Eq. (6-2) for E_d into Eq. (6-7).

$$V_Z = \frac{V_o}{A} + fV_o = V_o\left(\frac{1}{A} + f\right) \qquad (6\text{-}8)$$

Force Eq. (6-8) into a gain relation between the known zener voltage V_Z and V_o.

$$\frac{V_o}{V_Z} = \frac{A}{1 + fA} = \frac{1}{\dfrac{1}{A} + f} \qquad (6\text{-}9)$$

Equation (6-9) is in the classic feedback format, and shows that if A is very large, the term $1/A$ is negligible with respect to f, and V_o is related to V_Z by

$$V_o \cong \frac{V_Z}{f} = V_Z \frac{R_1 + R_2}{R_2} \qquad (6\text{-}10)$$

Note that we obtain the same results by equating V_Z and fV_o in Eq. (6-5), and by our intuitive analysis in Eq. (6-1).

A Physical Explanation for V_o

For a physical explanation of the establishment of V_o, assume power is turned on in Fig. 6-3. Op-amp and supply voltage on pin 11 cause some positive output voltage on pin 10 that causes V_o to go positive. Voltage across the zener goes positive and stops at $+5$ volts. At any time during this increase of V_o, the voltage on pin 5 will be more positive than voltage on pin 4. In other words, until the zener turns on we will have V_o on pin 5 and fV_o on pin 4. Once pin 5 is clamped to V_Z, V_o continues to rise until fV_o rises to within a few microvolts of V_Z (where $E_d = V_Z - fV_o$).

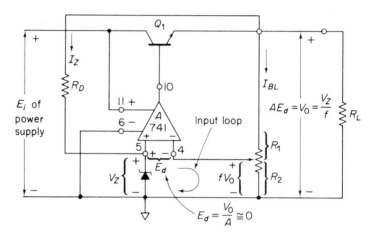

Figure 6-3

Voltage relationships for the feedback regulator.

EXAMPLE 6-1:

If $V_Z = 5V$, $A = 200,000$, and $R_1 = R_2$, what will be the value of V_o and E_d in Fig. 6-3?

SOLUTION:

From Eq. (6-4)

$$f = \frac{R_2}{R_1 + R_2} = \frac{R_2}{R_2 + R_2} = 0.5$$

From Eq. (6-9)

$$V_o = V_Z \frac{A}{1 + fA} = 5\frac{200,000}{1 + 0.5(200,000)} = 5 \times \frac{200,000}{100,001} = 10V$$

From Eq. (6-2)

$$E_d = \frac{V_o}{A_v} = \frac{10V}{200,000} = 50\mu V$$

Thus pin 5 is $50\mu V$ more positive than pin 4 in Fig. 6-3.

6-4 OUTPUT RESISTANCE

Output resistance R_o is defined as the resistance seen into the output terminal of the regulator. It is found from the ratio of a measured peak-to-peak change in load voltage to a resulting measured peak-to-peak change in load current. Usually the regulator is *loaded to one-half its rated current.* Then the load is varied sinusoidally (see Fig. 5-19a). Thus R_o is really an ac output resistance. But R_o can also be used to predict the approximate change in load voltage (ΔV_o) that will result from a change in current from no-load to full-load ($\Delta I_L = I_{Lmax}$). Thus R_o is also an estimate of load regulation.

Deriving R_o

To find R_o, assume R_L is adjusted for one-half the rated load. Apply a small ac test voltage V_T to the output terminals and employ the hybrid-pi model for Q_1 in Fig. 6-4. Assume R_L and $R_1 + R_2$ will draw negligible current from V_T because they are very large with respect to R_o. If we take a Thévenin equivalent

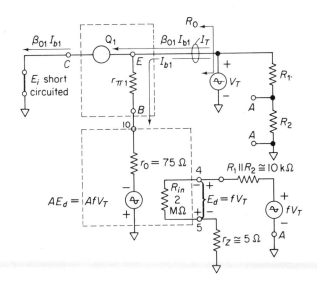

Figure 6-4

Small-signal model to find output resistance $R_o = V_T/I_T$.

of the feedback voltage across R_2, we see that fV_T divides among $R_1 \| R_2$, R_{in} of the op-amp, and r_Z. Since R_{in} is much larger than r_Z or $R_1 \| R_2$, almost all of fV_T is fedback as differential input voltage E_d. Compare the polarity of E_d in Fig. 6-4 with E_d in Fig. 6-2b to see that pin 10 will go negative. The op-amp output is modeled by dependent voltage source $AfV_T = AE_d$ in series with op-amp output resistance r_o. Data sheet value of r_o is listed as typically 75Ω.

Writing the I_{b1} loop equation:

$$AfV_T + V_T = I_{b1}(r_{\pi1} + r_o), \text{ solving for } \frac{V_T}{I_{b1}}$$

$$\frac{V_T}{I_{b1}} = \frac{r_{\pi1} + r_o}{1 + fA}$$

But $I_{b1} = I_T/(\beta_{o_1} + 1)$, substituting for I_{b1}

$$\frac{V_T}{I_T} = \frac{r_{\pi1} + r_o}{(1 + fA)(\beta_{o_1} + 1)} = R_o \qquad (6\text{-}11)$$

EXAMPLE 6-2:

In Fig. 6-4, $V_Z = 5V$, $f = 0.5$, and $\beta_{o_1} = 99$. What is output resistance R_o at 1/2 rated load?

SOLUTION:

Evaluate $r_{\pi1}$ at $I_{Lmax}/2 = 50\text{mA}$.

$$g_m = \frac{50\text{mA}}{25\text{mV}} = 2\text{U}, \qquad r_{\pi1} = \frac{99}{2} \approx 50\Omega.$$

From Eq. (6-11)

$$R_o = \frac{r_{\pi1} + r_o}{(1 + fA)(\beta_{o_1} + 1)} = \frac{50\Omega + 75\Omega}{(1 + 0.5 \times 200{,}000)(100)}$$

$$= \frac{125\Omega}{1 \times 10^7} = 12.5\mu\Omega$$

Practical Considerations of R_o

The outstanding low value of R_o predicted by Example 6-2 is not physically realizable. Thermal variations will vary V_Z, R_1, and R_2. Slight unbalances within the op-amp cause slight changes in V_o that will mask changes in V_o due to R_o. Even tiny heat changes due to changes in load current will yield real values for R_o of a few milliohms. The net effect will be manifested by a slow drift in V_o of a millivolt or so. In any case, load regulating performance of the op-amp regulator is remarkable.

The conclusion to be drawn from this section is that R_o is negligible. In addition, Eq. (6-11) and Fig. 6-4 lead us to conclude (1) any resistance added between op-amp pin 10 and base of Q_1 will be divided by $(1 + fA)(\beta_{o1} + 1)$ or 1×10^7; (2) any resistance added between emitter of Q_1 and V_T (top terminal of R_1) will be divided by $(1 + fA) = 1 \times 10^5$. These conclusions will be fully explored in section 6-9.

6-5 POWER REQUIREMENTS

Q_1 Power Requirements

Maximum power dissipation will occur in Q_1 under the full-load condition of Fig. 6-5. We will work with test data given for an unregulated power supply

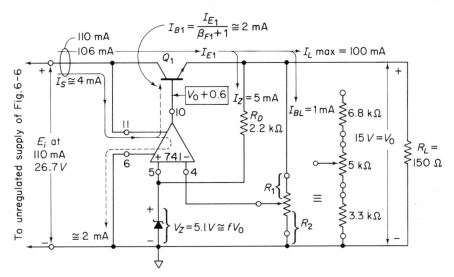

Figure 6-5

Circuit to calculated power dissipation.

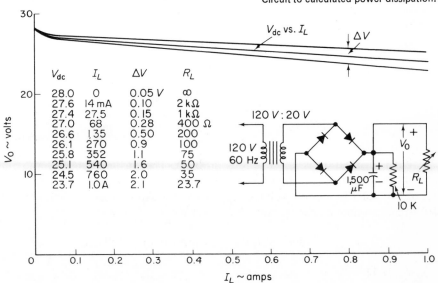

V_{dc}	I_L	ΔV	R_L
28.0	0	0.05 V	∞
27.6	14 mA	0.10	2 kΩ
27.4	27.5	0.15	1 kΩ
27.0	68	0.28	400 Ω
26.6	135	0.50	200
26.1	270	0.9	100
25.8	352	1.1	75
25.1	540	1.6	50
24.5	760	2.0	35
23.7	1.0 A	2.1	23.7

Figure 6-6

Power supply for Example 6-3 and Figure 6-5.

in Fig. 6-6. Maximum heat dissipation P_{D1max} in Q_1 is the product of collector current and collector-emitter voltage.

$$P_{D1max} = V_{CE1} \times I_{C1} = (E_i - V_o)(I_{Lmax} + I_Z + I_{BL}) \qquad (6\text{-}12a)$$

where E_i is dc output voltage of the unregulated supply when it delivers total current expressed in Eq. (6-12). Usually I_Z and I_{BL} add about 10mA and can be neglected to simplify Eq. (6-12a) to

$$P_{D1max} \approx (E_i - V_o)I_{Lmax} \qquad (6\text{-}12b)$$

Op-amp Power Requirements

From manufacturer's data sheets, 741 power comsumption is 50mW at no-load with $+V = -V = 15V$. At no-load the op-amp typically draws 50 mW/30V $\approx$ 1.7mA through pins 6 and 11 to set up internal bias currents. Assuming β_F of Q_1 is 49 in Fig. 6-5, the op-amp must supply 2mA of base current to Q_1, for $I_{Lmax} = 100$mA. The 2mA base current drive from pin 10 is in addition to the 1.7mA idle current. Op-amp supply current, I_S, must therefore equal about 4mA (2mA + 1.7mA).

We can estimate how much power is added to the op-amp (beyond the 50 mW idle power) by multiplying current $I_{B1} = I_{E1}/(\beta_{F1} + 1)$ times the voltage between pins 10 and 11, or

$$P_{\text{op-amp}} = 0.05 + \frac{(E_i - V_o - 0.6)(I_{E1})}{(\beta_{F1} + 1)} \text{ watts} \qquad (6\text{-}13a)$$

$$\cong 0.05 + \frac{P_{D1max}}{\beta_{F1} + 1} \text{ watts} \qquad (6\text{-}13b)$$

where E_i = unregulated supply voltage at $I_{E1} + I_S$ and

$$I_{E1} = I_{Lmax} + I_Z + I_{BL} \approx I_{Lmax}$$

Zener Power Requirements

Zener dissipation is not a serious problem. The smallest zeners are rated at 250mW. I_Z will be very stable and not vary with load. If we restrict I_Z to 10mA at the highest V_o (variable voltage regulator) and choose zeners from 5.1V to 10V, then maximum zener power dissipation will be

$$P_{Dmax} \text{ of zener} = I_Z V_Z = (10\text{mA})(10\text{V}) = 100\text{mW}$$

EXAMPLE 6-3:

Design a feedback regulator to supply a full-load current of 100mA at 15V. Assume $\beta_{F1} = 49$.

SOLUTION:

Since regulators are built for unregulated supplies and not vice versa, we begin with the power supply of Fig. 6-6. (a) Choose a 5.1V zener for good temperature stability and the cheapest $P_{D\max} = 250\text{mW}$. Assume $r_Z = 10\Omega$ and pick $I_Z = 5\text{mA}$. From Eq. (6-6)

$$R_D = \frac{(15 - 5.1)\text{V}}{5\text{mA}} \approx 2\text{k}\Omega$$

(b) Pick $I_{BL} = 1\text{mA}$; from Eq. (6-3)

$$R_1 + R_2 = \frac{V_o}{I_{BL}} = \frac{15\text{V}}{1\text{mA}} = 15\text{k}\Omega$$

(c) Find f from Eq. (6-10):

$$f = \frac{V_Z}{V_o} = \frac{5.1\text{V}}{15\text{V}} = 0.34$$

and R_2 from Eq. (6-4):

$$R_2 = f(R_1 + R_2) = 0.34(15\text{k}\Omega) = 5.1\text{k}\Omega$$

and

$$R_1 = (15 - 5.1)\text{k}\Omega = 9.9\text{k}\Omega$$

One satisfactory choice is the series combination of a 6.8kΩ and 3.3kΩ resistor with a 5kΩ potentiometer in Fig. 6-5. This eliminates the need for expensive precision resistors and allows adjustment of f.
(d) Power dissipation for Q_1 is found from Eq. (6-12b). Find

$$I_{C1} = 100\text{mA} + 5\text{mA} + 1\text{mA} = 106\text{mA}$$

From Fig. 6-6, E_i averages 26.7V at 106mA.

$$P_{D1} = (26.7\text{V} - 15\text{V})(106\text{mA}) = (11.7\text{V})(106\text{mA}) = 1.2\text{W}$$

Choose a medium power transistor heat-sinked for 2 to 3 watts. From Eq. (6-13), op-amp power is

$$P_{\text{op-amp}} = 0.050 + \frac{(26.7 - 15 - 0.6)0.106}{50} = 72\text{mW}$$

Zener power is $I_Z V_Z = 5\text{mA} \times 5.1\text{V} = 25\text{mW}$.

6-6 AUTOMATIC CURRENT LIMITING

Add an option to the basic regulater and incorporate *short-circuit protec-tion* with *current-limiting transistor* Q_S plus sensing resistor R_S in Fig. 6-7a. For load current below full-load, Q_S is cutoff and has no effect on operation of

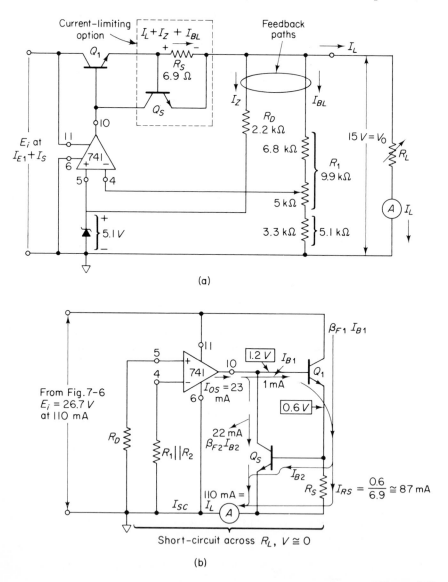

(a)

(b)

Figure 6-7

Current-limiting network Q_S and R_S in (a) hold overload and short-circuit currents to values in (b). Q_1 is a 2N3055 and Q_S a 40408.

the regulator. R_S is chosen by experiment to (1) establish a voltage drop of about 0.6V at the desired maximum emitter current of Q_1 in order to (2) forward-bias the base-emitter junction of Q_S so that (3) the collector of Q_S diverts base current drive away from Q_1. An equilibrium will be established where emitter current of Q_1 is constant and approximately equal to current through R_S of

$$I_{E1\max} \cong I_{RS} \cong \frac{0.6V}{R_S} \tag{6-14}$$

The R_S-Q_S option protects Q_1. It does not protect the op-amp against overload. The 741 was chosen because its internal circuitry is designed to automatically limit output current to a maximum of about $I_{o\max} = 25$mA, at an ambient temperature of 25°C.

Short-circuited Load Currents

If load R_L is short-circuited, V_o goes to zero in Fig. 6-7a and extends ground to the emitter of Q_S. The zener turns off and both amplifier inputs see the resistances modeled in Fig. 6-7b. (Note that there will be some small value of V_o at short circuit. All of it will appear at pin 5, but only fV_o will appear at pin 4. Therefore, output pin 10 must stay positive because pin 5 is more positive than pin 4.) In an actual short-circuit test, I_{os} self-limited to 23mA. R_S was set experimentally (by a 10Ω resistor in parallel with a 22Ω resistor) at 6.9Ω to limit I_L on short circuit at a 10% overload or 110mA. By inspection of Fig. 6-7b we can see the pattern of current limiting. Op-amp current is limited at 25mA and most of it is passed by the collector of Q_S to the short-circuit load. V_{BE} of Q_S and R_S limit current in Q_1 so that overload load current or short-circuit current I_{SC} is limited to

$$I_{SC} = I_{overload} = I_{O\max} + I_{Rs} \cong 23\text{mA} + \frac{0.6V}{R_S} \tag{6-15}$$

For the short circuit in Fig. 6-7b, $I_{SC} = 23$mA $+ 87$mA $= 110$mA.

Current-Limiting Action

R_L was varied in Fig. 6-7a to test current-limiting action. Op-amp output current at pin 10 and I_L were measured for each value of R_L and plotted versus V_o in Fig. 6-8. Observe that the onset of limiting occurs first in the op-amp. For example, at $R_L = 175$Ω, $I_o = 8.4$mA and $I_L = 85$mA. The base of Q_1 took 0.9mA from I_o and the remaining 7.5mA passed from collector to emitter of Q_S and contributed to I_L. Current through R_S was 85mA $-$ 7.5mA $= 77.5$mA (β_F at Q_1 measured at 85).

For $R_L \leqslant 140$Ω, I_{B1} stabilized at 1mA, I_{RS} at 87mA, I_o at 23mA, and I_L at 110mA. These values also pertain for $R_L = 0$, as in Fig. 6-7b.

Constant Current Power Supply

In Fig. 6-8 note the vertical constant current portion of the V_o vs. I_L curve. Note further that for all $R_L \leqslant 140\Omega$ our former regulated voltage supply now acts like, and is, a *regulated constant current* supply. It naturally follows that V_o is now determined by the product of constant current I_{SC} and load R_L. Not only has the simple addition of R_S and Q_S protected Q_1 against overload current, but it has also shown us the principle of a *constant current supply.*

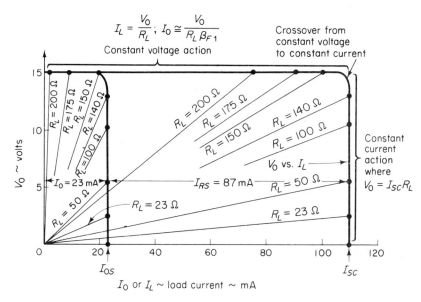

Figure 6-8

Voltage regulation curve for the circuit of Figure 6-7a.

6-7 MAXIMUM POWER DISSIPATION WITH SHORT-CIRCUIT PROTECTION

Power Dissipation in Q_1

Maximum heat must be dissipated by Q_1 under the short-circuit conditions of Fig. 6-7b. $P_{D1\max}$ will be the product of V_{CE1} and I_{C1}. Since I_{B2} will be small with respect to I_{RS}, the collector and emitter currents of Q_1 will approximately equal I_{RS}. The emitter of Q_1 is clamped at about 0.6V above ground by the base-emitter junction of Q_S, so $V_{CE1} = E_i - 0.6$. With short-circuited load

$$P_{D1} = V_{CE1}I_{C1} = (E_i - 0.6)I_{RS} \qquad (6\text{-}16)$$

where I_{RS} is found from Eq. (6-14). E_i is read from the unregulated supply voltage data at the current of I_{SC}, defined in Eq. (6-15).

Power Dissipation in Q_S

Power dissipation in Q_S is usually no problem. Collector current of Q_S essentially equals op-amp short-circuit current $I_{OS} \cong 23\text{mA}$. Base-emitter voltage drops of Q_S and Q_1 add to clamp the collector of Q_S at 1.2V. Thus $P_{DS\max}$ is evaluated from

$$P_{DS\max} = V_{CES}I_{CS} = 1.2(I_{OS}) \tag{6-17}$$

Power Dissipation in the Op-amp

Under short-circuit conditions, negligible currents will flow through pins 4, 5, and 6 of the op-amp. We can assume $I_{OS} \cong 22\text{mA}$ flows from pin 11 to pin 10. Pin 11 is at E_i and pin 10 is at 1.2V above ground. Thus maximum op-amp power $P_{O\max}$ is

$$P_{O\max} = (E_i - 1.2)I_{OS} \tag{6-18}$$

EXAMPLE 6-4:

Find short-circuit power dissipations for Q_1, Q_s, and op-amp in the circuit of Fig. 6-7. Assume $I_{O\max} = 23\text{mA}$.

SOLUTION:

From Eq. (6-15), $I_{SC} = 23\text{mA} + 0.6\text{V}/6.9\Omega = 110\text{mA}$. From Fig. 6-6, $E_i = 26.7\text{V}$ at 110mA. From Eq. (6-16)

$$P_{D1} = (26.7 - 0.6)(0.087) \cong 2.3\text{W}$$

From Eq. (6-17)

$$P_{DS\max} = (1.2\text{V})(22\text{mA}) = 27\text{mW}$$

From Eq. (6-18)

$$P_{O\max} = (26.7\text{V} - 1.2\text{V})(23\text{mA}) = 590\text{mW}$$

From the results of Ex. 6.4 we would probably choose a transistor heat-sinked for 3W or more for Q_1, and a small 300mW transistor for Q_S. Unfortunately, $P_{O\max}$ exceeds the maximum op-amp power rating of 500mW. There are two basic ways of transfering 100mW of heat power away from the op-amp (aside from installing a heat sink). In Fig. 6-9 op-amp current is limited to 23mA under short-circuit conditions. We must limit voltage at pin 11 to 500mW/23mA = 22V to protect the op-amp. In Fig. 6-9a the series zener is chosen to take up the voltage difference 26.7V − 22V = 4.7V and will absorb 4.7V × 23mA

110mW. In Fig. 6-9b the shunt zener fixes pin 11 at 22V. Resistor R is designed to pass 5mA to the zener and 23mA to the op-amp while dropping 4.7V. Thus $R = 4.7\text{V}/28\text{mA} = 170\Omega$.

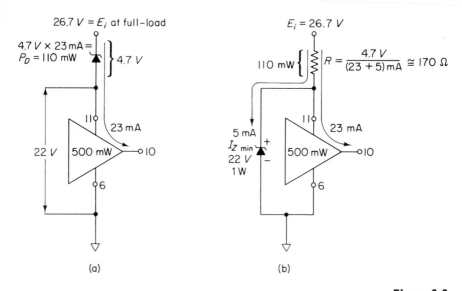

(a)   (b)

Figure 6-9

A series zener absorbs 110 mW in (a), and resistor R absorbs 110 mW in (b) to limit op-amp dissipation.

Over voltage protection of the op-amp may also be furnished by the circuit of Fig. 6-9a. For example, an unregulated supply with output voltage greater than 36V can be used to power Q_1, provided it is reduced below 36V to power the op-amp.

6-8 REMOTE VOLTAGE SENSING

Output resistance of the feedback regulator is so low, in the order of a few milliohms, that we should preserve its excellent performance wherever economically possible. In Fig. 6-10a load R_L is three feet away from the regulator's output terminals. Since wire resistance of #22 AWG is about 16.1 milliohms per foot, about 6 ft. $\times$ 16.1mΩ $\approx$ 100mΩ of resistance is added in series with the typically 5mΩ output resistance R_{OR} of the regulator. Output resistance seen by the load is $R_{OL} = 5\text{m}\Omega + 100\text{m}\Omega = 105\text{m}\Omega$. This is larger than output resistance at the regulator's output terminals by a factor of 21. Now a load current change of 1 ampere will change load voltage V_{OL} by 1A $\times$ 105m$\Omega \cong$ 0.1V, while regulator output voltage V_{OR} changes by only 1A $\times$ 5mΩ = 5mV. In those power supplies that feature remote sensing, the feedback loop is brought out to screw-type terminals. Heavy links are screwed down tight, closing the feedback loop to power output terminals designated $+$ and $-$. The regulator then regulates V_{OR}, not V_{OL}, to be constant.

To compensate for wire resistance voltage loss and achieve optimum regulation at the load terminals, *extend the feedback loop* to the load terminals as in Fig. 6-10b. Low-current sensing wires are shielded and connected between sense terminals and load *before* the links are removed and power turned on. Now

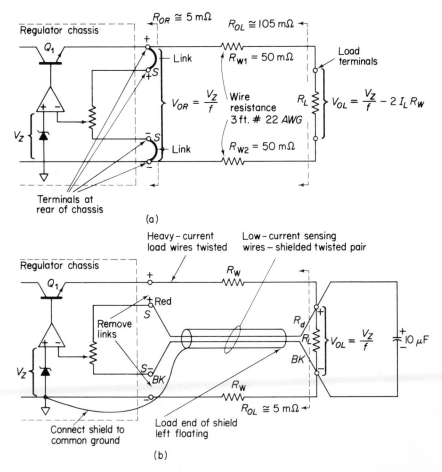

(a)

(b)

Figure 6-10

Remote load voltage sensing in (b) minimizes the effect of wire resistance in (a).

the regulator regulates V_{OL} to be constant and V_{OR} rises to whatever voltage is necessary to compensate for wire drop. The effect of wire resistance and current-sensing resistance on output resistance will be investigated in section 6-9.

If the red sense wire is opened, feedback voltage goes to zero and output voltage usually climbs to the unregulated supply value, which may be much higher than the regulators rated load voltage. If the black sensing lead is opened, we have a voltage-follower circuit and output voltage goes to V_Z. If both sensing leads are opened, all negative feedback is lost and the regulator may oscillate at high frequency. It is sound practice to connect a 10 to 50μF *decoupling* capacitor across the *load terminals* to eliminate possible oscillations caused by inductance in the sensing or line wires. Load wires should never be disconnected at the load terminals, or heavy load current will flow through the small gauge sensing wires. Remote voltage sensing is a desirable and relatively inexpensive option for feedback regulators.

6-9 EFFECT OF CURRENT-SENSING RESISTANCE ON OUTPUT RESISTANCE

Refer to Figs. 6-7a and 6-10b to draw a regulator schematic incorporating both current limiting and remote sensing. Assume R_L is set for one-half rated load so that Q_S is cutoff and has no effect on circuit operation. Both line wire

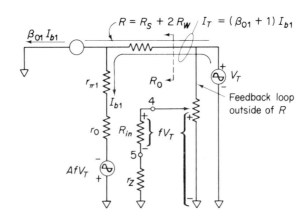

Figure 6-11

Circuit model to find effect of sensing resistor on output resistance.

resistors and sense-resistor R_S will be in series *inside* the feedback loop. To find an expression for resulting output resistance, R_O, apply test voltage V_T to the small signal circuit model in Fig. 6-11. (Compare with Fig. 6-4 to see the difference caused by R.) Resistor R models the effect of both R_S and both line wire resistors. Write the loop equation

$$AfV_T + V_T = I_{b_1}(r_{\pi 1} + r_o) + I_{b_1}(\beta_{o1} + 1)R$$

Substitute for $I_{b_1} = I_T/(\beta_{o1} + 1)$

$$V_T(1 + fA) = I_T\left(\frac{r_{\pi 1} + r_o}{\beta_{o1} + 1} + R\right)$$

and arrange to get output resistance $R_o = V_T/I_T$

$$\frac{V_T}{I_T} = R_o = \frac{r_{\pi 1} + r_o}{(1 + fA)(\beta_{o1} + 1)} + \frac{R}{1 + fA} \tag{6-19}$$

where $R = R_{SC} + R_{W_1} + R_{W_2}$.

Compare Eqs. (6-11) and (6-19) to see that R is still divided by $(1 + fA)$, but not by the transistor $(\beta_{o1} + 1)$. Note also that if the feedback loop were moved to the left of R in Fig. 6-11, there would be no reduction in R and R_o would be R plus the expression in Eq. (6-11).

EXAMPLE 6-5:

Show how output resistance in Example 6-2 would be affected by the addition of current limiting and remote sensing. Use 0.1Ω for $(R_{W_1} + R_{W_2})$ and 6.9Ω for R_S from Figs. 6-7a and 6-10b.

SOLUTION:

From Eq. (6-19) and data from Ex. (6-2)

$$R_o = \frac{50 + 75}{1 \times 10^7} + \frac{6.9 + 0.1}{100,000} = (0.0125 + 0.070)\text{m}\Omega = 0.0825\text{m}\Omega$$

While output resistance has been increased by addition of series resistance, it is still remarkably low, provided added resistance is kept inside the feedback loop.

6-10 REMOTE PROGRAMMING OF A REGULATED POWER SUPPLY

Remote programming is defined as a means of controlling regulated output voltage through a variable resistance or variable voltage, located at a distance remote from the regulator. This concept comes naturally from Eq. (6-10), where $V_o = V_Z/f$. It is possible to arrange a circuit to vary V_Z, or else we can vary f by controlling R_1 or R_2.

Controlling V_o with Remote Resistance

Voltage control terminals are brought out to a terminal strip on the regulator chassis for easy access in Fig. 6-12a. A rotary switch and resistor network, at a remote location, is connected via a shielded twist pair (don't use the shield as a wire) to the voltage control terminals. The switch can select any one of four values for R_2—in the order 2.5kΩ, 3.3kΩ, 5kΩ and 10kΩ—to give regulated

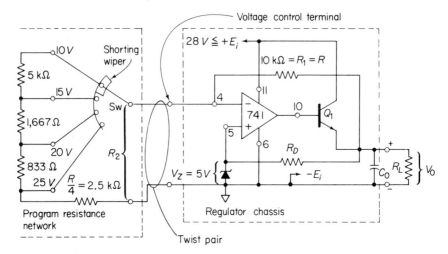

Figure 6-12

Remote programming of a power supply with resistance control of output voltage.

output voltages of 25, 20, 15, and 10V, respectively. Switch Sw has a shorting wiper so that R_2 will not be a momentary open ($R_2 = \infty$) while the wiper is transferring from one switch stator contact to the next. If desired, we could replace the top three resistors with a 7.5kΩ variable resistor and control V_o continuously from 10V to 25V. Other applications of remote voltage control by programming resistance are possible by placing R_1 rather than R_2 outside the regulator. If R_1 is a motor-driven resistor, the amplitude and frequency of V_o will be synchronized with the motor shaft position and speed respectively. Our regulator now acts as a tachometer and shaft position indicator.

EXAMPLE 6-6:

Calculate V_o for the two lower switch positions in Fig. 6-12.

SOLUTION:

From Eq. (6-10)

$$V_o = \frac{R_1 + R_2}{R_2}V_Z = \left(\frac{R_1}{R_2}\right)V_Z + V_Z = \frac{(10k\Omega)(5V)}{R_2} + 5V$$

For the 25V point

$$V_o = \frac{(10k\Omega)(5V)}{2.5k\Omega} + 5V = 25V$$

For the 20V point

$$V_o = \frac{(10k\Omega)(5V)}{3333\Omega} + 5V = 20V$$

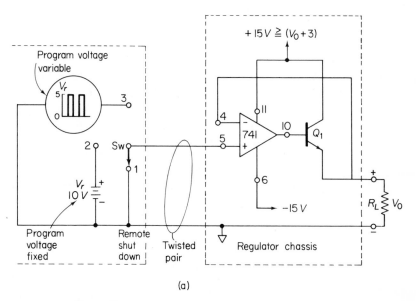

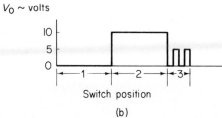

(b)

Figure 6-13

Voltage-follower programmed regulator in (a) has output voltages for each switch position in (b).

Controlling V_o with Remote Voltage

Up to now, for purposes of economy, we have derived reference voltage V_Z from the regulator itself. However, if we wish to program the regulator from a remote location with a voltage source, the voltage source will probably *not* be derived from the regulator. Figure 6-13 shows a special *follower-regulator* in which reference voltage V_r is applied to pin 5. Due to 100% negative feedback ($f = 1$), pin 4 is at the same potential (essentially) as pin 5. Since pin 10 is wired to pin 4, then output voltage V_o must equal or follow the program voltage. The advantage of this circuit is that only nanoamperes are drawn from V_r, and up to 100mA may be drawn from pin 10. Thus the follower-regulator performs as a power amplifier or as an *isolator* that almost completely isolates load R_L from low-power control voltage V_r. In Fig. 6-13, V_o is either at 0V or $+10$V when switch Sw is on positions 1 or 2 respectively. Position 1 also shows how to add *remote shutdown* to an IC regulator.

Switch position 3 connects a square wave from a multivibrator as reference voltage to show another application as an amplifier. Output decoupling capacitor C_o (used in Fig. 6-12) may have to be reduced or even removed for V_o (in Fig. 6-13) to be able to follow V_r. This is because C_o can discharge only through R_L, not back through Q_1, or the $-$ input. The time constant $C_o R_L$ (for lowest R_L) should be less than one-tenth of the time interval between pulses.

Remote programming can also be accomplished by eliminating R_D and the zener in Figs. 6-1 or 6-3 and bringing out a twist pair from ground and pin 5 to external program voltage V_r. V_o will than be related to V_r by Eq. (6-10), and if V_r can be adjusted to zero, V_o will go to zero. An alternate method of programming that will now be studied has an advantage whereby V_o cannot only be varied to zero, but can also be reversed in polarity.

6-11 NEGATIVE OR POSITIVE OUTPUT-VOLTAGE PROGRAMMING OF A FEEDBACK REGULATOR

In Fig. 6-14a the unregulated supply or full-wave rectifier with capacitor filter has been changed from our familiar bridge-type to a center-tapped transformer type. The reason for this change will be given in section 6-12.

Positive V_o with Negative Reference Voltage

There are three fundamental differences in the feedback regulator of Fig. 6-14a: (1) positive input 5 is grounded; (2) reference voltage V_r is connected via a resistor to pin 4 (instead of pin 5) and is *reversed* in polarity; (3) sensing resistors R_1 and R_2 no longer feed back a fraction of V_o, but now feed back a cur-

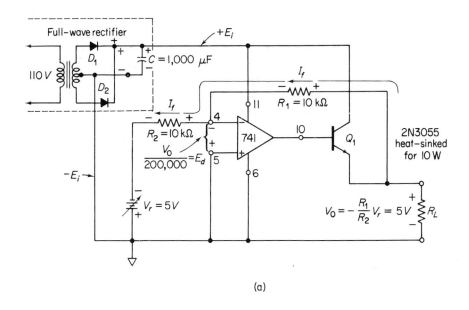

(a)

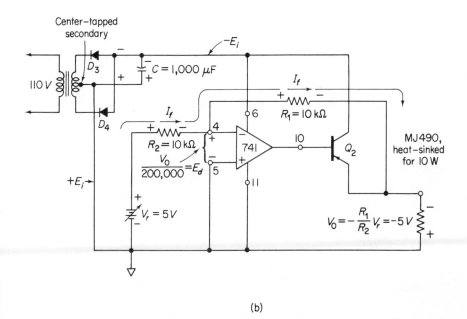

(b)

Figure 6-14

Programming voltage V_r gives positive output voltage control in (a) and
negative output voltage control in (b).

rent I_f. To understand what controls V_o, we reason that differential input voltage E_d must be very small ($V_o/A \leq 25\mu$V). Therefore, pin 4 is practically at the same potential as pin 5 or ground. Current I_f must be equal to the voltage across R_2 divided by R_2 or V_r/R_2. Since negligible current is drawn from pin 4 by the op-amp, I_f also flows through R_1 and is equal to voltage across R_1 divided by R_1 or V_o/R_1. V_r feeds the voltage drop across R_1 and E_d and sets the polarity of E_d, so that V_o must be positive or 180° out of phase with V_r. Calling current coming into pin 5 from R_1 positive, and current leaving pin 5 to R_2 negative

$$\frac{V_r}{R_2} = -I_f \quad \text{and} \quad I_f = \frac{V_o}{R_1}$$

Substituting for I_f in either equation

$$V_o = -\frac{R_1}{R_2}V_r \qquad (6\text{-}20)$$

For Fig. 6-14a, V_o is equal to $-V_r$ since $R_2 = R_1$. Gain can be obtained between V_o and V_R by choosing R_1 larger than R_2.

Negative V_o with Positive Reference Voltage

In Fig. 6-14b Q_1 is replaced by *pnp* transistor Q_2, and V_r is reversed in polarity along with the rectifier. Output voltage V_o can be varied from 0 to any negative voltage within the ratings of Q_2 and the 741 op-amp. This will be a maximum of approximately 0 to -30V.

Observe that V_o can never go positive in Fig. 6-14b. That is, if the polarity of V_r is reversed to drive pin 4 negative, V_o can only drop to 0 volts and can never go positive. This is because when pin 10 tries to go positive (assuming it could), Q_2 would be reverse-biased and shut off. The same conclusions apply in reverse to Fig. 6-14a: Any positive-going V_r will hold V_o at zero, and V_o can never go negative.

The Linear Half-wave Rectifier

This property is quite useful because the circuits of Fig. 6-14 have direct application as a low-voltage, *linear* half-wave rectifier. If V_r is replaced by a sinusoidal signal with peak amplitude of 0.5V, V_o will be a half-rectified sine wave. Only positive half-cycles will appear in the output of Fig. 6-14a, and negative half-cycles in the output of Fig. 6-14b. Peak values of either output will go to $+0.5$V or -0.5V respectively. Recall that with a normal half-wave rectifier using a silicon diode, the signal voltage must rise to about 0.6V before the diode becomes forward-biased. Thus any information or control data below 0.6V would be lost. In the linear rectifier application, Q_1 and Q_2 may be removed, and both R_L and R_1 are connected to pin 10. Figures 6-14a and b can also be combined into one circuit; the result will now be examined.

6-12 PROGRAMMING A REGULATOR AS AN AUDIO AMPLIFIER

In Fig. 6-15 (or Fig. 5-6), the same center-tapped secondary winding of the power transformer (corresponding to Figs. 6-14a and b) is used to develop two equal but opposite rectifier voltages. D_1 and D_2 conduct on alternate half-cycles to give positive dc voltage $+E_i$ with respect to the grounded center tap.

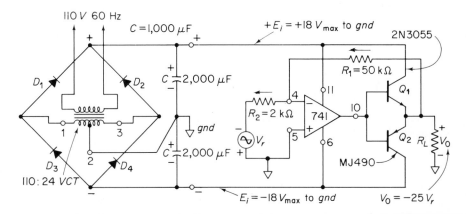

Figure 6-15

Both regulators of Figure 6-14 are combined to form this audio amplifier. V_o is 180° out of phase with V_r.

D_3 and D_4 conduct on alternate half-cycles to give negative dc voltage $-E_i$ with respect to the grounded center tap. Since the op-amp restricts maximum supply voltage between pins 11 and 6 below 36V, the peak transformer secondary voltage between terminals 1 and 3 cannot exceed 36V, or 36V/1.41 = 25V rms due to the center tap.

Positive or Negative Regulator Operation

Only one op-amp is required for either $+V_o$ or $-V_o$ in Fig. 6-15, because, for negative-going V_r as shown, the op-amp and Q_1 function almost exactly as they did in Fig. 6-14a. The difference is that there is a gain of $-R_1/R_2 = -25$ between V_r and V_o. Since the maximum value of V_o should be less than E_i by about 3 volts, $V_{omax} \cong 15V$. During positive-going V_r, the op-amp and Q_2 function exactly as they did in Fig. 6-14b to control negative-going V_o. When $V_r = 0$, $V_o = 0$ and no current flows through R_L, since both emitters are very close to ground potential.

We conclude that the circuit in Fig. 6-15 can regulate either positive or negative output voltages. Furthermore, if V_r is an audio signal, the circuit can also perform as an audio amplifier. Since there are no coupling capacitors, the amplifier will amplify dc input voltages at V_r.

We summarize our analysis of the regulator-amplifier by observing that V_o can be controlled continuously from 0 to either $+15$ or -15V. Gain can be changed easily by changing the value of either R_1 or R_2. In fact, gain can be varied by replacing R_2 with a variable resistor. Signal voltage V_r should have very low internal resistance, or its internal resistance R_g will add to R_2. Amplifier gain will be reduced to $V_o/V_r = R_2/(R_1 + R_g)$, unless we compensate by reducing R_1 by an amount equal to R_g. Input resistance to the amplifier equals R_2 since pin 4 is essentially at ground potential. Finally, when Q_1 drives V_o positive, Q_2 is cut off, and when Q_2 drives V_o negative, Q_1 is cut off. The circuit of Q_1 and Q_2 is known as *complementary symmetry* or *push-pull operation*.

EXAMPLE 6-6:

Revise either regulator in Fig. 6-14 to vary V_o between 5 and 30V. Use a center-tapped transformer and reference voltage of 5V.

SOLUTION:

To use a $+5$V reference voltage and get a positive V_o, wire $V_r = 5$V to $+$ input 5, as in Fig. 6-16a. From Eq. (6-10), for $V_o = 30$V and $V_r = V_z = 5$V, then $f = 5V/30V = 1/6$. If $V_o = 5$V, $f = 1$. Using Eq. (6-4), design a variable feedback resistor network to vary f from 1 to 6. From Eq. (6-4)

$$\frac{R_1 + R_2}{R_2} = \frac{1}{f} = 6 \quad \text{and} \quad R_1 = 5R_2$$

$$\frac{R_1 + R_2}{R_2} = \frac{1}{f} = 1 \quad \text{and} \quad R_1 = 0$$

Choose $R_2 = 2\text{k}\Omega$ and $R_1 = 0$ to $10\text{k}\Omega$ potentiometer.

EXAMPLE 6-7:

Employ Example 6-6 and Fig. 6-15 for guidance to make a regulator-amplifier with a gain of 26 and no phase shift between V_r and V_o.

SOLUTION:

A full-wave bridge and center-tapped transformer in Fig. 6-16b allow V_o to swing positive or negative for a maximum of ± 15V. Either a reference voltage or audio signal is applied between noninverting terminal pin 5 and ground. V_o will be in phase with V_r. Use Eqs. (6-10) and (6-4) to find $R_1 = 25R_2$. Pick $R_2 = 1\text{k}\Omega$ and $R_1 = 25\text{k}\Omega$.

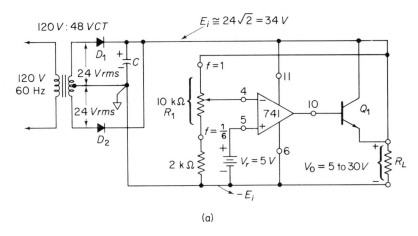

(a)

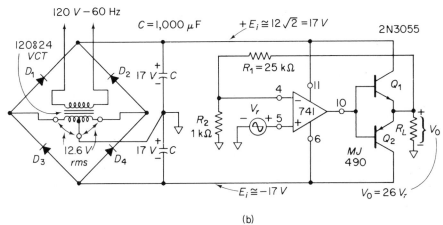

(b)

Figure 6-16

The positive output regulator in (a) is modified to form the audio amplifier in (b). V_o is in phase with V_r. In both circuits there are 34 V between op-amp pins 11 and 6.

The amplifier-regulator of Example 6-7 has the advantage of higher input resistance being presented to V_r than with the inverting amplifier of Fig. 6-15. If a battery is used for V_r, its useful life will approximately equal shelf life. Gain can be adjustable by making R_1 a potentiometer.

6-13 CURRENT-BOOSTING THE FEEDBACK REGULATOR

Load current capacity of the basic feedback regulator can be increased by adding a high-power *current-boost transistor*. In Fig. 6-17, boost-transistor

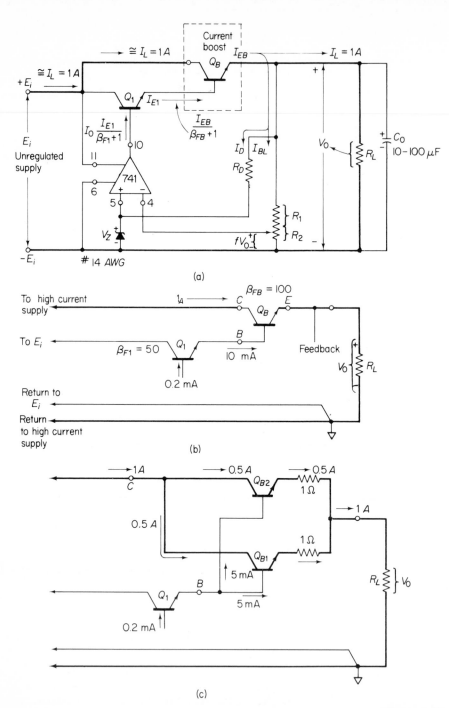

(a)

(b)

(c)

Figure 6-17

Current-boost transistor Q_B increases load current capacity in (a) and (b). Q_B is powered by a separate supply in (b), or paralleled in (c) to double its power-handling capacity.

Q_B multiplies emitter current of Q_1 by $(\beta_{FB} + 1)$. Therefore, the 100mA current capacity of the basic Q_1 op-amp regulator is increased by β_{FB} of Q_B. Maximum load current capacity is theoretically dependent on β_F of both transistors and the maximum output current from the op-amp, I_{omax}. Restricting I_{omax} to a comfortable 5mA and assuming typical $\beta_{F1} = 100$ and $\beta_{FB} = 50$, then I_{Lmax} is

$$\text{Theoretical } I_{Lmax} \cong I_{omax} \, \beta_{F1}\beta_{FB} = 5\text{mA} \times 100 \times 50 = 25\text{A}$$

This figure should be taken with a large grain of salt, as should *all maximum current* ratings. I_{Lmax} is set by heat dissipation limitations of Q_B to (1) an optimistic 10A, (2) a more realistic 5A if Q_B is to be on a practical size heat sink, or (3) a more practical 2.5A if our funds to pay for the transformer are limited.

Design considerations for R_1, R_2, V_Z, f, and E_i are identical with those previously developed. Since maximum power dissipation occurs at full-load current, maximum power relationships are developed in terms of I_{Lmax}.

Maximum Boost Transistor Dissipation P_{DB}

By inspection of Fig. 6-17a, maximum dissipation in Q_B is

$$P_{DB} = (E_i - V_o)(I_{Lmax}) \tag{6-21}$$

P_{DB} may be too high for one transistor to handle. For example, with E_i 30V and $V_o = 5$V, then $P_D = (30\text{V} - 5\text{V})(1\text{A}) = 25$W. One 2N3055 transistor heat-sinked for 15W would be destroyed. We can replace Q_B with two paralleled transistors to share the heat. One-ohm emitter resistors insure that Q_{B_1} and Q_{B_2} equalize their collector currents at one-half I_L. The resultant circuit is shown in Fig. 6-17c. Note that V_{CE} of *both* Q_{B_1} and Q_{B_2} is equal to Q_B, but current and consequently power is reduced by one-half for both Q_{B_1} and Q_{B_2}.

Maximum Pass Transistor Dissipation P_{D1}

The emitter of Q_1 is $V_o + V_{BEB} = (V_o + 0.6\text{V})$ volts above ground, and its collector is at E_i. Its collector and emitter currents are approximately equal, and also equal to the base current of Q_B. Neglecting the 1 in the current divider $(\beta_{FB} + 1)$

$$P_{D1} \cong [E_i - (V_o + 0.6\text{V})]\frac{I_{Lmax}}{\beta_{FB}} \tag{6-22a}$$

or

$$P_{D1} \cong \frac{P_{DB}}{\beta_{FB}} \tag{6-22b}$$

Maximum Op-amp Dissipation P_{Do}

Pin 10 is above ground by $V_o + V_{BEB} + V_{BE1} = V_o + 1.2V$. Pin 11 is at E_i. Current I_o flows between these two terminals to generate heat that is in addition to the 50mW idle current dissipation. Thus

$$P_{Do} \approx 50\text{mW} + [E_i - (V_o + 1.2V)]\frac{I_{Lmax}}{\beta_{FB}\beta_{F1}} \cong 50\text{mW} + \frac{P_{D1}}{\beta_{F1}} \qquad (6\text{-}23)$$

since $I_{omax} \cong I_{Lmax}/(\beta_{FB}\beta_{F1})$. E_i is the average dc output voltage of the unregulated rectifier supply, measured at I_{Lmax}.

Regulation of a Separate High-current Supply

Finally, it should be noted that the collector of Q_B can be wired to a separate high-current rectifier, where convenient, as in the simplified schematic of Fig. 6-17b. This change does not affect regulator performance provided the high-current supply voltage exceeds V_o by 3 volts.

EXAMPLE 6-8:

Revise Example 6-3 for a regulated supply of 1A at 15V. Assume $\beta_{F1} = 50$, $V_Z = 5.1V$, and employ the rectifier of Fig. 6-6.

SOLUTION:

(a) Since the solution circuit of Example 6-3 could furnish 100mA, we merely add a Q_B transistor with $\beta_F \geqslant 10$ and the design is completed, except for power considerations.

(b) Retain I_{BL}, I_Z, plus R_1 and R_2 of Example 6-3, and we have the same zener dissipation of 25mW.

(c) From Fig. 6-6, $V_{dc} = 23.7V$ at 1.0A for power calculations. Minimum E_i is $V_{dc} - \Delta V/2 = 23.7V - (2.1V/2) = 22.7V$. Our minimum input-output voltage differential is $22.7V - 15V = 7.7V$ and is satisfactory. (Note that V_o could be adjusted by f to a maximum of $22.7V - 2V \cong 20V$.)

(d) From Eq. (6-21), $P_{DB} = (23.7V - 15V)1A = 8.6W$. Select a 2N3055 heat-sinked for 12W. Assume $\beta_{FB} = 50$.

(e) From Eq. (6-22), $P_{D1} = (23.7V - 15.6V)(1.0A/50) = 8.1V \times 20\text{mA} = 0.16W$. Maximum current demand from Q_1 has been reduced from 100 mA to 20mA and its heat problems reduced accordingly, by a factor β_{FB}.

(f) From Eq. (6-23)

$$P_{Do} \cong 50\text{mW} + (23.7V - 16.2V)\frac{1.0A}{50 \times 50}$$

$$P_{Do} \cong 50\text{mW} + 3\text{mW} = 53\text{mW}$$

Finally, we construct an expression for R_o from inspection of Fig. 6-17a. Looking into the emitter of Q_1 we see $(r_o + r_\pi)/(\beta_o + 1)$ due to transistor action of Q_1. Then, looking into the emitter of Q_B, we see this resistance divided by $(\beta_{oB} + 1)$ plus $r_{\pi B}/\beta_{oB} + 1)$ due to transistor action of Q_B. Then the result is divided by $(1 + fA)$ because of the op-amp feedback loop. Thus R_o is very small at

$$R_o = \left[\left(\frac{(r_o + r_{\pi 1})}{\beta_{o1} + 1} \right) \left(\frac{1}{\beta_{oB} + 1} \right) + \frac{r_{\pi B}}{\beta_{oB} + 1} \right] \times \frac{1}{(1 + fA)} \qquad (6\text{-}24)$$

6-14 CURRENT-LIMITING THE CURRENT-BOOST TRANSISTOR

From the data sheets in Appendix A we know the op-amp will limit its own output current at about 25mA. If we install current-limiting on Q_1 alone, as in Fig. 6-18a, results are unsatisfactory. For example, assume we want to current-limit the supply of Example 6-8 to 1.0A. Then Q_1 must furnish 20mA to Q_B for full-load of 1.0A, and we would set R_{S1} to turn on Q_{1S} at 20mA. Under overload conditions, I_o can rise to 25mA and pass through Q_{1S}. This will add to I_{RS1} and furnish a maximum of 45mA to Q_B. Thus our supply will begin limiting at about 2.25A rather than 1.0A.

On the other hand, we could apply limiting to Q_B only in Fig. 6-18b. If we set I_{RS1} to 1A, then I_{B1} will only be $1A/50 = 20mA$. I_o will self-limit at 25mA and be multiplied by β_F of Q_1. The resultant 1.25A is passed through Q_{SB} to the load. Again limiting begins at 2.25A instead of 1.0A.

The defect in both circuits is that $I_{o\max}$ is multiplied by either β_F of Q_1 or β_F of Q_2. The solution is to current-limit both transistors and by-pass $I_{o\max}$ around both transistors. This procedure is shown in Fig. 6-18c. Protection for Q_B is established by choosing an overload current I_{OL} = short-circuit current I_{SC} of 1.1A. Then

$$R_{SB} = \frac{0.6V}{I_{SC}} = \frac{0.6V}{1.1A} = 0.55\Omega \qquad (6\text{-}25)$$

Pick R_{S1} to begin limiting at $I_{SC}/(\beta_{FB} + 1)$, or

$$R_{S1} = \frac{0.6V(\beta_{FB} + 1)}{I_{SC}} = \frac{0.6V(51)}{1.1A} = 20\Omega \qquad (6\text{-}26)$$

Resistance values obtained from these equations are approximate and should be modified by experiment.

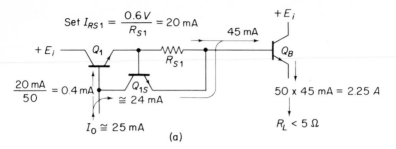

(a)

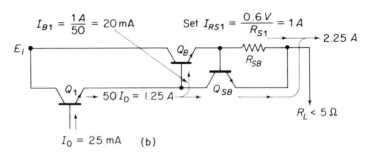

(b)

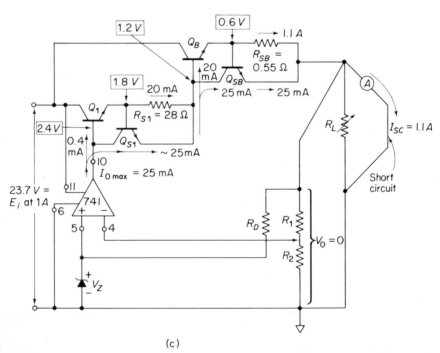

(c)

Figure 6-18

Current-limiting is applied incorrectly to the boosted supply in (a) and (b).
Correct limiting is shown in (c) and (d).

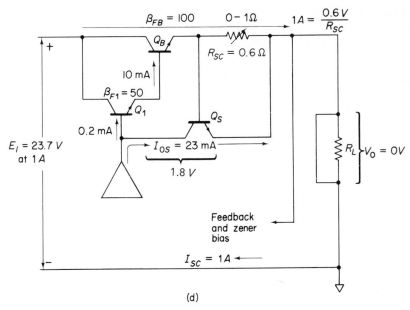

Figure 6-18 (Cont.)

Short-Circuit Power Limitations

Short-circuit power expressions are derived from an inspection of Fig. 6-18c. Allowing 0.6V for each base-emitter voltage drop, the voltage-to-ground for each transistor base is enclosed by a box. Boost transistor maximum-power dissipation is

$$P_{DB} = (E_i - 0.6\text{V})I_{SC} = (23.1\text{V})(1.1\text{A}) = 25.6\text{W} \qquad (6\text{-}27)$$

Equation (6-27) tells us that Q_B must dissipate as much heat as a small soldering iron. We could help Q_B by adding a collector resistor to absorb 5.7V and leave the collector of Q_B 3 volts above $V_o = 15$V at full-load. The resistor would absorb 5.7V $\times$ 1.1A = 6W and reduce transistor dissipation to 20W. An alternate solution would be to replace Q_B with two transistors in parallel to absorb 12W each. A one-ohm emitter resistor would be needed for each parallel Q_B to insure that they share the load.

Dissipation in Q_{SB} and Q_{S1} are equal at

$$P_{S1} = P_{SB} = 1.2\text{V}(I_{o\max}) = 1.2\text{V} \times 25\text{mA} = 30\text{mW} \qquad (6\text{-}28)$$

Dissipation in Q_1 is

$$P_{D1} = (E_i - 1.8\text{V})\left(\frac{I_{SC}}{\beta_{FB} + 1}\right) = 21.9\text{V} \times 20\text{mA} = 0.44\text{W} \qquad (6\text{-}29)$$

Dissipation in the op-amp is

$$P_{omax} = (E_i - 2.4V)I_{omax} = 21.3V \times 25mA = 530mW \qquad (6\text{-}30)$$

An alternate and more economical current-limiting option is shown in Fig. 6-18d. Q_S by-passes I_{OS} around *both* Q_1 and Q_B. R_{SC} determines short-circuit current I_{SC} from $0.6/R_{SC}$.

In practice, R_{SC} is a variable resistor and may be mounted on the front panel. It is adjusted as follows:

1. Set R_{SC} to zero.
2. Vary R_L to obtain maximum load current I_L, for example, 1A.
3. Increase R_{SC} until V_o just begins to drop, signifying onset of current-limiting.
4. Calibrate the setting of R_{SC} as 1A.

Other short-circuit current points can be calibrated on R_{SC} by repeating steps 1 to 4, but set I_L to the desired intermediate value in step 2, for example, 0.5A. The result is a calibrated, variable, current-limiting, or an adjustable constant current power supply.

6-15 INTEGRATED CIRCUIT VOLTAGE REGULATORS

Integrated circuit voltage regulators result from combinations of the latest advances in IC fabrication technology and circuit design. Teams of physicists, engineers, and technicians first design a circuit with the aid of computers, and then grow the circuit, in the sense that layer upon layer of semiconductor junctions, MOS capacitors, resistors, and interconnecting conductors are deposited by vacuum-diffusion technology. Internal circuitry is extremely complex. Improvements in fabrication technology are forced by improvements of existing design technology. Knowledge obtained during development, test, and manufacture of one IC points the way to better, more useful, and usually more complex ICs. Thus we should expect that, as we read about the characteristics of a device becoming available shortly, development has already been initiated on a new device that will hopefully be even better and perhaps cheaper.

How does a practicing technician, engineer, student, or instructor learn about the advantages of IC regulators? Rapid obsolescence of regulators does not furnish a partial answer. For example, if we purchase a number of price-reduced, discontinued IC regulators from a surplus outlet for experiment, we probably could have purchased more recently developed regulators for the same or lower cost. Furthermore, details of how ICs are built are not of real interest to those who want to use them. We must learn how to use these IC regulators (or any other IC) from the people who build them. Because of their relatively lengthy production time, most educational books are necessarily of little value in this respect. For this reason we will not discuss in detail how to apply representative IC regulators, but will merely furnish a partial list of excellent references in the bibliography. The remainder of this chapter is concerned with general guidance to the application of IC regulators.

Selection Guides

The first step in learning about IC regulators is to find out what is available and how much they cost. Send to manufacturers for a copy of their Selection Guide and price list for the family of devices. The selection guides show that the IC regulator family is classified into four branches: (1) Positive Voltage; (2) Negative Voltage; (3) Dual Voltage; and (4) Special Purpose Regulators.

Application Notes

After determining what type of regulator suits your needs, write to the manufacturer for his application notes on that type. The application notes and data sheets give vital information on maximum ratings of the device. For example, application guidance on positive regulators will give circuit diagrams showing the specifics of adding options to the basic regulator. The options of programming, current boost, and current limiting operate on the principles given in this chapter. Additional features, such as remote shutoff, digital programming, and floating regulator operation for higher voltage control, may be given for the more sophisticated regulators.

Data Sheets

Send for data sheets on the ICs of interest. Many of the newer data sheets contained essential data on mounting hardware, layout, and precautions. Some data sheets, particularly those for the special purpose regulators, perform double duty as application notes. For example, you can buy, for about $2.00, an excellent 15V-1A IC regulator that has only 3 terminals: input, output, and common. No adjustments of limiting current or output voltage are necessary because the IC contains all this circuitry, including the pass transistor and boost transistor. Because operation is virtually foolproof it does not require much application information. All you really need is the minimum input voltage and thermal resistance for heat-sinking. This means unregulated dc voltage can be distributed through a system and a local IC regulator can be placed on, for example, an op-amp card to deliver correct voltage to that card. A digital card requiring 5.0V would have its own regulator. Thus local regulators act as dc transformers, and save bussing different voltages as well as the need for extra transformers.

Typical IC Regulator Applications

Figure 6-19 illustrates flexibility and applications of National Semiconductor's IC regulators. Unregulated positive and negative voltage is supplied from the full-wave centertapped bridge rectifier of section 5-5. The LM 305 is a

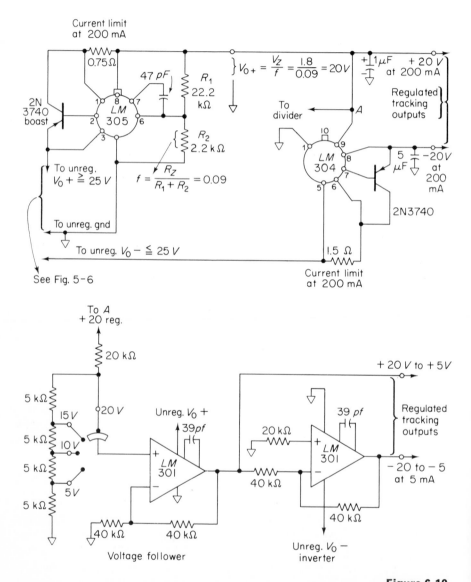

Figure 6-19

Tracking voltage regulator.

positive voltage regulator containing its own voltage reference of 1.8V, so that R_1 and R_2 are installed to set positive output voltage at $+20V$. Internal current limit circuitry is controlled by a 0.75Ω resistor to limit at 200mA. The LM 304 is a negative voltage regulator set up as a voltage follower to track and invert regulated positive voltage from the LM 305. Output of both regulators are boosted to 200mA by boost transistors 2N3740. Two low-cost op-amps are added to give adjustable, lower voltage, low-current, tracking, regulated $+$ and $-$ supply voltages. The 301 op-amp current limits itself to about 20mA.

One of the 3-terminal, fixed-voltage, positive regulators of Fairchild's μA 7800 series is illustrated in Fig. 6-20. The regulators are inexpensive and installation is as simple as connecting load and unregulated supply. They are current-limited by internal circuitry.

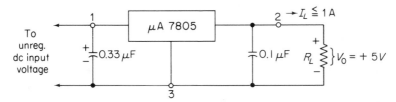

Figure 6-20

Three-terminal fixed-voltage regulator.

PROBLEMS

1. In Fig. 6-3, $V_Z = 5.0V$, $R_1 = 2R_2$, $A = 200,000$. Find V_o and E_d. What is the minimum value of E_i if a minimum input-output differential voltage of 3V must be maintained?

2. Repeat problem 1 for $R_2 = 2R_1$.

3. Given that $V_Z = 1.8V$ (as in some IC regulators), find V_o for each of the following feedback resistor combinations: (a) $R_1 = 5.55k\Omega$, $R_2 = 3.15k\Omega$; (b) $R_1 = 16.7k\Omega$, $R_2 = 2.3k\Omega$; (c) $R_1 = 22.2k\Omega$, $R_2 = 2.2k\Omega$ (Note: $R_1 \| R_2 = 2k\Omega$ for each combination.)

4. Find output resistance in Example 6-2 if we substitute a transistor with $\beta_F = 49$.

5. If the wiper arm of the $5k\Omega$ potentiometer were set to its upper limit in Fig. 6-5 and Example 6-3, find the new values of: (a) R_1 and R_2; (b) f; (c) V_o; (d) power dissipation of Q_1 at $I_{Lmax} = 100mA$; (e) op-amp power at $I_{Lmax} = 100mA$.

6. Repeat problem 5 for the wiper arm of the $5k\Omega$ potentiometer at the lower end of the $5k\Omega$ potentiometer.

7. What is the maximum and minimum possible V_o in Fig. 6-5 for $I_{Lmax} =$ 100mA?

8. Using Example 6-3 for guidance, design a feedback regulator to supply $I_{LFL} = 100$mA at 12V.

9. Find short-circuit current in Fig. 6-7 if $R_S = 6\Omega$ and $I_{OS} \cong 22$mA.

10. In the short-circuit conditions of Fig. 6-7, the op-amp self-heats and internal circuitry senses the heat increase to reduce I_{OS} to 17mA. Find the resulting device power limitations.

11. Redesign R_2 with a two-position switch to give regulated output voltages of 12V and 15V in Fig. 6-12.

12. Revise Example 6-6 for a -5V reference voltage.

13. Revise Fig. 6-15 for a gain of -10, but retain $R_1 = 50$kΩ.

14. Revise Fig. 6-16 b to obtain a gain of $+10$, but retain R_2.

15. When V_Z is changed to 2.5V in Fig. 6-12, what will be the new values of V_o for each switch position?

16. With sw on point 2 in Fig. 6-13, V_r replaced by a 9V transistor radio battery, and $R_L = 100\Omega$ to represent a typical small transistor radio, what current is drawn (a) by the radio and (b) from the transistor battery? (c) Estimate how long the battery would last.

17. The amplifier-regulators of Fig. 6-15 and Fig. 6-16b have about the same gain. Which circuit has a definite advantage over the other?

18. If $\beta_{FB} = \beta_{F1} = 50$ in Fig. 6-17a, $I_L = 2.0$A, $V_o = 15$V, and $E_i = 20$V: (a) what current is drawn from op-amp pin 10? (b) what approximate power will be dissipated in Q_B and Q_1?

19. With $R_{SC} = 1.0\Omega$ in Fig. 6-18d, evaluate emitter current of (a) Q_B, (b) Q_1, and (c) I_{B1}, (d) Find I_{SC} and power dissipation of Q_B.

20. Design a regulated supply with the following options: (1) output voltage variable from 10.0V to 20V; (2) maximum load current capacity of 1A; (3) short-circuit protection at 1A. Use the unregulated supply of Fig. 6-6, $V = 5.0$V, $\beta_{FB} = \beta_{F1} = 50$. Assume P_{Dmax} of available power transistors, with sinks, is 15W, and low-power transistors is 1W. Hints:
(a) Select R_D for $I_{BL} = 5$mA, at $V_o = 10$V.
(b) Select $R_1 + R_2 = 10$kΩ for $I_{BL} = 1$mA, when $V_o = 10$V. Select top and bottom stop resistors as in Fig. 6-7.

chapter seven
triggering devices

7-0 INTRODUCTION

A device is classified as a triggering device if it changes from one stable off-state to another stable on-state upon sensing a particular voltage level called the *trigger*, *firing*, or *peak-point* voltage. Triggering devices can do one of three basic jobs: (1) sense or detect a particular value of voltage; (2) determine the time interval between two events; or (3) control frequency of a relaxation oscillator.

In most applications, triggering devices sense voltage magnitude across a capacitor. When the capacitor voltage reaches the trigger voltage, the device turns on and provides a discharge path for the capacitor. What happens next depends on the circuit. We will learn how to control what happens next in the rest of this chapter. But before using the devices we must learn something about their stable on-state, stable off-state, and unstable in-between state called the *negative-resistance region*. The basic ideas, principles of operation, design, and analysis procedures will be introduced by an examination of the unijunction transistor (UJT).

After a few basic applications of the UJT are studied, we will turn to the two-transistor switch, not only for its own capabilities but also as an introduction to the *programmable unijunction transistor* (PUT). In sections 7-14 and 7-15 we look at *I-V* characteristics of other triggering devices to see how each device has its own range of trigger voltages.

Most triggering devices are designed to work with semiconductor switches such as the *silicon controlled rectifier* (SCR) or *triac*. These devices, along with their applications, are studied in Chapters 8 and 9.

7-1 INTRODUCTION TO THE UNIJUNCTION TRANSISTOR

In its simplest form the unijunction transistor (UJT) is made by diffusing *p*-type material about 70 percent of the distance from one end (B_1) of a bar-shaped *n*-type semiconductor. The resulting structure has one *pn*-junction as

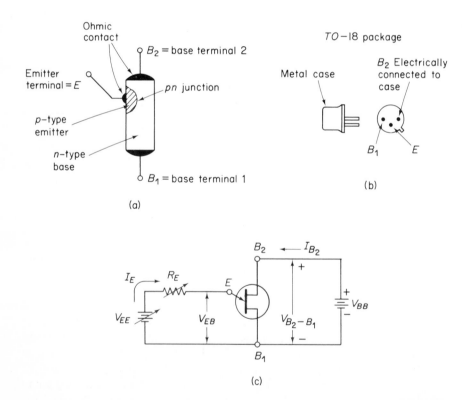

Figure 7-1

Structure in (a), packaging in (b), and circuit symbol in (c) of a unijunction transistor (UJT).

shown in Fig. 7-1a, with the typical package of Fig. 7-1b. Terminal, current, and voltage designations are shown in Fig. 7-1c, where the emitter arrow head points toward the *n*-type base material.

Interbase Resistance

From the circuit of Fig. 7-2a, *interbase resistance* r_{BB} is measured with emitter terminal open. This is the resistance of the *n*-type silicon between base-2 and base-1, and will range between 4kΩ and 10kΩ. Bar resistance from emitter junction to B_2 is called r_{B2}, and the remaining resistance is r_{B1}. Their sum is r_{BB}, or

$$r_{BB} = r_{B1} + r_{B2} \qquad (7\text{-}1)$$

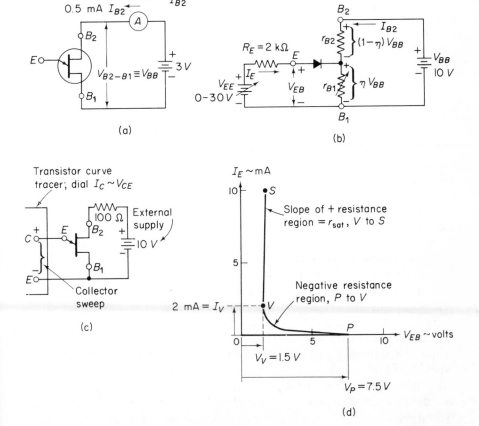

(a)

(b)

(c)

(d)

Figure 7-2

Interbase resistance r_{BB} is measured by the circuit in (a). Either test circuit in (b) and (c) can give the $I_E - V_{EB}$ characteristic in (d).

From measurements in Fig. 7-2a, $r_{BB} = 3V/0.5mA = 6k\Omega$. The UJT may be modeled by two resistors plus a diode in the test circuit of Fig. 7-2b. Let $V_{EE} = 0$ initially. I_{B2} develops a voltage drop across r_{B1} to reverse-bias the emitter diode. With $V_{EE} = 0$, the reverse-bias is equal to

$$\frac{r_{B1}}{r_{B1} + r_{B2}} V_{BB} = \eta V_{BB} = \frac{r_{B1}}{r_{BB}} V_{BB} \qquad (7\text{-}2)$$

Intrinsic Standoff Ratio

Voltage divider η (eta) is called *intrinsic standoff ratio*. Values of η are usually between 0.4 and 0.8. V_{EE} must be increased to a value large enough to forward-bias the emitter diode by $V_D \cong 0.5V$ before UJT triggering action is initiated. When V_{EE} is increased enough to make emitter-to-base 1 voltage V_{EB} equal to $\eta V_{BB} + V_D$, the UJT triggers. This value of V_{EB} is called *peak-point voltage* V_P where

$$V_P = V_D + \eta V_{BB} \qquad (7\text{-}3)$$

and $V_D \cong 0.5V$, $V_{BB} = V_{B2\text{-}B1}$.

UJT Switching Action

Once $V_{EB} = V_P$, the forward-biased emitter diode injects holes into the r_{B1} region, and electrons are drawn into the same region from terminal B_1. The increase in free current carriers lowers the value of r_{B1}. The decrease in r_{B1} sets off a regenerative switching action as follows: (1) Because r_{B1} drops, (2) emitter current increases, injecting more holes, causing (3) r_{B1} to drop. Of course V_{EE} divides between R_E and r_{B1} in Fig. 7-2b. Since neither V_{EE} nor R_E change, V_{EB} must necessarily drop. The collapse of r_{B1} is accompanied by an abrupt *increase* in emitter current and a *decrease* in emitter voltage. This collapse of r_{B1} is usually described as *negative resistance*. That is, when increasing current through a device causes increasing voltage across it, the device exhibits positive resistance. However, when increasing current causes decreasing voltage drops, the device behaves as a negative resistance. The decreasing value of r_{B1} stops when saturation occurs at a value called r_{sat}. Typical values for r_{sat} range between 5 and 25Ω.

The above description of UJT action is illustrated by the $I_E - V_{EB}$ emitter characteristic curve in Fig. 7-2d. The UJT is in its stable off-state from origin to point P, and in its stable on-state from valley-point V to point S. A minimum value of emitter current is required at peak-point P to initiate regenerative switching action. Manufacturers give this value as *peak-point* current I_P, and typical values are 0.5 to 50μA. Between peak-point P and valley-point V lies the unstable state along the negative resistance region of the $I_E - V_{EB}$ curve. Valley-point V is located on manufacturers' data sheets as *valley-current* I_V and ranges from 1 to 10mA, depending on the UJT. *Valley voltage* V_V is *not* found on data sheets, since it is circuit-dependent, but will usually be between 1 and 3 volts.

Other UJT Electrical Characteristics

Two other useful data points are given by manufacturers. A value is given for I_{B2} at a high-emitter current of 50mA, with $V_{B2\text{-}B1} = 10$V. This value is called $I_{B2\ \text{mod}}$ and is typically 12mA. A value is also given for $V_{EB\ \text{sat}}$ under the same test conditions ($I_E = 50$mA, $V_{B2\text{-}B1} = 10$V) and is usually 2 to 5V. This data is used to estimate r_{sat} from the slope of the positive resistance or saturation region (on-state)

$$r_{\text{sat}} = \frac{V_{EB\ \text{sat}} - V_V}{50\text{mA} - I_V} \simeq \frac{V_{EB\ \text{sat}} - V_V}{50\text{mA}} \tag{7-4}$$

EXAMPLE 7-1:

In the circuit of Fig. 7-2b, $I_{B2} = 1.66$mA when $I_E = 0$. V_{EB} is increased to 7.5V and the UJT triggers on. V_{EE} is then adjusted to 10V. Find: (a) r_{BB}; (b) η and V_P, assuming $V_D \cong 0.5$V; (c) r_{B1} and r_{B2}; (d) UJT emitter current when $V_{EE} = 10$V and $R_E = 2$kΩ; (e) r_{sat}.

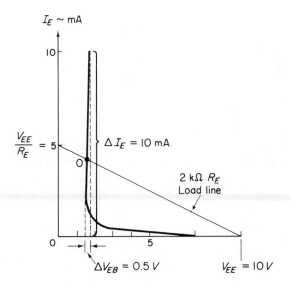

Figure 7-3

Solution to Example 7-1

SOLUTION:

(a) From Fig. 7-2a, $r_{BB} = V_{BB}/I_{B2} = 10V/1.66mA = 6k\Omega$.

(b) From Fig. 7-2d, $V_P = 7.5V = V_{EB}$ at trigger point. From Eq. (7-3), $V_P = V_D + \eta V_{BB}$, or $7.5V = 0.5V + \eta \times 10V$, or $\eta = 0.7$.

(c) From Eq. (7-2), $r_{B1} = \eta r_{BB} = 0.7 \times 6k\Omega = 4.2k\Omega$. From Eq. (7-1), $r_{B2} = r_{BB} - r_{B1} = 6k\Omega - 4.2k\Omega = 1.8k\Omega$.

(d) Draw an R_E load line on the emitter characteristic reproduced in Fig. 7-3, and locate operating point 0 at $I_E \cong 4mA$, $V_{EB} \cong 1.5V$.

(e) Slope of the $I_E - V_{EB}$ curve is found in the positive resistance region of Fig. 7-3 to be $\Delta V_{EB}/\Delta I_E = 0.5V/10mA = 50\Omega$.

7-2 OSCILLATOR ACTION WITH THE UJT

Addition of capacitor C_E and load resistor R_L in Fig. 7-4a forms a basic timer or oscillator, depending on the size of R_E. When switch sw is closed, capacitor voltage V_E charges toward V_{BB} at a rate determined by magnitudes of C_E and R_E. During charging time, the emitter diode is reverse-biased and the UJT is essentially disconnected from the charging circuit. In a time interval of one time constant equal to $R_E C_E$, V_E will charge from 0V to 63% of V_{BB}. Since η for many UJTs is approximately 0.6, then from Eq. (7-3), peak-point voltage $V_P \cong 65\%$ of V_{BB}. In other words, the time interval, T, required for C_E to charge up to a voltage equal to the trigger voltage V_P is approximately equal to the time constant

$$T = R_E C_E \tag{7-5}$$

Selection of Timing Resistor R_E

Once capacitor voltage V_E reaches V_P, the UJT triggers from its off-state to one of two possible states—either the stable on-state or the unstable negative resistance region, depending on the size of R_E. In Fig. 7-4b we *grossly* distort the $I_E—V_{EB}$ characteristic to show peak and valley-point coordinates on the same scale. If R_E is *below* some minimum value $R_{E\,min}$, C will discharge through the emitter, B_1 and R_L to develop an output voltage spike across R_L. R_L is usually small, from 10 to 100Ω. Therefore, the UJT will stabilize at an operating point located on its on-state curve. V_E will equal V_{EB} approximately at a value found by the procedure in Fig. 7-3. $R_{E\,min}$ is found in terms of circuit elements from Fig. 7-4b to be

$$R_{E\,min} = \frac{V_{BB} - V_V}{I_V} \simeq \frac{V_{BB}}{I_V} \tag{7-6}$$

If R_E is *larger* than some maximum value $R_{E\,max}$, current through R_E will be reduced below the minimum peak-point current I_P required to trigger the UJT. For example, suppose $R_E = 8M\Omega$ in Fig. 7-4a. When C charges to 6V,

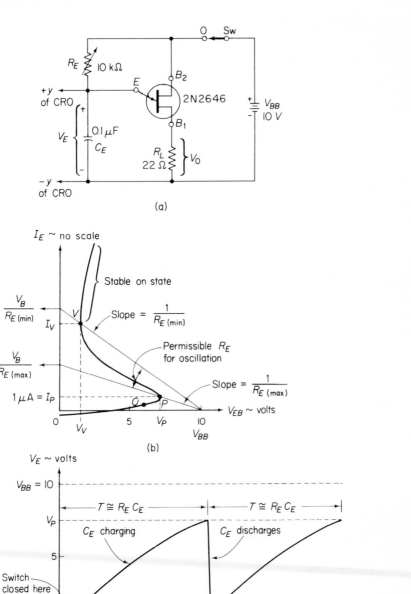

(a)

(b)

(c)

Figure 7-4

The UJT circuit in (a) will oscillate if R_E has a value between $R_{E\text{min}}$ and $R_{E\text{max}}$ in (b). A recurring triangular wave is developed across C_E in (c).

current through R_E will be limited to $(10 - 6)\text{V}/8\text{M}\Omega = 0.5\mu\text{A}$. UJT operation will stabilize at point O in Fig. 7-4b. Here the emitter conducts $0.5\mu\text{A}$ and none of the current through R_E is left to increase the charge on C. Thus capacitor voltage never reaches V_P and the UJT remains in the off-state. Since $R_{E\,max}$ will just barely furnish I_P at the peak-point voltage, define $R_{E\,max}$ from Fig. 7-4b as

$$R_{E\,max} = \frac{V_{BB} - V_P}{I_P} \tag{7-7}$$

The UJT will oscillate if R_E is selected to have a value *between* $R_{E\,min}$ and $R_{E\,max}$. Once V_E has charged to V_P the UJT triggers, collapsing r_{B1} toward its saturation value. C_E discharges through the now forward-biased emitter diode, R_L, and through r_{B1} to a voltage approximately equal to V_V. The time interval for discharge is in the order of 1 to $100\mu\text{seconds}$, depending directly on the size of C_E. Once C_E stops discharging, emitter current falls and r_{B1} increases in resistance to its high off-state value. This reverse-biases the emitter diode and C_E begins to recharge from an initial value of V_V toward V_P. Since V_V is often small with respect to V_P, the charging time of C_E will still be approximately given by Eq. (7-5). Since the UJT triggers once for every charge time interval T, the period is T and frequency of oscillation f is the reciprocal of the period, or

$$f \cong \frac{1}{T} = \frac{1}{R_E C_E} \tag{7-8}$$

EXAMPLE 7-2:

Given $C_E = 0.1\mu\text{F}$ in Fig. 7-4a. Typical values of $r_{BB} = 7\text{k}\Omega$, $I_P = 1\mu\text{A}$, $\eta = 0.6$, and $I_V = 6\text{mA}$ are taken from the data sheet of a 2N2646. No value is given for V_V. Find: (a) the permissible range for R_E to construct an oscillator; (b) minimum and maximum frequency of oscillation.

SOLUTION:

(a) Assume $V_V \cong 0\text{V}$ in Eq. (7-6), and $R_{E\,min} = 10\text{V}/6\text{mA} = 1.66\text{k}\Omega$. From Eq. (7-3), $V_P = 0.5\text{V} + 0.6 \times 10\text{V} = 6.5\text{V}$. From Eq. (7-7)

$$R_{E\,max} = \frac{(10 - 6.5)\text{V}}{1\mu\text{A}} = 3.5\text{M}\Omega$$

(b) From Eqs. (7-5) and (7-8)

$$T_{max} = R_{E\,max}C_E = 3.5 \times 10^6 \times 0.1 \times 10^{-6} = 0.35\text{sec}$$

$$f_{min} = \frac{1}{T_{max}} = \frac{1}{0.35\text{sec}} = 2.86\text{Hz}$$

$$T_{min} = R_{E\,min}C_E = 1.66 \times 10^3 \times 0.1 \times 10^{-6} = 0.166\text{ms}$$

$$f_{max} = \frac{1}{T_{min}} = \frac{1}{0.166\text{ms}} = 6.0\text{kHz}$$

UJT Oscillator Frequency Range

We conclude from Example 7-2 that the range in frequency of oscillation for a UJT is very large, over 1,000 to 1. Practical values for C_E are $0.001\mu F$ to $10\mu F$. The upper limit of oscillating frequency is about 100kHz because of the discharge time of C_E. The lower limit is about once per 5 minutes, due to leakage current of large electrolytic capacitors exceeding I_P. Because of leakage, low-leakage mylar capacitors should be used for long charge times.

7-3 MEASUREMENTS AND TEMPERATURE STABILIZATION FOR THE UJT

The simplest way to measure key UJT electrical characteristics η, V_P, V_V, I_V, and I_P is to build an oscillator, and measure the capacitor voltage waveform with a CRO. Measurement procedure is illustrated by an example.

EXAMPLE 7-3:

Design a UJT oscillator for operation at $T = 5$ms ($f = 200$Hz) with $V_{BB} = 10$V.

SOLUTION:

Based on experience gained in Example 7-2, pick a value for $R_E = 50k\Omega$ as a mean value between $R_{E\,min}$ and $R_{E\,max}$. From Eq. (7-8)

$$C_E = \frac{1}{R_E \times f} = \frac{1}{50,000 \times 200} = \frac{1}{10 \times 10^6} = 0.1\mu F$$

The circuit is built as in Fig. 7-5b, and measurements of $V_V = 1.5$V and $V_P = 7.5$V are obtained from the emitter voltage CRO picture in Fig. 7-5a. Next, R_E is increased until the UJT stops oscillating. This value of R_E is measured as $R_{E\,max} = 3M\Omega$. Then R_E is reduced to $R_{E\,min} = 2k\Omega$ where the UJT stops oscillating. Find η from Eq. (7-3)

$$\eta = \frac{V_P - V_D}{V_{BB}} = \frac{(7.5 - 0.5)V}{10V} = 0.7$$

I_V from Eq. (7-6)

$$I_V = \frac{V_{BB} - V_V}{R_{E\,min}} = \frac{(10 - 1.5)V}{2k\Omega} = 4.2mA$$

and I_P from Eq. (7-7)

$$I_P = \frac{V_{BB} - V_P}{R_{E\,max}} = \frac{(10 - 7.5)V}{3M\Omega} = 0.8\mu A$$

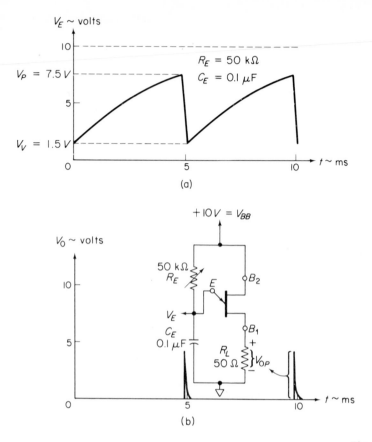

Figure 7-5

Solution to Example 7-3.

It should be noted in connection with Example 7-3 that if R_E is known and we measure T (or f) on a CRO, we can calculate C_E. Alternatively, if C_E is known we can calculate R_E from a measurement of T (or f). Thus, if we have a digital frequency meter to read T or f we can use a UJT to make a digital ohmmeter or a capacitance meter.

Temperature Dependence of Oscillator Frequency

If the UJT undergoes a temperature change, r_{BB}, η and V_D change. This will change V_P and oscillator frequency. It has been found experimentally that adding an ordinary carbon resistor with a positive temperature coefficient (increase resistance with increasing temperature) in series with B_2 can stabilize V_P against temperature changes. Temperature-stabilizing resistor R_T is shown in Fig. 7-6. Its value is *approximated* by

$$R_T \cong 0.015 V_{BB} r_{BB} \eta \tag{7-9}$$

For maximum oscillator stability, a trial value of R_T should be calculated from Eq. (7-9) and installed in the oscillator. Then oscillator frequency is monitored while temperature is cycled over the desired range. R_T is changed until frequency changes are minimized over the temperature range. For Example 7-3, R_T would be initially chosen at

$$R_T \cong 0.015 \times 10 \times 7{,}000 \times 0.7 \cong 730\Omega$$

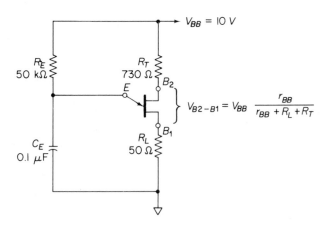

Figure 7-6

UJT oscillator with temperature compensator R_T.

Interbase voltage $V_{B2\text{-}B1}$ may differ significantly from V_{BB} because of R_L and R_T. Where *measured* values of V_D, V_V, R_E, C_E, η, and R_{BB} are known, a more exact formula for T is given by

$$T = R_E C_E \ln \frac{V_{BB} - V_V}{V_{BB} - V_D - \eta V_{B2\text{-}B1}} \tag{7-10}$$

EXAMPLE 7-4:
Using the data of Example 7-3, find T from Eq. (7-10). Let $r_{BB} = 7\text{k}\Omega$ and $V_D = 0.5\text{V}$.

SOLUTION:
From Fig. 7-6, $V_{B2\text{-}B1} = (10\text{V})(7{,}000\Omega/7{,}780\Omega) = 9\text{V}$

From Eq. (7-10)

$$T = 50 \times 10^3 \times 0.1 \times 10^{-6} \ln \frac{10 - 1.5}{10 - 0.5 - 0.7 \times 9}$$

$$T = 5\text{ms} \ln \frac{8.5}{3.2} = 5\text{ms} \ln 2.65$$

$$T = 5\text{ms} \times 0.974 = 4.86\text{ms}$$

Comparing Examples 7-3 and 7-4, we see the margin of error (5ms approximate versus 4.86ms exact) is small when using the approximation formula of Eq. (7-8) and the exact formula of Eq. (7-10).

7-4 UJT OUTPUT VOLTAGES

Sawtooth output voltage waveforms are available between emitter terminal and ground, as in Figs. 7-4c and 7-5a. However, positive-going pulses are available at B_1 if a small load resistor R_L is added. When the UJT triggers and C_E discharges through R_L, a spike output voltage V_O is developed as in Fig. 7-5b. The peak value of V_O or V_{OP1} is difficult to predict, since it depends on V_{BB}, C_E, R_L, and how fast r_{B1} of the UJT collapses. Peak values of V_{OP1} were measured for three different capacitors as R_L was varied from 0 to 50Ω, with $V_{BB} = 20$V, in Fig. 7-7. Note that pulse amplitude doubled for the same R_L as C_E was changed by a factor of 100 from 0.001μF to 0.1μF. For example, when $R_L = 20\Omega$, $V_{BB} = 20$V, V_{OP1} increased from about 4 to about 8V.

V_{OP1} was also measured for the same capacitors and same variation in R_L, but with V_{BB} halved to 10V. For $R_L = 50\Omega$, $C = 0.1\mu$F, we observe $V_{OP1} \cong 10.4$V for $V_{BB} = 20$V, and $V_{OP1} = 4.3$V for $V_{BB} = 10$V. We conclude that pulse height roughly doubles as supply voltage doubles, and that capacitors below 0.001μF do not generate much of an output pulse and should not be used except in special circumstances.

Base 1 Pedestal Voltage

During UJT off-time, V_{BB} divides among R_T, r_{BB}, and R_L in Fig. 7-6 to establish a small dc voltage across R_L. This voltage forms a pedestal from which V_O rises to its peak value when the UJT triggers. As will be shown in Chapters 8 and 9, the UJT pulse will be used to trigger a semiconductor switch such as the SCR or triac. The dc pedestal voltage must be much lower than the voltage required to trigger a semiconductor switch. For this reason, R_L must be restricted to a small value, so if we increase R_L to increase V_{OP1} we must check that the pedestal voltage is not too high for our needs.

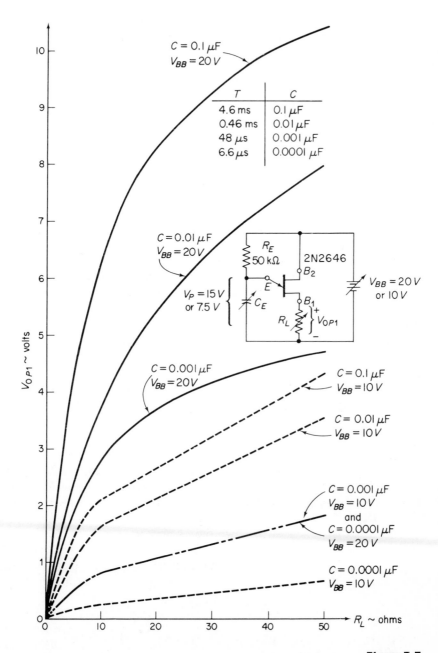

T	C
4.6 ms	0.1 μF
0.46 ms	0.01 μF
48 μs	0.001 μF
6.6 μs	0.0001 μF

$C = 0.1\,\mu F$
$V_{BB} = 20\,V$

$C = 0.01\,\mu F$
$V_{BB} = 20\,V$

$V_P = 15\,V$
or 7.5 V

$C = 0.001\,\mu F$
$V_{BB} = 20\,V$

$C = 0.1\,\mu F$
$V_{BB} = 10\,V$

$C = 0.01\,\mu F$
$V_{BB} = 10\,V$

$C = 0.001\,\mu F$
$V_{BB} = 10\,V$
and
$C = 0.0001\,\mu F$
$V_{BB} = 20\,V$

$C = 0.0001\,\mu F$
$V_{BB} = 10\,V$

R_E
50 kΩ 2N2646
B_2
E $V_{BB} = 20\,V$
C_E or 10 V
B_1
R_L V_{OP1}

$V_{OP1} \sim$ volts

$R_L \sim$ ohms

Figure 7-7

Variation of V_{OP1} with C_E, R_L, and V_{BB}.

EXAMPLE 7-5:

Assuming $r_{BB} = 7k\Omega$ in Fig. 7-6, what is the value of: (a) pedestal voltage? (b) off-state voltage from B_2 to ground? (c) I_{B2} off-state?

SOLUTION:

(a)

$$V_{B1} = \frac{R_L}{R_L + r_{BB} + R_T}V_{BB} = \frac{50}{50\Omega + 7k\Omega + 730\Omega}(10V) \cong 64mV$$

(b)

$$V_{B2} = \frac{r_{BB} + R_L}{R_L + r_{BB} + R_T}V_{BB} = \frac{7,050\Omega}{7,780\Omega}(10V) \cong 9.1V$$

(c)

$$I_{B2} = \frac{V_{BB}}{R_L + r_{BB} + R_T} = \frac{10V}{7,780\Omega} = 1.29mA$$

In Fig. 7-6, I_{B2} abruptly increases when the UJT triggers because of the collapse of r_{B1}. Increased interbase current I_{B2} increases the voltage drop across R_T, and B_2 goes negative. For example, if I_{B2} increased abruptly to 3.27mA in Ex. 7-5, voltage drop across R_T would increase from 1.29mA $\times$ 730Ω = 0.94V to 3.27mA $\times$ 730Ω = 2.39V. B_2 goes negative from +9.1V to (10V − 2.39V) = 7.6V for a negative-going pulse of 1.5V peak.

7-5 CONSTANT-CURRENT CHARGING OPTION FOR THE UJT OSCILLATOR

Capacitor voltage rises exponentially and does not give a pure triangular wave. Replacing R_E with a *constant-current charging* circuit will charge C at a constant rate. In Fig. 7-8a, I_{B2} flows through r_{BB} to forward-bias three silicon diodes. The diodes form a forward-bias voltage of $N \times 0.6V$ across R_E and base-emitter junction of BJT Q_2. N is the number of diodes. The 0.6V base-emitter voltage drop uses up one diode drop, so emitter current is set by R_E and $(N − 1) \times 0.6V$. That is, voltage across two diodes appears across R_E. Since emitter current I_E and collector current I_C are about equal, and I_C is the capacitor charging current I, C_E is charged by a constant current

$$I = \frac{(N − 1)(0.6)}{R_E} \tag{7-11a}$$

Voltage V_C across C_E rises *at a constant rate* and is determined by capacitor size C and charge Q where

$$V_C = \frac{Q}{C_E} \tag{7-11b}$$

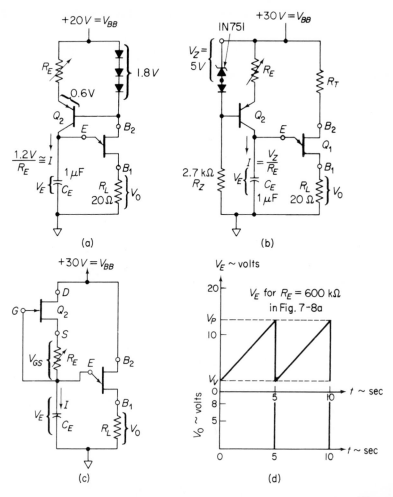

Figure 7-8

Constant current source Q_2, in (a), (b), and (c) converts the UJT oscillator into a linear sawtooth wave or ramp generator. Emitter voltage output ramp is shown in (d).

But Q equals current I (amps per second) times time t in seconds

$$Q = It \qquad (7\text{-}11c)$$

Now the UJT will trigger when V_C rises from V_V to V_P in time interval $t = T$, or

$$t = T = \frac{Q}{I} = \frac{(V_C)(C_E)}{I} = \frac{(V_P - V_V)C_E}{I} \qquad (7\text{-}12)$$

Control of Constant-Current Charge with a Zener

In Fig. 7-8b the diode temperature-stabilizes zener voltage and its voltage drop equals V_{BE} of Q_2. Resistor R_Z insures that the zener and diode are turned on. Zener voltage V_Z is established across R_E and charging current I is constant with magnitude of

$$I = \frac{V_Z}{R_E} \qquad (7\text{-}13)$$

Charging time interval T is found from Eq. (7-12) and frequency of oscillation from

$$f = \frac{1}{T} \qquad (7\text{-}14)$$

Supply and zener or total diode voltage must be chosen to insure that Q_2 of Figs. 7-8a and b remains out of saturation until V_E reaches V_P. Thus, for Figs. 7-8a and b respectively, allow $V_{CE2} = 1\text{V}$ as a guard against saturation, and minimum supply voltage V_{BB} must be, for Fig. 7-8a:

$$V_{BB} \geqslant V_P + 1 + (N - 1)0.6 \qquad (7\text{-}15a)$$

for Fig. 7-8b:

$$V_{BB} \geqslant V_P + 1 + V_Z \qquad (7\text{-}15b)$$

JFET Constant-Current Generator

An N-channel FET or constant-current diode is used in Fig. 7-8c as a constant-current source. Charge current I is set by gate-source voltage V_{GS} and R_E from

$$I = \frac{V_{GS}}{R_E} \qquad (7\text{-}16)$$

The magnitude of I is determined by R_E and the drain current vs. drain-source voltage curves of the FET, as in section 3-2. A simpler procedure is to build the circuit with R_E as a potentiometer and adjust R_E for the desired time interval.

The emitter voltage wave in Fig. 7-8d is triangular and quite linear. The UJT oscillators in Fig. 7-8a, b, and c are often called *ramp* generators because the sawtooth emitter voltage waveform looks like a ramp.

EXAMPLE 7-6:

Given $V_D = 0.5\text{V}$, $I_P = 1\mu\text{A}$, $\eta = 0.65$, $I_V = 4\text{mA}$, and $V_V = 2\text{V}$ in Figs. 7-8a and b. (a) What value of R_E is required for $I = 2\mu\text{A}$ in Fig. 7-8a, and what are the corresponding oscillator frequency and charge time interval? (b) Repeat for Fig. 7-8b.

SOLUTION:

(a) For Fig. 7-8a, find V_P from Eq. (7-3).

$$V_P = V_D + \eta V_{B2\text{-}B1} = 0.5V + 0.65(20 - 1.8)V = 12.3V$$

From Eq. (7-11a)

$$2 \times 10^{-6} = \frac{(3 - 1)0.6}{R_E}, \quad R_E = 0.6M\Omega$$

From Eq. (7-12)

$$T = \frac{(V_P - V_V)C_E}{I} = \frac{(12.3 - 2)V \times 1\mu F}{2\mu A} = 5.1s$$

From Eq. (7-14)

$$f = \frac{1}{5.1s} \cong 0.2Hz$$

For Fig. 7-8b, neglect R_T, and from Eq. (7-3)

$$V_P = 0.5V + (0.65)(30V) = 20V$$

From Eq. (7-13)

$$R_E = \frac{V_Z}{I} = \frac{5V}{2\mu A} = 2.5M\Omega$$

From Eq. (7-12)

$$T = \frac{(20 - 2)V \times 1\mu F}{2\mu A} = 9s, f = \frac{1}{9s} = 0.11Hz$$

Calibration of Charge Current

Charge current $I = 2I_P$ in Example 7-6 and is the slowest practical rate of charging C. Since T depends directly on I and R_E, T is directly proportional to R_E. This means we can use a potentiometer for R_E and calibrate R_E in terms of time, frequency, or charge current. For example, in Fig. 7-8b we could mark R_E at its point of 1.25MΩ (half the value of R_E in Example 7-6b) with the calibration $I = 4\mu A, T = 4.5sec, f = 0.22Hz$. There are many possible applications based on this observation; a few are illustrated in the next section.

7-6 MEASUREMENTS WITH THE BASIC UJT TIMER-OSCILLATOR

The ramp generator of Fig. 7-8b will be modified as required for a variety of applications. The lowest permissible value of R_E corresponds to a charge current I equal to I_V and is found from Eq. (7-13) to be

$$R_{E\,\min}\frac{V_Z}{I_V} = \frac{5V}{4mA} = 1.25k\Omega$$

Corresponding time interval T and frequency of oscillation f are found from Eq. (7-12) and Example 7-6 to be

$$T - \frac{(20-2)V \times 1\mu F}{4mA} = 4.5ms, f = \frac{1}{4.5ms} = 222Hz$$

(If convenient, T can be reduced and f increased by a factor of 10 if C_E is reduced by a factor of 10 to 0.1μF.) End points $R_{E\,\min}$, $I_{\max}$ and $R_{E\,\max}$, $I_{\min}$ are located in Fig. 7-9 and connected to form a calibration curve. Log-log scales are necessary to get the data on one piece of paper. This data will be used in the applications that follow.

Resistance or Temperature Measurement with the UJT Oscillator

An unknown resistor can be connected in place of R_E in Fig. 7-8b and can be measured by reading f or T on a CRO or frequency counter. Thus we have a

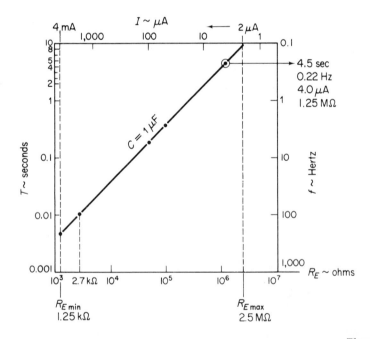

Figure 7-9

Calibration curve for Figure 7-8b relating charging time T and current I to R_E and oscillating frequency f.

resistance to frequency converter. For example, when f reads 100Hz we find from Fig. 7-9 that the unknown resistance is 2,700Ω. Since thermistors are resistors that vary with temperature, replace R_E with a thermistor to obtain a temperature to frequency converter. If a thermistor has a resistance of 2,700Ω at 200°C and 1,250Ω at 100°C, a frequency meter will read 100 and 222Hz, respectively, for a change of 2Hz per °C. See also Fig. 7-14.

Light Intensity, Distance, and Size Measurement with a UJT Oscillator

R_E can be replaced with a light sensitive resistor (cadmium sulphide cell) or LSR whose resistance decreases with increasing light. Typically, dark resistance of an LSR is 2.5MΩ, and resistance in bright sun is about 1kΩ. With an LSR replacing R_E in Fig. 7-8b, a frequency meter would read from 0.1Hz to 200Hz. A more convenient frequency range of 1 to 2,000Hz could be obtained by changing C_E to 0.1μF (or 10 to 20,000Hz with $C_E = 0.01\mu$F). See also Fig. 7-14.

A light source can be aimed at an object so that reflected light falls on the LSR. As the object is moved farther away, reflected light intensity falls and LSR resistance increases. Now object distance is translated into light intensity and the UJT converts light intensity to a frequency count, so we have a distance to frequency converter. Experimentally, it is easiest to calibrate the frequency counter in terms of distance.

If objects of different size are held at the same distance from a light source, we can calibrate the UJT-frequency counter to measure their size—a small object on a conveyor belt interrupts or reflects less light than a large object on the same conveyor belt. In any of these applications logic circuitry can be set to act on a particular frequency and actuate a control. For example, a large object causes a low-frequency reading that is sensed by a logic circuit to divert the large object to another conveyor heading toward a packaging machine.

Capacitor Size or Leakage Current Measurement

Large electrolytic capacitors have leakage currents exceeding I_P. If an electrolytic capacitor replaces C_E in Fig. 7-8b, we can increase R_E until the UJT just stops oscillating and measure leakage current. For example, if R_E is increased to 500kΩ before oscillation ceases, leakage current is 10μA. Now all of the charging current leaks through the capacitor and cannot charge it to V_P.

If R_E is set to 2.7kΩ, then $f = $ 100Hz in Fig. 7-9, and frequency will vary if C_E is varied from Eq. (7-12). For example, if C_E is changed to 0.1μF, f will equal 1kHz, and capacity can be read directly from the frequency meter by means of

$$C_E \text{ (in } \mu\text{F)} = \frac{100}{f\text{(in Hz)}} \tag{7-17}$$

Voltage-to-Frequency Conversion

Other related applications of the UJT, such as the event timer, voltage-controlled oscillator, and voltage level sensing, will be studied with the programmable unijunction transistor at the end of this chapter.

7-7 OTHER OPTIONS FOR THE UJT OSCILLATOR

Q_D is added in Fig. 7-10a to discharge C_E to almost zero volts, or $V_{CE\text{ sat}}$ of Q_D. While C_E charges, Q_D is cut off. When the UJT triggers a pulse through R_L, Q_D is driven into saturation and C_E discharges through Q_D as well as through

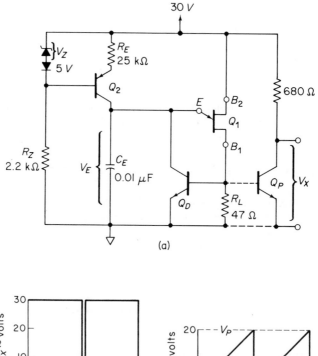

(a)

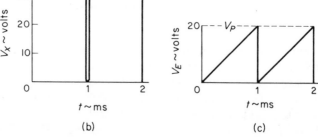

(b)   (c)

Figure 7-10

Q_D lowers valley voltage to almost $0V$ in (a) and (c). Q_P generates an output pulse in (b).

the emitter of the UJT. C_E discharges to a few tenths of a volt, lowering the valley voltage to roughly zero. Since the UJT does not have to discharge all of the charge on C_E, it turns off faster, and fall time of the V_E ramp voltage is faster. By connecting Q_P instead of Q_D in Fig. 7-10a, we make a pulse generator with a negative-going pulse duration of about 5 to 10μs duration. Q_P operates like Q_D, being saturated during the discharge of C_E. Output voltage V_X of Q_P is shown in Fig. 7-10b.

EXAMPLE 7-7:

What is the frequency of oscillation if Q_D is added in Fig. 7-10a? Assume $V_P = 20$V.

SOLUTION:

From Eq. (7-13), $I = V_Z/R_E = 5$V$/25$k$\Omega = 0.2$mA. Modify Eq. (7-12) to

$$T = \frac{V_P C_E}{I} = \frac{20 \times 10^{-8}}{2 \times 10^{-4}} = 10 \times 10^{-4} = 1\text{ms}$$

$$f = \frac{1}{T} = 1,000\text{Hz}$$

V_E is shown in Fig. 7-10c.

Minimizing Oscillator Loading Effects

When a load resistance is connected across C_E, oscillator timing is changed because the load draws current that would otherwise charge C_E. In the basic circuit of Fig. 7-6, any load resistor across C_E would form a voltage divider with R_E. V_{BB} would divide between R_E and the load resistor. Usually, voltage drop across the load is less than V_P, so C_E cannot charge to V_P and the UJT never fires. In Fig. 7-11a, Q_3 and Q_4 are added between UJT emitter and load. Assuming $\beta_3 = \beta_4 = 100$, any emitter resistance at Q_4 will be multiplied by $\beta^2 = 10,000$. For a 1kΩ volume control, load on Q_1 emitter is 10MΩ. V_E reaches 20V $= V_P$, where base current of Q_3 is only $(20 - 1.2)$V$/10$M$\Omega = 1.8\mu$A. For charging currents above 20μA there will be no significant change in oscillator performance by adding the Darlington screened load. Voltage at the emitter of Q_4 will be 1.2V below and follow the shape of V_E. Added capacitor C_C charges to the average value of $(1/2)(V_E - 1.2$V$)$, giving V_o as a sawtooth wave, centered on zero volts. In Fig. 7-11b, V_o is shown for full volume control setting.

Triangular Wave Generator

A triangular wave generator is made in Fig. 7-12a by adding the option of diode D and discharge resistor R_D. As shown in Fig. 7-12b, C_E charges toward V_{TH} from V_V. While C charges, diode D is forward-biased and the UJT is off.

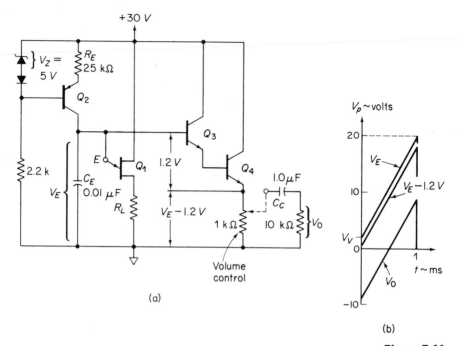

Figure 7-11

Darlington $Q_3 - Q_4$ minimize loading effects on the sawtooth voltage at V_E.

V_{TH} must exceed V_P or the UJT will not trigger. Charging time interval, T_{CH}, for C_E to charge from V_V to V_P is

$$T_{CH} = (R_E \| R_D)C_E \ln \frac{V_{TH} - V_V}{V_{TH} - V_P}, \quad V_E = V_V \text{ to } V_P \qquad (7\text{-}18)$$

when V_E reaches V_P, the UJT triggers. As shown in Fig. 7-12c, the emitter voltage drops to roughly V_V. R_E must lie between $R_{E\,min}$ and $R_{E\,max}$ to hold the UJT on in the negative resistance region. Diode D is now reverse-biased and C_E discharges through R_D from V_P to V_V, where diode D becomes forward-biased. D connects C_E back to the UJT emitter, where C_E draws current away from the emitter to trigger the UJT off and reset the charging circuit of Fig. 7-12b. Discharge time T_{DCH} is

$$T_{DCH} = R_D C_E \ln \frac{V_P}{V_V}, \quad V_E = V_P \text{ to } V_V \qquad (7\text{-}19)$$

EXAMPLE 7-8:

Given $V_P = 15V$, $V_V = 4V$, $R_{E\,max} = 3M\Omega$, $R_{E\,min} = 2k\Omega$, $C_E = 1.0\mu F$ in Fig. 7-12. (a) What is the minimum possible value for R_D (neglecting

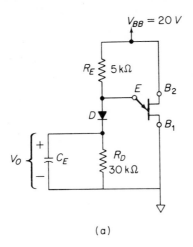

$V_{BB} = 20\,V$

R_E $5\,k\Omega$

E B_2

D

B_1

V_O $+$ C_E R_D
  $-$ $30\,k\Omega$

(a)

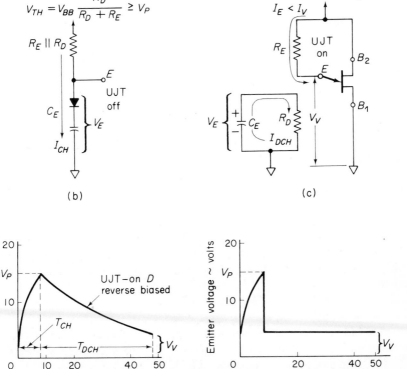

$V_{TH} = V_{BB} \dfrac{R_D}{R_D + R_E} \geq V_P$

$R_E \parallel R_D$

E

UJT off

C_E V_E

I_{CH}

(b)

$V_{BB} = 20\,V$

$I_E < I_V$

R_E UJT on B_2

E

B_1

V_E $+$ C_E R_D V_V
  $-$ I_{DCH}

(c)

(d)

$V_O \sim$ volts

20

V_P

10

UJT-on D reverse biased

T_{CH}

T_{DCH} $\}V_V$

0 10 20 40 50

$t \sim$ ms

(e)

Emitter voltage $\sim$ volts

20

V_P

10

$\}V_V$

0 20 40 50

$t \sim$ ms

Figure 7-12

Option R_D and D in (a) convert the UJT to a triangular wave generator.
Charge current in (b) and discharge current in (c) give the waveshapes in
(d) and (e).

277

diode voltage) if $R_E = 5\text{k}\Omega$? (b) Select $R_D = 2\,R_{D\,\text{min}}$ and estimate the waveshape of V_o.

SOLUTION:

(a) From Fig. 7-12b

$$15\text{V} = V_P \leq \frac{R_D}{R_D + R_E}V_{BB} = \frac{R_D}{R_D + 5\text{k}\Omega} \times 20\text{V}$$

$$R_D \geq 15\text{k}\Omega = R_{D\,\text{min}}$$

(b) Select $R_D = 30\text{k}\Omega$, and $R_D \| R_E = (5\|30)\text{k}\Omega = 4.28\text{k}\Omega$, and $V_{TH} = (30\text{k}\Omega/35\text{k}\Omega)(20\text{V}) = 17.1\text{V}$. From Eq. (7-18)

$$T_{CH} = (4{,}280)10^{-6}\,\ln\frac{17.1 - 4}{17.1 - 15} \simeq 8\text{ms}$$

From Eq. (7-19)

$$T_{DCH} = 20 \times 10^3 \times 10^{-6}\,\ln\frac{15}{4} \simeq 40\text{ms}$$

The resulting waveshapes of V_o and emitter voltage are shown in Figs. 7-12d and e.

Variable-Frequency, Variable-slope, Ramp Generator

Our next UJT application is shown in Fig. 7-13. Capacitor C_o charges at a linear rate from constant-current generator Q_2. R_R varies charging current and consequently rate of increase of V_o. Thus R_R is a variable slope control. C_o stops charging when Q_2 saturates to a maximum value of $V_{o\,\text{max}} = V_{BB} - V_Z$. UJT oscillator frequency is set by R_E and C_E according to Eq. (7-6), so R_E is a variable frequency control. When the UJT fires, it drives Q_D into saturation, and Q_D discharges C_o to almost 0V to reset the ramp. Circuit operation is explored in an example.

EXAMPLE 7-9:

(a) Find the period of the variable-frequency, variable-slope ramp generator in Fig. 7-13a if $R_E = 50\text{k}\Omega$. (b) Find $V_{o\,\text{max}}$. (c) Derive an expression for the ramp's slope in volts per second. (d) Show V_o for values of R_R at 2.5$\text{k}\Omega$, 5$\text{k}\Omega$, 10$\text{k}\Omega$, and 25$\text{k}\Omega$.

SOLUTION:

(a) From Eq. (7-8), $T = R_E C_E = (50\text{k}\Omega)(1\,\mu\text{F}) = 50\text{ms}$.
(b) $V_{o\,\text{max}} = V_{BB} - V_Z = 30\text{V} - 5\text{V} = 25\text{V}$.

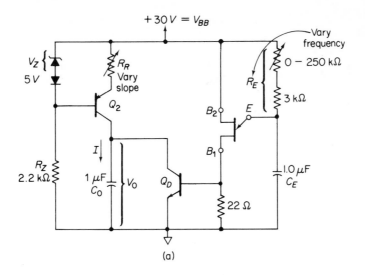

(a)

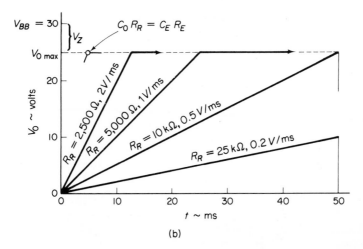

(b)

Figure 7-13

Variable-frequency variable-slope ramp generator in (a) has output voltage
waveshapes in (b) for $R_E = 50\ k\Omega$.

(c) From Eq. (7-13), $I = V_Z R_R$ for Fig. 7-13a. From Eq. (7-12)

$$V_{Co} = V_o = \frac{T}{C_o} \times I = \frac{T}{C_o} \times \frac{V_Z}{R_R} \quad \text{or} \quad \frac{V_o}{T} = \frac{V_Z}{C_o R_R}$$

(Note: If $R_R C_o = C_E R_E = T$, then V_o rises at V_Z volts per second.)

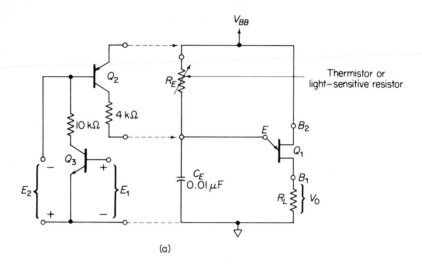

(a)

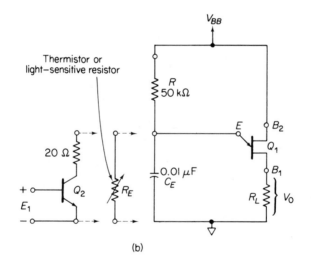

(b)

Figure 7-14

Series control of UJT frequency in (a) and parallel control in (b).

(d) Substituting in (c) for V_Z and C_E

$$\frac{V_o}{T} = \text{Volt/s} = \frac{5}{(1 \times 10^{-6})R_R} = \frac{5 \times 10^6}{R_R}$$

For plotting convenience convert to volts per millisecond:

$$\frac{V_o}{T} = \frac{\text{volts}}{\text{ms}} = \frac{5,000}{R_R}$$

Tabulating calculations

R_R in ohms	2,500	5,000	10,000	25,000
V_o/T in volts/ms	2	1	0.5	0.2

Resultant ramp voltages of V_o are shown in Fig. 7-13b.

Thermistor and LSR Control of Oscillator Frequency

Our final representative options are shown in Fig. 7-14. Timing resistor R_E represents a thermistor or light sensitive resistor that can make oscillator frequency vary with light or temperature variations. In Fig. 7-14a oscillations will stop when $R_E < R_{E\,min}$ or $R_E > R_{E\,max}$. In Fig. 7-14b oscillation stops when R_E becomes low enough to reduce its own voltage drop below V_P. From the voltage division of V_{BB}

$$\frac{R_E}{R_E + R}V_{BB} \leqslant V_P \cong \eta V_{BB}; \frac{R_E}{R_E + R} \leqslant \eta$$

we see that $R_E \leqq R/(1 - \eta)$ to stop oscillation.

On-off Control of Oscillators

Replacing R_E with Q_2 or the $Q_2 - Q_3$ combination in Fig. 7-14a allows a positive (E_1) or negative (E_2) voltage to control oscillator frequency by controlling the equivalent series resistance between emitter and base of Q_2. In Fig. 7-14b, Q_2 can replace R_E to vary frequency or act as a shutoff control, as determined by voltage E_1.

7-8 THE TWO-TRANSISTOR REGENERATIVE SWITCH

In Fig. 7-15a, *pnp* transistor Q_P and *npn* transistor Q_N are interconnected to form a regenerative switch. Only three terminals are brought out and labeled: G for *gate*, A for *anode*, and K for *cathode*. (Significance of these labels will become apparent when we study the *programmable unijunction transistor* in section 7-9). *Program voltage* V_S is analagous to ηV_{BB} of the UJT in Fig. 7-2b. Anode voltage V_{AK} is less than V_{GK}, so the base-emitter junction of Q_P is reverse-biased and Q_P is cut off. Since Q_P is essentially an open circuit, Q_N has no forward-bias and is also cut off. The two-transistor switch is in its stable off-state. (Note: V_S must be less than 6 to 7V or reverse-voltage rating on V_{BE} of Q_P may be exceeded.) We conclude that, for all negative values of V_{AA}, or positive values less than or equal to V_S (shown in Fig. 7-15a): (a) the switch is off; (b) no currents flow between terminals G, A, and K; (c) $V_{GK} = V_S$ and $V_{AK} = V_{AA}$.

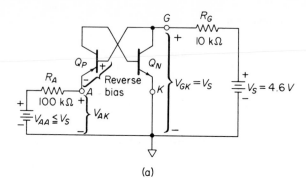

(a)

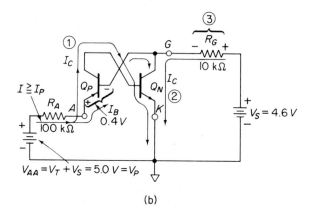

(b)

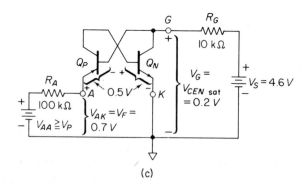

(c)

Figure 7-15

The two-transistor regenerative switch has a stable *off-state* in (a), unstable *transition* state in (b), and stable *on-state* in (c).

Turn On

In order to be changed from its stable off-state to its stable on-state, the switch must go through an unstable regenerative switching state. To begin regenerative switching action, anode voltage V_{AK} must be increased to a *peak-point voltage V_P* that will exceed program voltage V_S by enough voltage to forward-bias the base-emitter junction of Q_P. The required forward-bias is called *offset voltage V_T* (analagous to diode voltage V_D in the UJT) and is about 0.5V. Trigger or peak-point voltage can now be expressed by

$$V_P = V_T + V_S \qquad (7\text{-}20)$$

Regenerative Switching Action

Regenerative switching action is shown in Fig. 7-15b. Once Q_P becomes forward-biased, its collector current enters the base of Q_N and is amplified by β of Q_N as collector current of Q_N to be conducted through R_G and lower gate potential V_{GK}. As V_{GK} drops, Q_P becomes more forward-biased to accelerate the increase in collector currents of both Q_P and Q_N. This is a regenerative switching action, with Q_P and Q_N *driving one another very fast* from cutoff into saturation. *Peak-point current I_P* is the minimum current that must flow into the anode at $V_{AK} = V_P$ to initiate triggering.

When both Q_P and Q_N reach saturation, the regenerative switch is in its stable on-state. As shown in Fig. 7-15c, gate voltage is close to ground potential at $V_{CEN \, sat} = 0.2$V. Ground at cathode K clamps both base of Q_N and collector of Q_P at 0.5V. Forward-bias on base-emitter junction of Q_P, plus $V_{CEN \, sat}$, clamp anode A to about $0.5 + 0.2 = 0.7$V above the reference cathode ground. This on-state voltage between anode and cathode is called *forward voltage V_F*.

Sweep Measurements of Switching Action

All electrical characteristics of a two-transistor switch can be measured with a CRO, filament transformer, and dc voltage supply from the sweep circuit in Fig. 7-16c. The transformer-bridge arrangement is adjusted to supply a varying anode voltage V_{AA} whose peak value exceeds trigger voltage V_P. Program voltage V_S is set at 4.6V. Using cathode K as reference, V_{AA} is measured with a CRO (on internal sweep) and its waveform is shown in Fig. 7-16a. Anode voltage V_{AK} is seen, in Fig. 7-16a, to equal V_{AA} until V_{AK} rises to reach trigger voltage $V_P = 5.0$V. Regenerative switching from off- to on-state occurs at about 3.5ms, and V_{AK} drops to its on-state value of forward voltage $V_F = 0.7$V. The CRO shows that gate voltage V_{GK} equals program voltage V_S until triggering occurs, whereupon V_{GK} drops to the low 0.2V saturation voltage of Q_N.

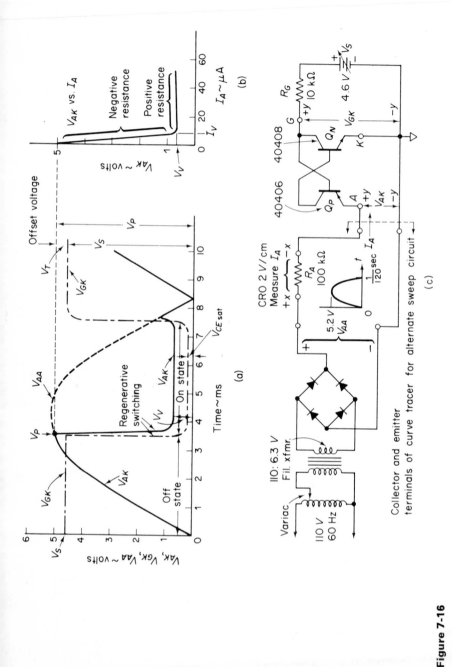

Figure 7-16

The sweep circuit in (c) shows gate and anode voltages with respect to time in (a), and anode $V_A - I_A$ characteristic in (b). (Note high resistance [10MΩ] probes must be used to avoid loading by CRO.)

Dial for external sweep on the CRO and wire its x-inputs to display anode current I_A on the horizontal axis. Anode $V_{AK} - I_A$ characteristic is shown on the CRO as in Fig. 7-16b. As expected, anode voltage V_{AK} goes down as anode current goes up, at trigger voltage $V_P = V_{AK}$. Thus the two-transistor switch exhibits a negative resistance region just like the UJT. However, regenerative switching action of the transistors is faster than the collapse of r_{B1} in the UJT. Valley voltage V_V and valley current I_V are measured from the CRO display, as shown in Figs. 7-16a and b. Offset voltage V_T is measured from the difference between V_P and V_S.

Turn-off

As sweep voltage V_{AA} drops in Fig. 7-16a to about 1.7V at $t \cong 7.5$ms, anode current I_A drops to

$$I_A = \frac{V_{AA} - V_{AK}}{R_A} = \frac{(1.7 - 0.7)\text{V}}{100\text{k}\Omega} = 10\mu\text{A} = I_V \tag{7-21}$$

This small current is not enough to hold the transistors in saturation, and the regenerative switching process is reversed, with both transistors driving one another swiftly through their active regions into cutoff. The minimum value of current, necessary to hold the transistors in saturation, is called *valley current* I_V. We conclude that anode current greater than I_V will maintain the regenerative switch in its stable on-state, and anode current less than I_V will force the switch to return swiftly to its stable off-state.

Voltage-Sensitive Switch

All conclusions, measurements, and definitions developed in this section are directly transferable to characteristics of the PUT described in the next section. The most important observation in both sections in that V_S can be *varied* or *programmed* to program trigger voltage V_P. Thus we have been studying a *voltage-sensitive switch* that looks for a particular voltage before it turns on. Furthermore, we can control the voltage. Applications for the two-transistor switch are identical with those of the programmable unijunction transistor, and will be introduced in the following sections.

7-9 INTRODUCTION TO THE PROGRAMMABLE UNIJUNCTION TRANSISTOR

The PUT or *programmable unijunction transistor* is *not* a single-junction semiconductor device like the UJT, but is a triode thyristor constructed from four alternating layers of p-type and n-type material. The I-V anode characteristic of a PUT is similar to the *I-V* emitter characteristic of a UJT. Therefore, although the PUT is actually a regenerative switch whose trigger voltage is programmable (established by an external resistance network), it was named for the UJT. Principal advantages of the PUT over the UJT and two-transistor switch are: (1) higher breakdown voltage; (2) low-voltage operation capability; (3) higher output voltage pulses; (4) programmable trigger voltage; and (5) most important, low cost and small size.

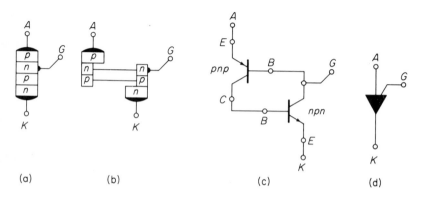

Figure 7-17

PUT construction in (a), *pnp-npn* transistor analogy in (c) and symbol in (d).

Construction of the PUT is pictured in Fig. 7-17a to derive the two transistor analogies in Figs. 7-17b and c. The schematic symbol is shown in Fig. 7-17d, where anode, gate, and cathode terminals are brought out from three of the four layers and labeled *A*, *G*, and *K*, respectively. Compare Figs. 7-17c and 7-15a to see that the PUT is a two-transistor switch in one package, and the *A*, *G*, *K* designations correspond to one another.

7-10 MEASURING CHARACTERISTICS OF A PUT

Replacing the two-transistor switch in the test circuit of Fig. 7-16c with a 2N6028 PUT gives the PUT characteristics in Fig. 7-18. A negative resistance trace is absent because V_{AK} cannot collapse directly from V_P to V_V due to capacitance from the CRO. However, observe the marked similarity between characteristics of the PUT and the two-transistor switch.

A simpler test set, shown in Fig. 7-19a, will be useful for other types of thyristors. Program voltage V_S is set at 20V, and then high-impedance voltmeter V_2 is shifted to read V_{GK}.

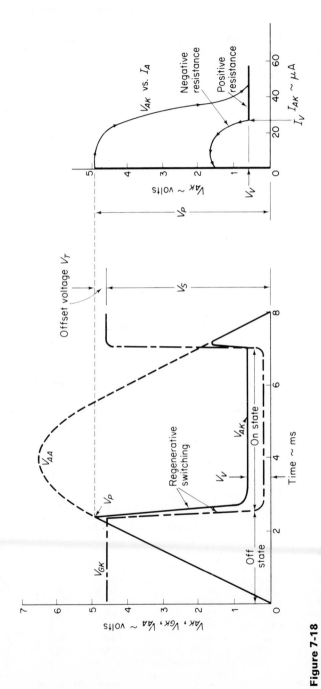

Figure 7-18

PUT characteristics obtained from the test circuit of Figure 7-15c.

Measurement of V_P and I_P

With $R_A = 200\text{k}\Omega$, increase V_{AA} slowly until V_{AK} and V_{GK} abruptly trigger to their on-state values of about 0.7V and 0.2V respectively. The highest values of V_A and I_A before triggering are V_P and I_P respectively.

Measurement of I_V

Once the PUT is on, I_A is determined by R_A and V_{AA}. (R_A was selected to restrict $I_{A\,\text{on}}$ below $120\mu\text{A}$ to protect the microammeter, yet allow I_P to flow at $V_{AK} = V_P$.) Reduce V_{AA} *slowly* and watch I_A decrease to a minimum *holding* or *valley current* where: (1) I_A jumps to zero; (2) V_{GK} jumps to V_S; (3) V_{AK} jumps to the low value of V_{AA} to (4) signify a switch to the off-state. The lowest value of I_A noted before PUT switched to its off-state is I_V.

Measurement of V_V and V_F

Increase V_{AA} until the PUT triggers to its on-state. Anode voltage V_{AK} will remain constant at 0.6 to 0.8V. This measurement is *valley voltage* V_V. Reduce R_A or increase V_{AA} until I_A is set at *forward current* $I_F = 50\text{mA}$. Measure *forward voltage* V_F between anode and cathode. V_F will be between 0.8 and 1.5V to locate an operating point well into the on-state for comparison of PUTS.

Offset Voltage V_T

V_T is found from measurements of V_P and V_S with the aid of Eq. (7-20). Test data was measured with the circuit of Fig. 7-19a and plotted in Fig. 7-19b to show how valley current I_V depends on program voltage V_S and gate resistor R_G. Data points are summarized in Table 7.1 to show I_V increases with increasing V_S and decreasing R_G. Manufacturers often give the variation in I_V as a curve plot for $R_G = 10\text{k}\Omega$ and $1\text{M}\Omega$.

Table 7.1

Measurement of PUT Parameters from Figure 7-18

R_G (ohms)	R_A (ohms)	V_P (volts)	V_s (volts)	I_V (μA)	$V_{G\,\text{off}}$ (volts)	$V_{G\,\text{on}}$ (volts)	$V_{A\,\text{on}} \cong V_V$ (volts)
10k	200k	20.2	20	63	20	0.3	0.6
100k	400k	20.2	20	15	20	0.3	0.6
1M	400k	20.2	20	7	20	0.3	0.6
10k	200k	10.2	10	44	10	0.3	0.6
100k	300k	10.2	10	10	10	0.3	0.6
1M	400k	10.2	10	4	10	0.3	0.6
10k	100k	5.3	5.0	28	5.0	0.3	0.6
100k	100k	5.3	5.0	8	5.0	0.3	0.6
1M	100k	5.3	5.0	3	5.0	0.3	0.6

Note: Characteristic curves for type 2N6027, 8 PUTs are given in Appendix 2.

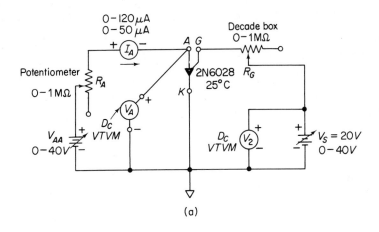

(a)

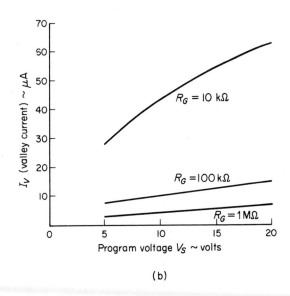

(b)

Figure 7-19

Effect of program voltage and gate resistance on valley current in (b) is
measured in the test circuit of (a).

7-11 PUT OSCILLATOR ACTION

Oscillator action of the PUT is similar to that of the UJT discribed in section 7-2. However, peak-point currents are slightly lower (between 0.1 to $10\mu A$)
with $1\mu A = I_P$ as a reasonable value. Program voltage V_S is set by R_1 and R_2

in Fig. 7-20a and is found from the gate's Thévenin equivalent in Fig. 7-20b to be

$$V_S = \eta V_B = V_B \frac{R_1}{R_1 + R_2} = \frac{27\text{k}\Omega}{16\text{k}\Omega + 27\text{k}\Omega}(20\text{V}) = 12.5\text{V} \qquad (7\text{-}12)$$

where $\eta = R_1/(R_1 + R_2) = $ *programmable intrinsic standoff ratio* and Thévenin resistance in series with the gate is

$$R_G = R_1 \parallel R_2 = 16\text{k}\Omega \parallel 27\text{k}\Omega = 10\text{k}\Omega \qquad (7\text{-}22)$$

Capacitor C charges to $V_P = V_T + V_s$ and triggers the PUT if anode current at V_P is greater than I_P. To insure triggering, the maximum value of R_A is found by reference to Eq. (7-7) as

$$R_{A\,\text{max}} \leqslant \frac{V_{BB} - V_P}{I_P} \qquad (7\text{-}23)$$

Refer to the discussion leading to Eq. (7-6) and Fig. 7-20 to find the minimum value of R_A necessary to keep the PUT in its negative resistance and maintain oscillation. $R_{A\,\text{min}}$ is found from

$$R_{A\,\text{min}} \geqslant \frac{V_B - V_V}{I_V} \qquad (7\text{-}24)$$

It is sound practice to choose R_A between the limits of $2R_{A\,\text{min}}$ and $1/2$ $R_{A\,\text{max}}$ to insure some margin of safety.

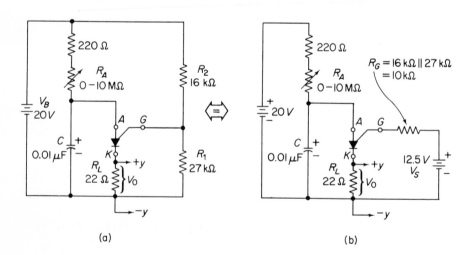

(a)

(b)

Figure 7-20

Basic oscillator circuit in (a) and equivalent gate circuit in (b).

EXAMPLE 7-10:

In Fig. 7-20, pulse outputs at V_o cease when R_A is increased to 7MΩ. Find I_P if offset voltage is 0.5V.

SOLUTION:

From Eq. (7-20), $V_P = V_T + V_S = 0.5\text{V} + 12.5 = 13.0\text{V}$. From Eq. (7-23)

$$I_P = \frac{V_B - V_P}{R_{A\,\text{max}}} = \frac{(20 - 13)\text{V}}{7\text{MΩ}} = 1\,\mu\text{A}$$

Example 7-10 illustrates how I_P can be measured precisely with an oscillator.

7-12 TIMING AND OSCILLATOR FREQUENCY OF THE PUT

Time T for capacitor C in Fig. 7-20 to charge from V_V to trigger or peak-point voltage V_P is given by

$$T = R_A C \ln \frac{V_B - V_V}{V_B - V_P} \tag{7-25a}$$

Equation (7-25a) is simplified considerably *if V_V is negligible compared to V_B* and if $V_P \cong V_s \cong \eta V_B$. Under these conditions, Eq. (7-25a) simplifies to

$$T = R_A C \ln \frac{R_1 + R_2}{R_2} \tag{7-25b}$$

for $V_V \ll V_B$ and $V_P \cong \eta V_B$. Furthermore, if we make $(R_1 + R_2)/R_2 = e = 2.71$

$$\frac{R_1 + R_2}{R_2} = 2.71 \quad \text{or} \quad R_1 \cong 1.7R_2$$

then in Eq. (7-25b), $\ln 2.71 = 1$ and it simplifies to

$$T = R_A C, \quad \text{for} \quad R_1 = 1.7R_2 \tag{7-25c}$$

When the PUT functions as timer, its charge time and time interval are given by Eq. (7-25). When it functions as an oscillator, oscillating frequency f is given by

$$f = \frac{1}{T} \tag{7-26}$$

Limits of oscillating frequency are explored in the next two examples.

EXAMPLE 7-11:

(a) Find valley current, valley voltage, peak-point voltage, and η for the circuit of Fig. 7-20. (b) Find maximum and minimum values of R_A to allow oscillation, if $I_P = 1\mu A$. Use test data in Figs. 7-18 and 7-19b.

SOLUTION:

(a) Enter $V_S = 12.5V$ in Fig. 7-19b to the intersection with $R_G = 10k\Omega$, and read $I_V = 50\mu A$. From Fig. 7-18 read $V_V = 0.7V$. (In the absence of such data assume $V_V \cong 1V$.)
From Eq. (7-21)

$$\eta = \frac{R_1}{R_1 + R_2} = \frac{27k\Omega}{27k\Omega + 16k\Omega} = 0.628$$

$V_s = \eta V_B = 0.628 \times 20V = 12.5V$. From Eq. (7-20), $V_P = V_T + V_s = 0.5V + 12.5V = 13V$.
(b) From Fg. (7-24)

$$R_{A\,min} \geqslant \frac{(20 - 1)V}{50\mu A} = \frac{19}{50} \times 10^6 = 380k\Omega$$

(c) From Eq. (7-23)

$$R_{A\,max} \leqslant \frac{(20 - 13)V}{1\mu A} = 7M\Omega$$

From Eqs. (7-25) and (7-26) we learn that high oscillating frequencies require a small capacitor and small resistance in the timing circuit. A practical lower limit for C is 1000pF ($0.001\mu F$). Since $R_{A\,min}$ is controlled primarily by valley current, we would select devices or circuitry to obtain high I_V.

For low oscillating frequencies, we must select large capacitors with leakage currents less than I_P. Size and cost factors usually limit our selection to glass or mylar capacitors with maximum capacitance of $10\mu F$.

EXAMPLE 7-12:

In Fig. 7-20, what is the range of oscillating frequencies for (a) $C = 1,000pF$ and (b) $C = 10\mu F$ if R_A is varied between limits of $500k\Omega$ and $5M\Omega$?

SOLUTION:

Resistors R_1 and R_2 were deliberately selected from standard resistor sizes to give $R_G = 10k\Omega$ and fulfill stipulations of Eq. (7-25c) by making $R_1 \cong$

$1.71R_2$. That is, $27k\Omega = 1.71 \times 16k\Omega = 27k\Omega$. Using Eqs. (7-25c) and (7-26):

(a) For $R = 500k\Omega$, $T = R_A C = 5 \times 10^5 \times 1,000 \times 10^{-12} = 50ms$ and $f = 1/(5 \times 10^{-4}) = 2,000Hz$. For $R = 5M\Omega$, $T = 5 \times 10^6 \times 1,000 \times 10^{-12} = 5ms$ and $f = 1/(5 \times 10^{-3}) = 200Hz$.

(b) For $R = 500k\Omega$, $T = 5 \times 10^5 + 10 \times 10^{-6} = 5sec$, $f = 1/5 = 0.2Hz$. For $R = 5M\Omega$, $T = 5 \times 10^6 \times 10 \times 10^{-6} = 50sec$, $f = 1,150 = 0.02Hz$.

Example 7-12 illustrates that the approximate frequency range of a basic PUT oscillator is from 0.02Hz to roughly 2kHz. With changes in circuitry, these limits can be extended. However, the PUT is well-suited for long interval timers and low-frequency oscillators.

7-13 APPLICATION OF THE PUT

Versatility of the PUT can be illustrated by a careful selection of representative applications. Keep in mind that you can select a particular value of gate resistance to obtain a maximum I_V for short charge time intervals or a minimum

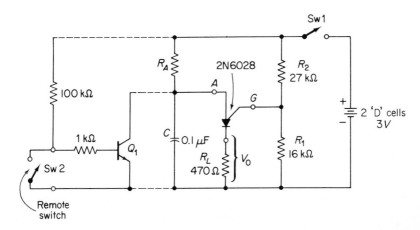

Figure 7-21

Low-voltage timer or oscillator circuit.

I_P for long charge time intervals. Even more important is the fact that you can select or vary η by controlling V_S through R_1, R_2, or V_B. Each of the following design-analysis examples focuses attention on one advantagous characteristic of the PUT. Our first example shows a low supply voltage oscillator.

EXAMPLE 7-13:

The circuit of Fig. 7-21 was selected to show a low-supply voltage application for the PUT. Evaluate R_G, η, and V_P from

$$R_G = R_1 \| R_2 = \frac{(27k\Omega)(16k\Omega)}{27k\Omega + 16k\Omega} = 10k\Omega,$$

$$\eta = \frac{R_1}{R_1 + R_2} = \frac{16k\Omega}{43k\Omega} = 0.37$$

$$V_P = V_T + \eta V_B = 0.5V + 0.37(3V) = 1.6V$$

Note that program voltage $V_S = \eta V_B = 1.1V$ was chosen low because of V_T. Since $V_S = 1.1V$ and $R_G = 10k\Omega$, I_V is estimated to be $20\mu A$ from Fig. 7-19b, and I_P is $1\mu A$.

Oscillator Application

Picking R_A between limits imposed by Eq. (7-23)

$$R_{A\,(\text{max})} \leqslant \frac{(3 - 1.6)V}{1\mu A} = 1.4M\Omega$$

and Eq. (7-24)

$$R_{A\,(\text{min})} \geqslant \frac{(3 - 0.6)V}{20\mu A} = 120k\Omega$$

gives about 1 to 10 range for R_A to permit oscillation. Since V_V is not small with respect to V_B, we must use Eq. (7-25a). For $R_{A\,(\text{min})} = 120k\Omega$

$$T = 0.12 \times 10^6 \times 0.1 \times 10^{-6} \ln\frac{3.0 - 0.6}{3.0 - 1.6} = 0.012 \ln 1.71 = 0.012(0.536)$$

$$T = 6.4\text{ms}$$

and $f = 1/T = 1/(6.4 \times 10^{-3}) \cong 150\text{Hz}$. For $R_{A\,(\text{max})} = 1.4M\Omega$, $T = 1.4 \times 10^6 + 0.1 \times 10^{-6} \times 0.536 = 75\text{ms}$, and $f = 1/(75 \times 10^{-3}) = 13\text{Hz}$.

Figure 7-21 can also perform as a timer by choosing R_A for *any* value less than R_{max}. After closing sw 1, capacitor C charges from zero volts. Equation (7-25a) must be modified to account for this fact by setting $V_V = 0$, and for $R_{\text{max}} = 1.4M\Omega$ the maximum time interval would be modified to

$$T = 0.14 \ln\frac{3.0}{1.4} = 0.14 \ln 2.12 = 0.14(0.75) = 105\text{ms}$$

Transistor Q_1 may be added to start timing instead of switch sw$_1$. Assume sw$_1$ is closed. With sw$_2$ open, Q_1 is in saturation, placing the anode of the PUT at essentially ground potential. When sw$_2$ is closed, Q_1 cuts off and timing begins. When the PUT is triggered and C discharges through R_L, V_o is an output pulse and can be used to trigger an SCR or triac into its stable on-condition to record the end of timing, or initiate a sequence of events. In our next example we program the anode voltage.

Programming Anode Voltage

EXAMPLE 7-14:

In Fig. 7-22 gate resistance is a minimum of 100kΩ to give a low I_P of, typically, $0.1\mu A$. This low I_P arrangement is optimum for long interval timers. When switch sw is closed to begin timing, the gate goes to $+24V$ because C is uncharged. Program voltage is now placed on the anode and is equal to $V_B \times R_2/(R_1 + R_2) = 10V$. G is more positive than A, so the PUT is initially in its off-state.

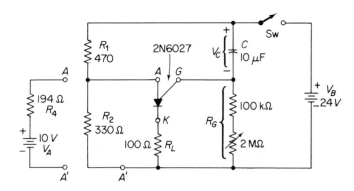

Figure 7-22

Timer with anode program voltage.

C charges from zero volts toward V_B. But when V_G becomes $V_T = 0.5V$ less than the voltage across R_2, the anode becomes more positive than the gate to trigger the UJT on. Charging time is given by

$$T = R_G C \ln \frac{R_1 + R_2}{R_2} \qquad (7\text{-}27)$$

With R_G adjusted for 1.1MΩ, the PUT triggers in

$$T = 1.1 \times 10^6 \times 10 \times 10^{-6} \ln \frac{470 + 330}{330} = 11 \ln 2.43 = 10\text{sec}$$

Here the PUT is used as a *solid state relay*, or to replace an SCR-UJT pair.

After triggering, anode current equals load current through R_L, and neglecting V_F gives

$$I_{AK} = \frac{V_A}{R_A + R_L} = \frac{10V}{(194 + 100)\Omega} = 34mA$$

Values of I_{AK} up to 150mA are permissible, so load R_L can be switched on with a single PUT. R_L could be replaced with a relay to control even heavier currents.

Voltage Controlled Oscillator

For our last design example we select a voltage-controlled ramp generator to: (1) learn how to add a constant-current charging source to the PUT, and (2) learn how the PUT can sense different values of voltage levels on the gate.

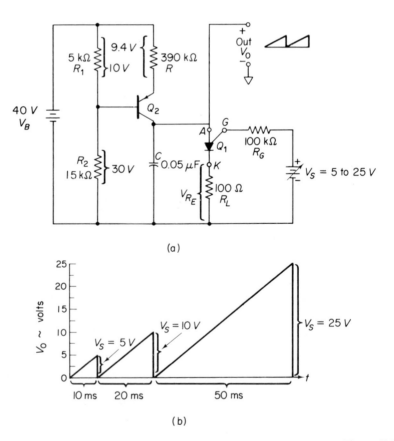

(a)

(b)

Figure 7-23

Program voltage V_S controls trigger voltage V_P of the PUT in (a). Period of the output ramp voltage in (b) is controlled by V_S.

EXAMPLE 7-15:

Resistors R_1 and R_2 in Fig. 7-23a set a constant voltage across R of 9.4V to establish a constant collector-charging current from Q_2 of

$$I = \frac{9.4V}{R} = \frac{9.4V}{390\text{k}\Omega} \cong 25\mu\text{A}$$

I must be restricted between $I_P = 2\mu\text{A}$ and $I_V = 50\mu\text{A}$. Charge time interval is found from Eq. (7-12)

$$T = \frac{V_S C}{I} = V_S \times \frac{0.05\mu\text{F}}{25\mu\text{A}} = \left(2\frac{\text{ms}}{\text{volt}}\right) V_S$$

Note the unit conversion:

$$\frac{\text{farads}}{\text{amps}} = \frac{\text{farads}}{\text{volts/ohms}} = \frac{\text{farads} \times \text{ohms}}{\text{volts}} = \frac{\text{seconds}}{\text{volts}}$$

Thus, if V_S is set to 5V, trigger voltage V_P is approximately 5V and capacitor C will charge to V_P and trigger the PUT in $T = (2\text{ms/volt}) \times (5 \text{ volt})$ $= 10\text{ms}$. Frequency of oscillation will be $f = 1/10\text{ms} = 100\text{Hz}$. The highest permissible value for V_S is approximately equal to the voltage drop across R_2, so that Q_2 stays out of saturation.

7-14 THE DIAC TRIGGER

The *diac* is a three-layer, two-terminal semiconductor device designed to trigger triacs or provide protection against over-voltages. It is made like an *npn* transistor without an external base connection, but has an *I-V* characteristic similar to that of a neon glow lamp. *Bidirection trigger diode* is a more descriptive name for the diac, and two alternate circuit symbols are shown in Fig. 7-24. Two opposite-pointing arrowheads in the symbol mean the diac can conduct in either direction. It would be meaningless to call either terminal anode or cathode, so they are numbered 1 and 2. Arbitrarily using terminal 2 as reference, we increase positive voltage at terminal 1 in the sweep circuit of Fig. 7-24. Diac voltage $V+$ equals applied sweep voltage, and only leakage current ($< 50\mu\text{A}$) flows until $V+$ reaches *breakover voltage*, V_{BR}, of 22V. Diac voltage then abruptly drops, accompanied by an increase in diac current to exhibit the trigger characteristic of a negative resistance. The change in voltage, ΔV, that occurs just after V_{BR} is circuit-dependent and is specified at a diac current of 10mA. Slope of the *I-V* characteristic never goes positive, so the concept of holding or valley current does not apply. As sweep voltage is reduced from 30V below V_{BR}, the *I-V* characteristic retraces its original negative resistance path to V_{BR} and abruptly becomes nonconducting from V_{BR} to zero. For negative sweep voltages $(V -)$, the diac has symmetrical *negative breakover voltage* (usually within a few volts), although not necessarily symmetrical ΔV.

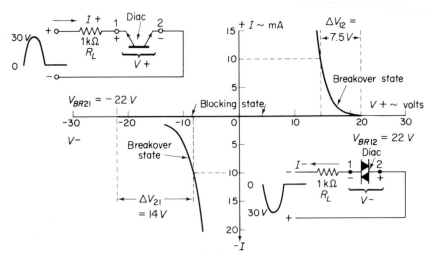

Figure 7-24

Current-voltage characteristics of a diac.

Diac Applications

The diac can be employed as a relaxation oscillator since it has a characteristic like the UJT. However, for phase shift control of a triac, the diac monitors rising voltage across a capacitor. When capacitor voltage reaches V_{BR}, the diac triggers to discharge the capacitor by ΔV and conduct discharge current into the triac's gate. Typical peak discharge current is roughly $1A/0.1\mu F$. Such applications will be explored when we study the triac in Chapter 9.

7-15 OTHER TRIGGER DEVICES

Trigger devices vary in appearance and trigger voltage level but share one thing in common—a negative resistance region somewhere on their I-V characteristic. Once trigger voltage and current plus valley voltage and current are specified, all equations developed in preceding sections apply. A representative selection of common trigger devices will be examined.

The SUS, Silicon Unilateral Switch

The *Silicon Unilateral Switch*, *SUS*, is an integrated circuit containing a resistor, 6.8V zener diode, *pnp* and *npn* transistors. Its circuit, symbol, equivalent circuit, and anode-cathode $I_A - V_{AK}$ characteristic are shown in Figs. 7-25 a, b, and c, respectively. Comparison of the two-transistor switch of Fig. 7-15 with the SUS model in Fig. 7-25 shows their similarity. The SUS is actually a two-transistor switch with a built-in program voltage $V_s = V_Z$. Offset voltage is

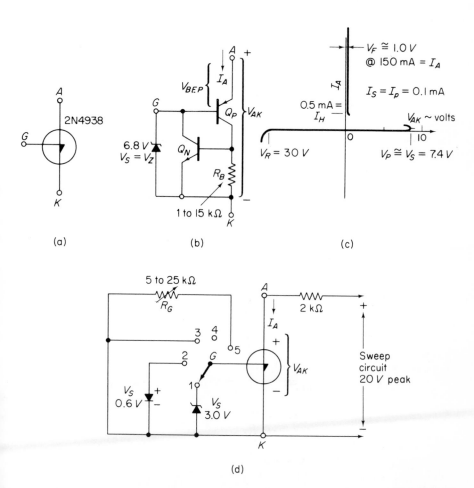

(a)   (b)   (c)

(d)

Figure 7-25

The SUS has the symbol in (a), model in (b), and characteristics in (c).
Trigger voltage can be changed by any of the arrangements in (d).

$V_T = V_{BEP}$. So the SUS will trigger on when anode voltage goes more positive than $V_P = V_Z + V_{BEP} = V_S + V_T$. The SUS can be used as a two-terminal device (A to K) that will trigger on about 7.4V. For example, with switch on point 4 in Fig. 7-25d, the characteristic in Fig. 7-25c is shown on a CRO to measure $V_P \cong V_S = 7.4$V. Switching to point 3, the SUS is a three-terminal device with $V_S = 0$ and triggers on about 0.6V. Point 2 and 1 show how trigger voltage can be modified to 0.6V + 0.6V = 1.2V and 3.0V + 0.6V = 3.6V, respectively. On point 5 an external gate resistor will modify V_P by an amount that may be determined experimentally.

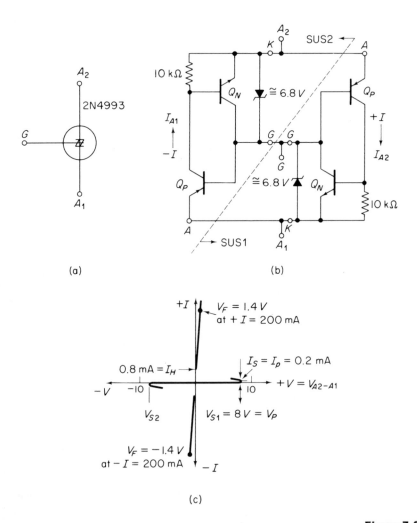

(a)

(b)

(c)

Figure 7-26

The SBS symbol in (a) is modeled in (b) and has symmetrical IV characteristics in (c).

The SBS, Silicon Bilateral Switch

Arrowheads pointing in both directions signify current conduction in both directions (bilateral or bidirectional) for the *SBS, Silicon Bilateral Switch* in Fig. 7-26a. Its equivalent circuit in Fig. 7-26b shows that it will act like one two-transistor switch with built-in program voltage, connected in inverse parallel with another identically programmed, two-transistor switch. There is no anode or cathode because of bidirectional current flow, so main current carrying terminals are simply labeled A_1 and A_2. The *I-V* characteristic is symmetrical in Fig. 7-26c for either direction of current flow. For special purposes, trigger voltages can be reduced by connecting lower value zeners between gate and either or both anodes. By adding a zener in series with a SBS we have an *asymmetrical silicon bilateral* switch, ASBS, with characteristics shown in Fig. 9-11b.

The Shockley Diode

The *Shockley* diode is a two-terminal, four-layer *npnp* trigger device with an *I-V* characteristic similar to the SUS in Fig. 7-25c. However, no gate terminal is used, and its symbol is identical to that of the SUS with gate terminal removed. A four-layer diode such as the M4L3054 has a typical trigger voltage (V_{BR}) of 10V and an on-voltage of $V_F \cong 1V$. Note the similarity in *I-V* characteristics for the Shockley diode, SUS, PUT in Fig. 7-18 and two-transistor switch in Fig. 7-16.

The Tunnel Diode

In Fig. 7-27 a tunnel diode, IN3712, has a negative resistance region in its *I-V* characteristic. The tunnel diode is a *pn* diode so heavily doped that its junction is in reverse breakdown with no applied voltage. When reverse-biased, as in quadrant 3 of Fig. 7-27, it acts like a diode in reverse breakdown. In quadrant 1, forward-bias exceeding about 0.4V cancels the internal breakdown due to doping, causing the tunnel diode to act like a normal junction diode. In the process of moving the junction out of breakdown, a negative resistance region is encountered between trigger voltage $V_P = 60mV$ and valley voltage $V_V = 0.38V$.

Neon Glow Lamps

Neon glow lamps such as the NE83 or NE51 have a bidirectional *I-V* characteristic like a high voltage diac. Their trigger or breakdown voltages are typically between 60 and 100V and sensitive to light and temperature. I_P is small, at $1\mu A$ or less, and neon lamps can deliver current pulses of up to 15mA into a 20Ω load when discharging a $0.1\mu F$ capacitor. In Fig. 7-28 trigger voltages are symmetrical at $78V = V_P$. Once the lamp is triggered it needs a *sustaining* volt-

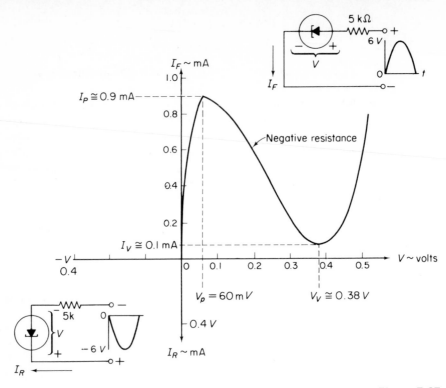

Figure 7-27

Measured *IV* characteristic of a IN 3712 tunnel diode.

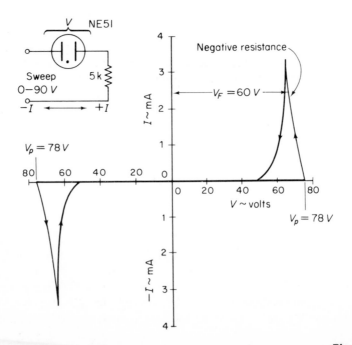

Figure 7-28

Trigger *IV* characteristic of an NE 51 neon lamp.

age less than V_P but greater than about 50V to remain in the on-state. Neon lamp triggers have a relatively short operating lifetime ($\cong$ 5,000 hours) compared to semiconductors, but have the important advantages of low cost and visible indication of triggering.

When obtaining the I-V characteristic of a neon glow lamp, use a very high impedance probe ($\geqslant$ 10MΩ), or else breakdown voltages will be lowered.

PROBLEMS

1. In Fig. 7-1c, $I_{B2} = 2$mA, $\eta = 0.6$, and $V_{BB} = 10$V. If the UJT has not been triggered, find (a) r_{BB} and (b) r_{B1}.

2. In Fig. 7-2b, $V_{EE} = 10$V, and the UJT is triggered into the positive resistance region. If $V_B \cong 2$V, find I_E.

3. In Example 7-1, η was measured at 0.7. What is V_P if V_{BB} increased to 20V?

4. In Example 7-2 and Fig. 7-4a, V_{BB} is doubled to 20V. Find: (a) permissible R_E for oscillation, and (b) minimum and maximum frequency of oscillation. (c) Does increasing V_{BB} raise the maximum possible oscillator frequency?

5. A UJT is selected with characteristics idential to those in Example 7-2, except $I_V = 2$mA. How does this change affect f_{max}?

6. In Fig. 7-4, $R_E = 50$kΩ, $C_E = 0.01\mu$F. Find (a) oscillating frequency and (b) peak value of V_o. (Reference: Example 7-3 and Fig. 7-7.) (c) How does decreasing C_E by a factor of 10 affect f?

7. If we replace the UJT in Example 7-4 with the same characteristics given, except $\eta = 0.5$: (a) What is the new value of T? (b) Does f increase with a decrease in η?

8. If C_E began charging from 0V instead of V_V, what is the charge time T in Example 7-4?

9. What peak voltages are available for $C = 0.01\mu$F, $V_{BB} = 20$V in Fig. 7-7 for R_L of: (a) 10Ω? (b) 27Ω? (c) 33Ω? (d) 47Ω?

10. (a) If $r_{BB} = 7$kΩ and $R_T = 0\Omega$, what are the base 1 pedestal voltages in problem 9? (Reference: Example 7-5.) (b) If the dc pedestal voltage cannot exceed 100mV, what is the maximum R_L—10, 27, 33, or 47Ω?

11. Given $V_D = 0.5$V, $I_P = 2\mu$A, $I_V = 2$mA, $\eta = 0.65$, $R_T = 0\Omega$, and $V_V = 2$V for the UJT in Fig. 7-8. What is the (a) maximum and (b) minimum value of R_E to just maintain oscillation? Hint: Find R_E to maintain charge current between limits of I_P and I_V.

12. What are the minimum and maximum charging time intervals and corresponding oscillator frequencies for problem 11 if $C_E = 0.01\mu$F?

13. If oscillation stops when R_E is increased to 1MΩ in problem 11, what is the leakage current of C_E?

14. If $R_E = 25\text{k}\Omega$ in the circuit of problem 11, and $f = 1,110\text{Hz}$, what is the value of C_E?

15. (a) Find oscillating frequency in Example 7-9 and Fig. 7-13 if R_E is reduced to 25kΩ. (b) Sketch the shape of V_o for $R_R = 2.5\Omega$, 5kΩ, 10kΩ, and 25kΩ.

16. Evaluate minimum and maximum possible oscillator frequency at $R_{A\,min}$ and $R_{A\,max}$ in Example 7-12 (a) and (b).

17. If both R_1 and R_2 are increased by a factor of 10 in Fig. 7-20 and Example 7-12, what are the new values of $R_{A\,min}$ and $R_{A\,max}$? Assume I_P remains at $1\,\mu\text{A}$.

18. What new minimum and maximum possible oscillator frequencies result from the increased values of R_1 and R_2 in problem 17? Compare your results with those of problem 16 and draw a conclusion on the effect of increasing R_G on minimum and maximum oscillating frequency. Assume I_P remains at $1\,\mu\text{A}$.

19. What value of R is required in Fig. 7-23a to generate a ramp voltage rise of 5V in 15ms? Hint: Assume $V_s = 5\text{V}$ and refer to Example 7-15.

20. What is the approximate shape of V_{RL} in Fig. 7-20 if $V_{AK\,on} \cong 1\text{V}$?

chapter eight
silicon controlled rectifier

8-0 INTRODUCTION

Multilayer semiconductor devices that are intended primarily for use as on-off switches, have two stable states—conducting (on) or nonconducting (off), and change states swiftly because of internal regenerative switching action, are classified into the *thyristor* family. The thyristor family is subdivided into two main branches: (1) those able to conduct current through their switch terminals in one direction only—*unidirectional* or *unilateral* (such as the PUT), and (2) those able to conduct current through their switch terminals in both directions— *bidirectional* or *bilateral* (such as the diac). The *silicon controlled rectifier* or SCR and *triac* are the most popular examples, respectively, of unidirectional and bilateral thyristors. This chapter will be concerned with learning how to use the SCR; Chapter 9 will cover the triac.

8-1 INTRODUCTION TO THE SILICON CONTROLLED RECTIFIER

As shown by the simplified construction model of Fig. 8-1a, the SCR is a four-layer, three-terminal semiconductor device. A single arrowhead on its symbol in Fig. 8-1b shows conventional current flows in one direction only between switch terminals anode and cathode. Once the SCR is turned on, usually by a small suitable signal between gate and cathode, it acts like a closed switch,

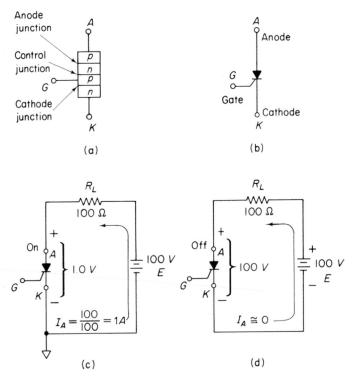

Figure 8-1

Simplified construction of an SCR in (a) and circuit symbol in (b). Stable on- and off-states are shown in (c) and (d), respectively

as in Fig. 8-1c. The gate can only turn the SCR on, and then exercises no further control over it. The circuit must force load or anode current I_A to practically zero (below holding current I_H) to turn the SCR off. When off, SCR switch terminals A-K act like an open switch, as in Fig. 8-1d. In summary, the SCR normally conducts only when its anode is positive with respect to its cathode, and simultaneously the gate is made sufficiently positive with respect to the cathode. The circuit must turn the SCR off by interrupting anode current (and removing gate drive so that the SCR will not turn back on when anode voltage is restored.)

Two-transistor Analogy

A two-transistor analogy of the SCR is developed in the sequence of Figs. 8-2a to c. It is similar to the PUT and two-transistor switch in Chapter 7 except for the gate connection. To simplify explanation of regenerative switching ac-

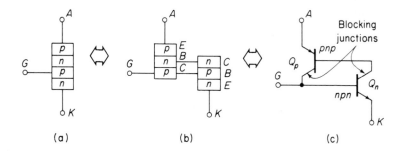

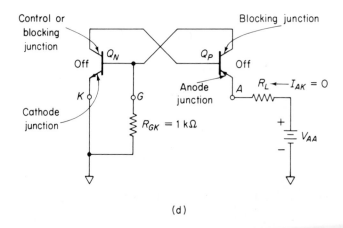

(d)

Figure 8-2

A two-transistor analogy of the SCR is developed in (b), (c), and (d) from the construction in (a). The SCR is in its stable off-state.

tion, we redraw transistors in the flip-flop arrangement of Fig. 8-2d. Here the SCR is off or in a *forward blocking condition* where V_{AA} reverse-biases the blocking junction of Q_N. Resistor R_{GK} is connected to give a low resistance path between gate and cathode for low-current SCRs. This path by-passes internal leakage currents and external noise signals away from the cathode junction to minimize uncontrolled turnon. If V_{AA} were reversed in polarity, both cathode and anode junctions would be reverse-biased while the control junction would be forward-biased. This condition is known as *reverse blocking*.

8-2 TURNING ON THE SCR

To change the SCR's state from off to on, we must inject or create enough free holes or electrons in the cathode junction to initiate triggering or regenerative switching action. Making the gate positive with respect to the cathode is

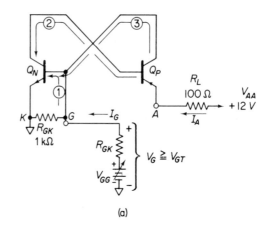

(a)

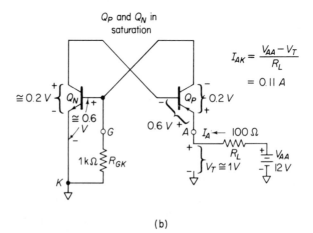

(b)

Figure 8-3

Turn-on regenerative switching action in (a) results in the stable on-state in (b).

the most common method of triggering an SCR. In Fig. 8-3a, V_{GG} is increased to raise gate voltage V_G up to a minimum value required to initiate triggering. This value of V_G is called *gate trigger voltage* V_{GT} and will be in the order of 0.6V, to forward-bias the silicon *pn* junction.

SCR Regenerative Switching Action

When V_G equals V_{GT}, gate current flows into the base of Q_N and is amplified into collector current of Q_N to drive the base of Q_P, where it is amplified into collector current of Q_P to drive the base of Q_N. Thus once Q_N begins to conduct, both transistors drive one another rapidly into saturation in a *regenerative switching action*. As a result of the switching action, anode current I_A abruptly changes to a value determined by R_L and V_{AA} in Fig. 8-3b. Voltage across switch terminals A-K of the on-SCR is appropriately called V_{AK-ON} or V_T, and is typically 1V.

Turnoff is accomplished by: (1) opening the gate; (2) maintaining V_{AA} constant at about 7V; (3) increasing R_L to reduce I_{AK} to holding current I_H. At about $I_H = I_{AK} = 5$mA there is insufficient current to maintain both transistors in saturation. As soon as one transistor comes out of saturation, each drives the other rapidly into cutoff and the SCR switch is in its stable off-state.

There are four other, less common, methods of triggering an SCR into its on-state:

1. *Forward break over turnon.* In Fig. 8-2d, V_{AA} can be increased to a value where reverse-bias on the control junction accelerates leakage-current minority carriers to a velocity sufficient to ionize semiconductor atoms. These additional free charges will flow through the cathode junction as a small emitter current for Q_N. This initiates regenerative switching action in the same fashion, as if triggering were initiated by a gate current pulse. Specification sheets list this value as *Repetitive Forward Blocking Voltage* V_{DRM} or V_{FXM}. SCRs are made with minimum V_{FXM} ratings between 15V and 800V. Once V_{FXM} is exceeded, the SCR turns on and remains in its stable on-state. Resistor R_L should be increased to limit anode current below the maximum current and power dissipation ratings. Typically, $R_L = 50\text{k}\Omega$.

2. *Rate effect turnon.* When power is first connected by a switch to an SCR circuit, the anode-cathode voltage can rise very quickly. That is, anode-cathode voltage change dV_{AK} will change in a short interval of time dt. For example, in Fig. 8-2d, if $V_{AA} = 150$V and a switch connected V_{AA} in 1 microsecond, V_{AK} would change from 0 to 150V in 1 microsecond, or

$$\frac{dV_{AK}}{dt} = \frac{150\text{V}}{1\,\mu\text{s}} = 150\text{V/microsecond}$$

Since all *pn* junctions have capacitance, a charging current can flow into the emitter junction and initiate a false triggering sequence. Thus the SCR would give a false turnon.

To control the rate of change of V_{AK}, we use the fact that maximum rate-of-change of voltage across a capacitor occurs when it just begins to charge. When voltage V_{AA} is applied in Fig. 8-4, capacitor voltage aims for V_{AA}. During the first time constant τ, the capacitor charges to 63 percent of V_{AA}. The *maximum* rate-of-change of voltage, which is also approximately equal to the maximum dV_{AK}/dt, is the initial slope

$$\frac{dV_C}{dt} = \frac{dV_{AK}}{dt} = \frac{V_{AA}}{\tau} \tag{8-1}$$

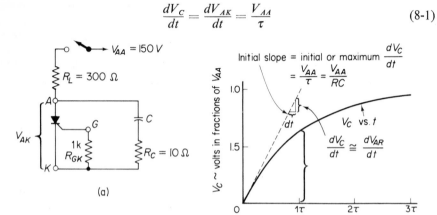

(a)

(b)

Figure 8-4

Capacitor C in (a) limits dV_{AK}/dt according to the time constant in (b).

where $\tau = (R_L + R_c)C$. Manufacturers specify maximum dV/dt and a typical maximum dV/dt is 10 volts per microsecond. Resistor R_C limits capacitor discharge current when the SCR fires.

EXAMPLE 8-1:

In Fig. 8-4a, V_{AA} represents the maximum voltage that could be present when a switch connects a 105V ac line voltage ($105 \times 1.4 \cong 150V$) to an SCR. Find C to limit dV/dt below 10V/microsecond.

SOLUTION:

From Eq. (8-1), $\tau = (R_L + R_c)C = (300\Omega + 10\Omega)C = 310\Omega C$, and

$$\frac{V_{AA}}{\tau} = \frac{150}{(310)C} \leqq \frac{10V}{1 \times 10^{-6}sec} = \frac{dV_{AK}}{dt}$$

$$C \geqq \frac{150 \times 10^{-6}}{310 \times 10} = 0.0485 \times 10^{-6}F \cong 0.05\mu F$$

3. *Temperature turnon.* Since leakage currents increase with temperature increase, the resultant increase in minority carriers through the cathode junc-

tion acts as a gate current and can initiate regenerative switching. The presence of R_{GK} will by-pass a portion of these leakage currents and prevent false turnons.

4. *Radiation energy turnon.* Radiant energy, impinging on a semiconductor, will rupture covalent bonds if the frequency of the radiant energy is within the semiconductor's spectral bandwidth. The ruptured bonds release current carriers that act like a gate current and initiate a turnon. A window may be installed in the top of an SCR package and allow light energy to fall on the cathode and control junctions. Any light pulse of sufficient magnitude will then initiate turnon. This device is called the *light activated silicon controlled rectifier* *LASCR*. The LASCR also may be turned on by any of the above methods, including gate control.

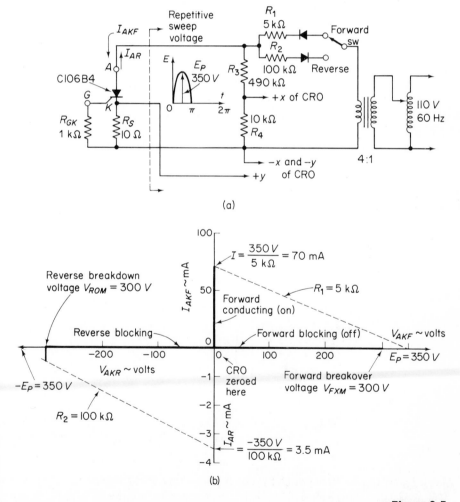

(a)

(b)

Figure 8-5

Sweep circuit in (a) measures forward breakover voltage or reverse breakdown voltage by the CRO display of (b). *E* adjusted for 350 *V* peak.

8-3 CHARACTERISTIC CURVES AND ELECTRICAL
MEASUREMENTS OF THE SCR

Either a curve tracer or the sweep circuit of Fig. 8-5 can measure repetitive forward breakover V_{FXM} or reverse breakdown V_{ROM} voltages. Resistors R_3 and R_4 divide the sweep voltage by 50 so that CRO x-sensitivity can be set for 1V/cm. The x-axis is then calibrated at 50V/cm. R_1 limits forward anode current I_{AKF} to 70mA, although I_{AKF} can be swept to over an ampere without exceeding any maximum rating. With switch Sw in the reverse position, negative-going sweep voltage is applied to the anode and breakdown current is limited by R_2. CRO y-sensitivities are set at 0.1V/cm for 10mA/cm and 0.01V/cm for 1mA/cm calibrations with Sw in forward and reverse positions, respectively. Observe that the first-quadrant I_{AK}-V_{AK} characteristic is the same as a switch (on this scale).

Minimum Gate Trigger Current I_{GT} and
Minimum Gate Trigger Voltage V_{GT}

In Fig. 8-6, V_{GG} is increased from zero while watching both V_{AK} on the CRO and gate current I_G on the microammeter. The highest value of I_G observed just before V_{AK} goes from off to on is I_{GT}. In one test on a C106B4, V_{GG} measured 14V,

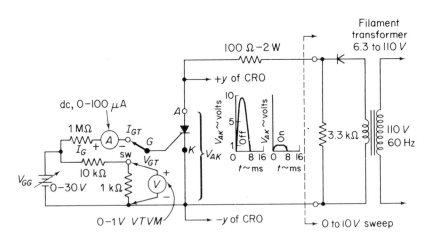

Figure 8-6

Measuring minimum gate trigger voltage and current (C106B4).

with $I_{GT} = 13\mu A$. With switch Sw down, V_{GG} is raised and we record the highest value of gate voltage V_G as V_{GT}, which occurs just before the SCR switches on. The half-wave sweep circuit automatically shuts off the SCR (non-conducting between anode and cathode) during the second half-cycle. In one test on the same C106B4, V_{GT} measured 0.54V. Accordingly, the minimum gate power required to trigger this SCR is a mere (0.54V)(13μA) = 7μW.

Gate Characteristic

A curve tracer or sweep circuit measurement of forward gate current versus forward gate voltage is used to insure that maximum gate ratings are not exceeded. For example, maximum *average gate power* $P_{G(AV)}$ for small SCRs is specified at 0.1W, and maximum *peak gate power* P_{GM} is 0.5W. Maximum *peak gate current* is 0.2A and maximum *peak gate reverse* voltage is 6V. Typically, the gate would be operated with an average dissipation of 0.01W. In order to show the gate trigger point on the I_G-V_G characteristic of Fig. 8-7, we must use scales the are too small to plot maximum ratings. The highest gate power shown is only 0.002W and illustrates the wide latitude of selection for trigger points.

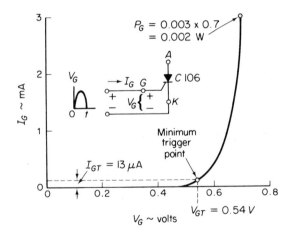

Figure 8-7

Typical low-power SCR gate characteristic.

In higher power or less sensitive SCRs there may be a gate-to-cathode resistor included within the SCR case. The gate characteristic curve will have a slope equal to this resistor on the forward characteristic between $V_G = 0$ and $V_G \cong 0.4$V, and also on the reverse gate characteristic.

Anode Characteristic

The test circuit of Fig. 8-6 is modified in Fig. 8-8a to display anode-to-cathode current I_{AK} versus anode-to-cathode voltage V_{AK}. Resistors R_1 and R_{GK} divide the sweep voltage by 10. Since $V_{GT} = 0.5$V, the gate will trigger the

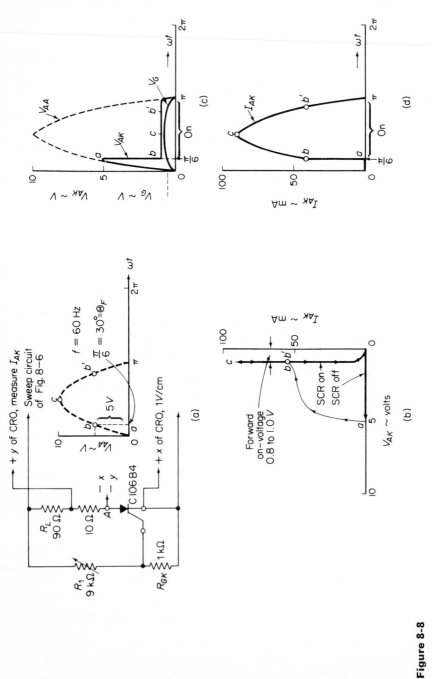

Figure 8-8

The sweep circuit in (a) plots I_{AK} vs. V_{AK} in (b). V_{AK} vs. time in (c). and I_{AK} vs. time in (d).

SCR into conduction when the sweep voltage rises to 5.0V. In Figs. 8-8b through d, the SCR is off from 0 to point (a), where $V_{GT} = 0.5V$ triggers the SCR to its on-state. Anode current rises to about 40mA while anode voltage drops to about 1.0V at point (b). For the time interval (b) to (c) to (b') V_{AK} is held to 1V and V_G is clamped at about 0.7V, while I_{AK} goes sinusoidally through a maximum of 90mA. Finally, all CRO traces return from (b') to 0 where the SCR turns off. To show how the display information can be used to calculate peak currents, we employ the following example.

EXAMPLE 8-2:

Based on an estimate that $V_{AK(on)} = 1V = V_F$, $V_{G(on)} = 0.8V$, and $V_{GT} = 0.5V$, calculate: (a) the conduction angle or on-time for the SCR in Fig. 8-8; (b) peak anode current; and (c) peak gate current.

SOLUTION:

(a) The open circuit gate voltage is related to V_{AA} by

$$\frac{R_{GK}}{R_{GK} + R_1} V_{AA} = V_G \text{ or } V_G = \left(\frac{1}{1+9}\right)(10 \sin 2\pi \, 60t) = 1 \sin \omega t$$

Since the gate triggers to $V_G = V_{GT} = 0.5V$, solve for ωt at $V_G = 0.5V$. $0.5 = \sin \omega t$, or $\omega t = \sin^{-1} 0.5 = 30° = \pi/6$. The conduction angle is $180° - 30° = 150°$, and since one half-cycle of a 60Hz wave lasts $(1/2)$ $(1/60) = 8.33$ ms, set up the proportion

$$\frac{150°}{180°} = \frac{\text{on-time}}{8.33\text{ms}}, \text{ on-time} = 5.6\text{ms}$$

The SCR conducts for 5.6ms out of every 16.66ms.

(b) Peak anode current is found from Ohm's law when V_{AA} is at its peak of 10V from

$$I_{AK \text{ peak}} = \frac{V_{AA \text{ peak}} - V_{AK \text{ on}}}{90\Omega + 10\Omega} = \frac{(10 - 1)V}{100\Omega} = 90\text{mA}$$

(c) The peak Thévenin gate voltage V_{GG} of

$$V_{GG} = \frac{R_{GK}}{R_{GK} + R_1} V_{AA \text{ peak}} = \frac{1}{1+9} \times 10 = 1V$$

is in series with Thévenin resistance

$$R_{GG} = R_1 \parallel R_{GK} = 1\text{k}\Omega \parallel 10\text{k}\Omega = 910\Omega$$

and peak gate current is

$$I_{G \text{ peak}} = \frac{V_{GG} - V_{G(on)}}{R_{GG}} = \frac{(1 - 0.8)V}{910\Omega} \cong 220\mu\text{A}$$

Holding Current i_H

When gate current is removed or drops below I_{GT}, there is a minimum value of anode current below which an SCR cannot remain on, but must revert to its off-state. This value of anode current is called *holding current* I_H or I_{HX}. In Fig. 8-9, reduce R to zero. Touch the jumper wire to the gate to trigger the SCR. VTVM V is used as a burn-out proof milliameter. A one volt reading corresponds to an anode current of 1mA. Increase R until V drops abruptly to zero. The lowest current read just before the SCR turns off is I_H, where $I_H = V/1,000$ amps $= V$ milliamps.

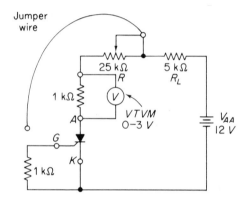

Figure 8-9

Circuit to measure anode holding current I_{HX}.

From a study of Example 8-2 and Fig. 8-8, it is evident that average current through any load in series with the anode can be varied quite easily. We can pick any ratio of R_1 to R_{GK} to make the gate trigger voltage occur any place during the first 90° (from *a* to *c*). Very large load powers can be varied with relative small amounts of gate power. We now turn our attention to understanding how to control the on-time or conduction angle through techniques known as *phase control*.

8-4 AVERAGE AND RMS CURRENT VERSUS FIRING ANGLE

Rectifiers and SCRs conduct current in only one direction, so they are rated in terms of *average current* as read by a dc ammeter. However, the effective heating power of a resistive load is determined by the *rms load current or rms load voltage*. Measuring rectified (but not filtered) average load current or average load voltage with dc meters is simple, but the product of their readings is *not*

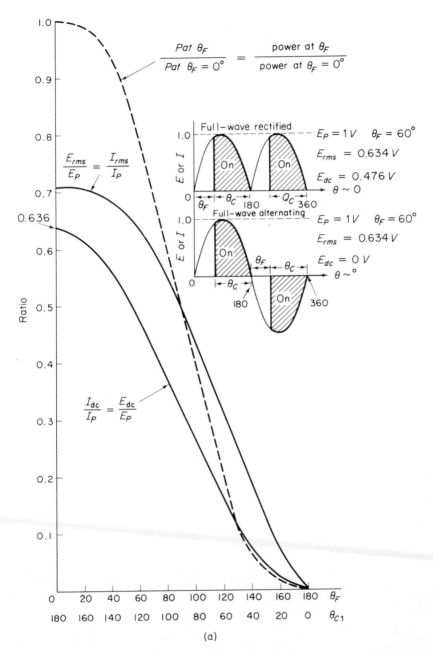

$$\frac{P \text{ at } \theta_F}{P \text{ at } \theta_F = 0°} = \frac{\text{power at } \theta_F}{\text{power at } \theta_F = 0°}$$

Full-wave rectified $\quad E_P = 1V \quad \theta_F = 60°$

$E_{rms} = 0.634\,V$

$E_{dc} = 0.476\,V$

$\theta \sim 0$

Full-wave alternating $\quad E_P = 1V \quad \theta_F = 60°$

$E_{rms} = 0.634\,V$

$E_{dc} = 0\,V$

$\theta \sim °$

$$\frac{E_{rms}}{E_P} = \frac{I_{rms}}{I_P}$$

$$\frac{I_{dc}}{I_P} = \frac{E_{dc}}{E_P}$$

Ratio

(a)

Figure 8-10

Ratios of $E_{dc}/E_p = I_{dc}/I_p$, $E_{rms}/E_p = I_{rms}/I_p$, and ratio of power at each conduction angle to full-wave, full-conduction power for full-wave rectified or alternating sine waves in (a) and half-wave in (b).

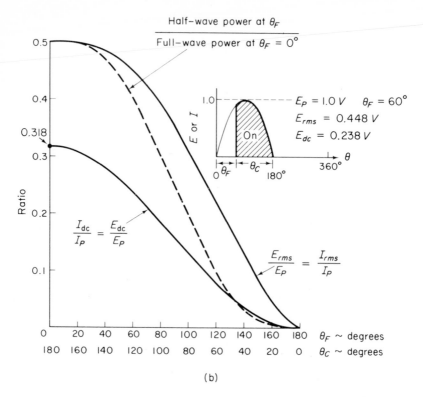

Figure 8-10 (Cont.)

the load power. We must convert from dc to rms values in order to predict or control load power. The easiest way to perform the conversions is to express *average load voltage* E_{dc} in terms of *peak voltage* E_p. Then express *rms voltage* E_{rms} in terms of *peak voltage* E_p. But firing of an SCR can be delayed by a firing angle Θ_F, so that conduction occurs over a conduction angle Θ_C. For a 60Hz line voltage,

$$\Theta_F + \Theta_C = 180°, \text{ time} = 8.3\text{ms} \tag{8-2}$$

Our conversions between E_{dc} and E_p as well as between E_{rms} and E_p must therefore include the effects of firing angle Θ_F. Furthermore, the SCR may control either a half sine wave or a rectified full sine wave. The ratio of E_{dc} to E_p and E_{rms} to E_p is expressed for half-wave operation by

$$\frac{E_{dc}}{E_p} = \frac{1}{2\pi} \int_{\Theta_F}^{\pi} \sin \Theta d\Theta = \frac{1 + \cos \Theta_F}{2\pi} \tag{8-3a}$$

$$\frac{E_{rms}}{E_p} = \sqrt{\frac{1}{2\pi} \int_{\Theta_F}^{\pi} \sin^2 \Theta d\Theta} = \sqrt{\frac{2(\pi - \Theta_F) + \sin 2\Theta_F}{8\pi}} \tag{8-3b}$$

and for fully rectified sine waves by

$$\frac{E_{dc}}{E_p} = \frac{1}{\pi} \int_{\Theta_F}^{\pi} \sin \Theta d\Theta = \frac{1 + \cos \Theta_F}{\pi}, \text{ full-wave} \qquad (8\text{-}4a)$$

$$\frac{E_{rms}}{E_p} = \sqrt{\frac{1}{\pi} \int_{\Theta_F}^{\pi} \sin^2 \Theta d\Theta} = \sqrt{\frac{2(\pi - \Theta_F) + \sin 2\Theta_F}{4\pi}} \qquad (8\text{-}4b)$$

Where Θ_F = firing angle, measured with a CRO
          E_p = peak load voltage measured on a CRO
          E_{dc} = load voltage measured with a dc voltmeter
          E_{rms} = ac load voltage measured with a true rms voltmeter

Note that the usual VTVM or multimeter reads true rms only for full sine waves or 360° of conduction, and is grossly incorrect for conduction angles less than 330°. Rather than use the cumbersome formulae in Eqs. (8-3) and (8-4), refer to Fig. 8-10, where the voltage ratios are plotted. For resistive loads the same ratios apply to average, peak, and rms currents.

Load Power versus Firing Angle

We normally calculate power in a load resistor R_L from either an alternating or rectified full sine wave by $I_{rms}^2 R$ or V_{rms}^2/R. Specifically, our experience in fundamentals of electricity is based on a reference power for a full sine wave (either rectified or alternating) that conducts for 180° during both half-cycles ($\Theta_F = 0°$). We should therefore compare power at smaller conduction angles for either half- or full-wave conduction with *full-wave full-conduction* power. The ratio of *full-wave power* at any firing angle Θ_F to *full-wave full-conduction* power is plotted as a dashed line in Fig. 8-10a. The ratio of *half-wave power* at any conduction angle to *full-wave full-conduction* is the dashed line in Fig. 8-10b.

EXAMPLE 8-3:
Power (brightness) of a 100W-110V tungsten lamp is to be varied by controlling firing angle of an SCR in a half-wave circuit supplied with 110V ac. (a) What rms voltage and current is developed in the lamp at $\Theta_F = 0°$? (b) What average load voltage and current will be read on a dc voltmeter and ammeter at $\Theta_F = 0°$? (c) What actual heating (rms) power is developed in the lamp at $\Theta_F = 0°$? (d) For comparison, repeat each part of this problem for $\Theta_F = 60°$ or $\Theta_C = 120°$.

SOLUTION:
(a) From Fig. 8-10b, $E_{rms} = 0.5E_p$ at $\Theta_F = 0°$. Find E_p from 1.41 × 110V = 156V, and $E_{rms} = 0.5 × 156V = 78V$. Find lamp resistance by assum-

ing (for simplicity) it is a fixed resistance from

$$R_L = \frac{V^2}{P} = \frac{(110V)^2}{100W} = 121\Omega$$

And $I_{rms} = E_{rms}/R_L = 78V/121\Omega = 0.64A$, $I_p = E_p/R_L = 156V/121\Omega = 1.29A$.

(b) From Fig. 8-10b, $E_{dc} = 0.318E_p$ at $\Theta_F = 0°$, or $E_{dc} = 0.318 \times 156V = 49.6V$, $I_{dc} = E_{dc}/R_L = 49.6V/121\Omega = 0.41A$.

(c) Under normal full-wave full conduction, rms power is 100W. From Fig. 8-10b, half-wave power at $\Theta_F = 0°$ is 0.5 × full-wave power, or $P_{half-wave}$ at $\Theta_F = 0° = 0.5 \times 100W = 50W$.

Repeat (a) Fig. 8-10b, $E_{rms} = 0.45 \times E_p$ at $\Theta_F = 60°$, or $E_{rms} = 0.45 \times 156V = 70V$, $I_{rms} = 0.45 \times I_p = 0.45 \times 1.29A = 0.58A$.

(b) Fig. 8-10b, at $\Theta_F = 60°$, $E_{dc} = 0.238 \times E_p = 0.238 \times 156V = 37.2V$, $I_{dc} = 0.238 \times 1.29A = 0.30A$.

(c) Fig. 8-10b, at $\Theta_F = 60°$, $P_{half-wave} = 0.405 \times P_{full-wave}$, $\Theta_F = 0°$, or $P = 0.405 \times 100W = 40.5W$.

Example 8-3 shows that, for a fixed load and supply voltage, load power, load current or load voltage decreases with decreasing conduction or increasing firing angle. This method of controlling power is very efficient because other methods, such as added series resistance, waste much power in the added control element. Since the SCR is a unidirectional device, we usually measure rms load power by a dc measurement. First, measure firing angle with a CRO and *average* dc load voltage with a dc voltmeter across R_L. Second, convert dc load voltage to peak and then to rms with the aid of Fig. 8-10. Finally, find load power from $(V_{rms})^2/R_L$.

Peak Currents and Average Currents

Because *average* load or anode current is our reference in an SCR circuit, we will examine further in Example 8-4 the relation between peak current and average current.

EXAMPLE 8-4:

In a half-wave rectifier, what peak load currents will occur if we measure an average dc load current of 1.0A at conduction angles of 0° to 180° in 30° steps?

SOLUTION:

From Eq. (8-3a)

$$I_p = \frac{2\pi I_{dc}}{1 + \cos\Theta_F} = \frac{2\pi \times 1}{1 + \cos\Theta_F} = \frac{6.28}{1 + \cos\Theta_F}$$

Tabulate results:

Θ_C (degrees)	180	150	120	90	60	30	0
Θ_F (degrees)	0	30	60	90	120	150	180
Cos Θ_F	1	0.87	0.5	0	−0.5	−0.87	−1
$1 + \cos \Theta_F$	2	1.87	1.5	1	0.5	0.13	0
$I_p \approx$ A	3.14	3.36	4.18	6.28	12.56	46.8	∞

Figure 8-11 shows how dramatically peak current values must increase at low conduction angles to maintain the same average load current. Perhaps unexpected is the *increase in rms* current value with *decreasing* conduction angle for the same *average dc current,* as shown in the next example.

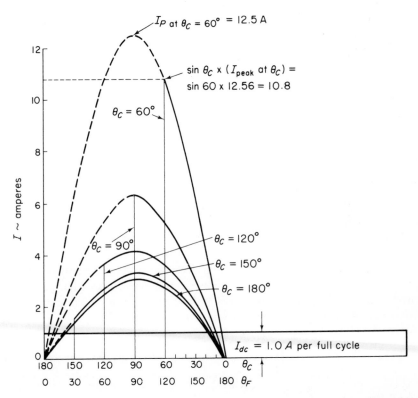

Figure 8-11

All current waves have the same average value. Solution to Example 8-4.

EXAMPLE 8-5:

For the same conduction angles and average current of 1A in Example 8-4, find the corresponding rms current (heating effect) value.

SOLUTION:

Multiply each peak current value in Example 8-4 by the corresponding ratio of E_{rms}/E_p from Fig. 8-10b. Tabulate results:

I_p from Ex. 8-4 (amperes)	3.14	3.36	4.18	6.28	12.56	46.8
E_{rms}/E_p, Fig. 8-10b	0.5	0.49	0.45	0.35	0.23	0.09
Θ_C (degrees)	180	150	120	90	60	30
I_{rms} (amperes)	1.57	1.6	1.9	2.2	2.9	4
I_{dc} (amperes)	1.0	1.0	1.0	1.0	1.0	1.0

Example 8-5 shows that, from the same 1.0A dc measurement, about three times as much rms load current is present (4A) for a 30° conduction angle as for a 180° conduction angle (1.57A). This is because, as shown in Eq. (8-3b), each instantaneous current value is squared before it is averaged to find the rms value. As far as load power is concerned, the difference is even greater, because the rms value is squared again. For example, with a 1 ohm load in Example 8-5, load power is $(1.57)^2 \times 1\Omega \cong 2.5$W at $\Theta_C = 180°$, $I_{dc} = 1$A. But at $I_{dc} = 1$A and $\Theta_C = 30°$, load power is $(4A)^2 \times 1\Omega = 16$W, or almost 7 times as much power. We conclude that an average current reading means little unless conduction angle is known.

8-5 SCR HEAT LIMITATIONS

The SCR is heated during conduction, and this heat is a power loss that must be dissipated. Since forward voltage ($V_{AK\,on}$) drop is fairly constant, and anode current is usually some portion of a sine wave, we cannot simply calculate SCR power by an I-V product. Manufacturers provide information (as in Appendix 3) on how SCR power dissipation varies with conduction angle Θ_C (not firing angle Θ_F). Some SCRs are intended for use without heat sinks, so maximum allowable *ambient* temperature is specified for various average currents. Larger SCRs are intended for use only with heat sinks, and maximum allowable *case* temperature is specified for various average currents.

Maximum Junction Temperature and Forward Current

To simplify our study of SCR heat problems, focus attention on the fundamental limitations of *maximum forward current* $I_{T(rms)}$ and maximum junction temperature T_{Jmax}. For example, an SCR rated at 4A rms and $T_{Jmax} = 110°$C means that neither of these limits may be exceeded at any time. This 4A rms and 110°C limitation must be examined carefully with respect to how the SCR is to be used. In each of the following examples, assume $\Theta_{JC} = 2°$C/w, $\Theta_{CS} = 0$, $\Theta_{JA} = 50°$C/w for the SCR, as in Appendix 3.

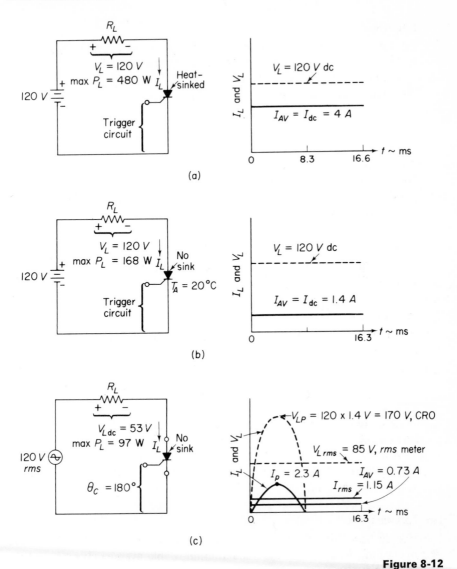

(a)

(b)

(c)

Figure 8-12

Circuits for Examples 8-6 to 8-8. SCR is MCR 4-6-4. V_L and I_L are
dc instrument readings.

EXAMPLE 8-6:

In Fig. 8-12a the SCR conducts dc current. From Fig. 4 of Appendix 3,
6.3W average power P_F is dissipated at 4A = $I_{F(AV)}$ for dc. (a) under what
conditions can this SCR handle 4A dc? and (b) what maximum load power
will it control? Assume an ambient of 20°C.

SOLUTION:

(a) From $T_{j\,max} = T_A + \Theta_{JC}P_F + \Theta_{CA}P_F$ or $T_C = T_{j\,max} - \Theta_{JC}P_F = T_A + \Theta_{CA}P_F$, find maximum allowable case temperature from $T_C = T_{j\,max} - \Theta_{JC}P_F = 110° - 2°C/W \times 6.3W \cong 97°C$. Verify this rating in Fig. 1 of Appendix 3 at $I_{F(AV)} = 4A$, $T_C = 97°$. Find required Θ_{CA} from

$$\Theta_{CA} = \frac{T_C - T_A}{P_F} = \frac{(97 - 20)°C}{6.3W} = \frac{77°C}{6.3W} = 12°C/W$$

Since required Θ_{CA} is less than SCR's $\Theta_{CA} = 50 - 2 = 48°C/W$, a heat sink is required with thermal resistance of $12°C/W$.

(b) With a 120V source we can control $120V \times 4A = 480W$ with 6W power loss in the SCR.

EXAMPLE 8-7:

With no heat sink, how much power can the SCR control in the circuit of Fig. 8-12b in an ambient of 20°C?

SOLUTION:

From junction temperature limitation find maximum SCR power dissipation

$$P_F = \frac{T_{j\,max} - T_A}{\Theta_{JA}} = \frac{(110 - 20)°C}{°C/W} = \frac{90°C}{50°C/W} = 1.8W$$

Now we must enter Fig. 4 of Appendix 3 at 1.7W to the intersection with line dc and read $I_{F(AV)} = 1.4A$. (Check Fig. 2 of Appendix 3 to see that at $1.4A = I_{F(AV)}$ with dc conduction the ambient must be equal to 20°C or lower.) With a 120V source we can control up to $120V \times 1.4A = 168W$, without a heat sink.

EXAMPLE 8-8:

For no heat sink, $T_A = 20°C$, and a conduction angle of 180° (half-wave operation), what is: (a) the maximum allowable average current, and (b) the maximum power we can control? The circuit is shown in Fig. 8-12c.

SOLUTION:

(a) From Fig. 2 of Appendix 3, at $T_{A\,max} = 20°C$ and $\Theta_C = 180°$, read $I_{AV} = 0.73A$. From Fig. 8-10b, convert to peak from $I_p = I_{dc}/0.318 = 0.73A/0.318 = 2.3A$, and then to rms with $I_{rms} = 0.5I_p = 0.5 \times 2.3A = 1.15A$. Rms value of the half-wave load voltage is $(120V/0.707) \times 0.5 = 84.5V$, so maximum controllable power is $84.5V \times 1.15A = 97W$. Minimum R_L is $84.5V/1.15A \cong 75\Omega$.

SCRs in Inverse Parallel

EXAMPLE 8-9:

Two SCRs rated at 4A rms are operated in inverse parallel, as shown in Fig. 8-13a. If $\Theta_C = 180°$, $T_A = 20°C$, and with no heat sink, what is the maximum load power they can control?

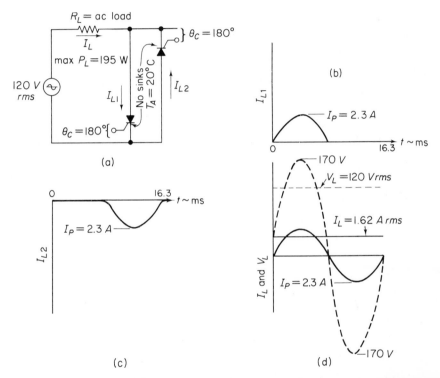

(a)

(b)

(c)

(d)

Figure 8-13

Two SCRs conduct on alternate half-cycles in (b) and (c) to give full-wave full conduction load current in (d). Solution to Example 8-9.

SOLUTION:

From Example 8-8, each SCR carries a maximum of 0.73A average current, or $0.73A/0.318 = 2.3A = I_p$ and $2.3A \times 0.5 = 1.15A = I_{rms}$, as shown in each half-cycle of Fig. 8-13b. Line current is shown in Fig. 8-13b for a full cycle to have an rms value of $I_p \times 0.707 = 2.3A \times 0.707 = 1.62A = I_{rms}$. (Note that I_{rms} of a full cycle equals 1.41 times I_{rms} of a half-cycle.) Rms full cycle load voltage is 120V rms rather than 84.5V as in Example 8-8. Maximum load power is doubled by adding the second SCR to $120V \times 1.62A = 195W$.

Maximum rms and dc Current Rating

At this time it should be noted that in Fig. 1 of Appendix 3 the maximum dc current rating of 4A is also by definition a maximum rms current rating of 4A. Verify this important point that the end point for $I_{AV} = 2.54$A at $180°$ in Fig. 1 is equivalent to an rms rating from Fig. 8-10b of

$$I_{rms} = I_{AV} \times 0.5/0.318 = 1.57I_{AV}, \Theta_C = 180° \qquad (8\text{-}5)$$

or $I_{rms} = 1.57 \times 2.54$A $= 4$A. Thus all the end points of each curve correspond to 4A rms. Now in Fig. 2 of Appendix 3 the dc maximum rating of 1.4A tells

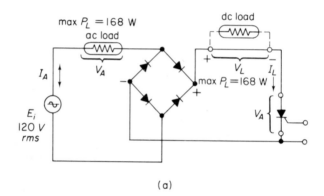

(a)

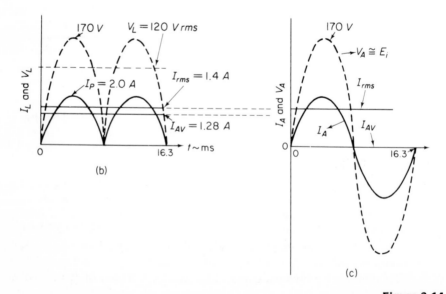

(b)

(c)

Figure 8-14

Solution to Example 8-10.

us that 1.4A rms is also the current limit for this SCR with no heat sink. If we employ Eq. (8-5) to find I_{AV} at $\Theta_C = 180°$ in Fig. 2, we get $I_{AV} = 1.4A/1.57 = 0.88A$. But in Fig. 2 the lowest ambient shown is $T_A = 20°C$, and 0.88A would cause T_J to exceed 110° at this ambient. So we must use the manufacturer's data of $I_{AV} = 0.72A$ for the maximum average current at $T_A = 20°C$.

SCR Conduction for 360°

Data is not normally given for conduction angles of 360°. In the final power example we use one SCR to conduct during *both* half-cycles. Assuming the SCR can conduct a maximum of 1.4A rms, we translate via Fig. 8-10a from full-wave full-conduction rms to I_{AV}:

$$I_{rms} = I_{AV} \times 0.707/0.638 = 1.1 I_{AV}, \Theta_C = 360° \tag{8-6}$$

to get $I_{AV} = 1.4A/1.1 = 1.28A$. Therefore, a dc reading of 1.28A means the load (either ac or dc) carries 1.4A rms.

EXAMPLE 8-10:
Employ the foregoing to calculate maximum controllable power, $\Theta_C = 360°$, $T_A = 20°C$, no sink, in the flexible circuit of Fig. 8-14a. Either an ac load *or* a dc load can be connected.

SOLUTION:
With $I_{L\,max} = 1.4A$ rms, average dc current through the SCR would measure $1.4A/1.1 = 1.28A$. Maximum load power is $1.4A \times 120V = 168W$. Dc and ac load currents are shown in Figs. 8-14b and c respectively.

8-6 SCR SWITCH OR RELAY REPLACEMENT

Figure 8-15 illustrates how the SCR may be applied to replace a switch or relay. With anode switch on point ac, the SCR will conduct half-wave current pulses (if turned on by a gate voltage) and reset itself to the off-state (as soon as I_A drops below I_H) once each cycle. With anode switch on dc, the SCR operates once and latches on (like a latching relay) and remains on until the anode switch is flipped to off.

When the gate control switch is on points 1 or 2, low-cost switches, located at a remote location, can transmit low-voltage ac or dc gate signals to turn the SCR on. In this type of application the SCR operates as a power control switch or solid state relay. No arcing will occur when the SCR turns load currents off or on, as would occur with relay contacts. With gate control switch in position

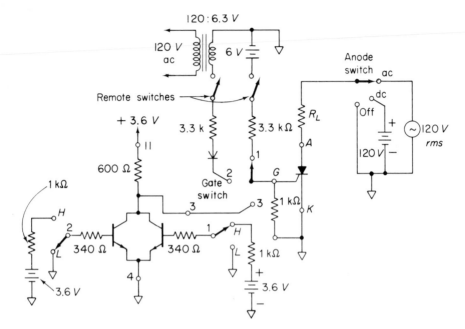

Figure 8-15

Representative applications of the SCR as a power switch or relay.

3, a two-input, integrated circuit NOR gate (such as the MC724P) gives logic control over the SCR. Both inputs must be low (L) before the SCR turns on. With either or both inputs high, the SCR will not turn on.

8-7 HALF-WAVE POWER CONTROL APPLICATIONS

A low-cost lamp dimmer circuit is featured in Fig. 8-16a. When the switch is closed, full brilliance occurs because the SCR is by-passed. Power control is taken over by the SCR when the switch is opened. For large values of R, V_C cannot charge up to the neon lamp's trigger voltage, and the lamp remains off. For small values of R, V_C charges fast enough to trigger the neon lamp, and in turn the SCR, early in the positive-going input cycle. When the SCR turns on, current is conducted through the lamp load, and terminals B and C are short-circuited by the SCR's anode and cathode. The SCR can be turned on at any firing angle between approximately 15° and 165°. In Fig. 8-16b the neon lamp trigger is replaced by a silicon unilateral switch (or four-layer diode). Timing resistor R is larger because of the lower trigger voltage (see Chapter 7). In Fig. 8-16c the simplest phase-control is a resistance divider that divides the input

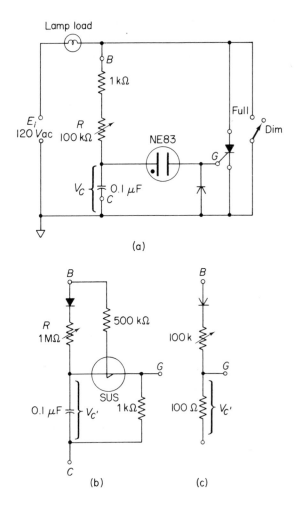

Figure 8-16

Half-wave lamp dimmer.

positive half-cycle down to V_{GT}. Firing angles of only $10°$ to $90°$ can be obtained. If V_{GT} is not reached by $90°$ (peak value of E_i), V_c must be less than V_{GT} during $90°$ to $180°$.

Motor Speed Control

The motor load in Fig. 8-17 is inserted in series with the cathode rather than anode in order to increase motor power, as motor speed slows down under load. For example, at no-load, motor speed will decrease as the wiper of potentiometer R is lowered. This is because V_G is lowered to trigger the SCR later in the cycle, applying shorter current pulses to the motor and, consequently, less power. For a given setting of R, gate and cathode voltages stabilize st some value. When the motor is loaded (for example, when bearing down on a drill), its back emf increases to raise cathode voltage so that capacitor C retains more

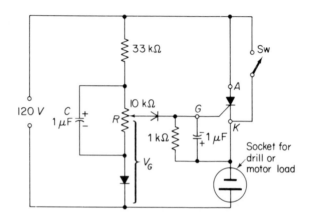

Figure 8-17

Half-wave speed control for universal series motor or brush-type motor
up to 1A rating.

of its charge after the SCR triggers. Now the capacitor does not have to charge as much to reach trigger voltage on subsequent cycles, and fires the SCR earlier in the cycle to increase power when it is needed.

8-8 EVENT-TIMER APPLICATION

Either a PUT or UJT can be used for Q_1 in the event timer of Fig. 8-18a. Any of the timing circuits in Chapter 7 can replace this representative application of a time delayed trigger pulse that will latch on an SCR. When switch Sw is thrown to time position, C_E charges at a rate determined by C_E and R_E to trigger the UJT. A peak output pulse, developed across R_L, turns the SCR on to energize the load. Peak gate current is found from an R_L-V_{op} load line drawn on the SCR's gate characteristic in Fig. 8-18b. R_L should be chosen: (1) small so that its IR drop before firing is less than V_{GT}; (2) large enough to furnish $V_{op} >$ V_{GT} upon firing; and yet (3) not so large as to exceed peak gate current ratings. Typical circuit values are $R_E = 1\text{k}\Omega$ to $2\text{M}\Omega$, $C_E = 0.001\mu\text{F}$ to $100\mu\text{F}$, $R_L = 10$ to 47Ω.

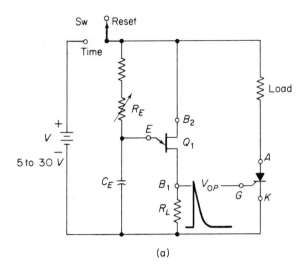

(a)

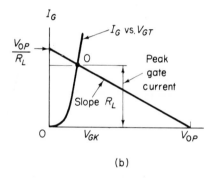

(b)

Figure 8-18

Event timer.

8-9 UJT-SCR APPLICATIONS

In some applications a UJT may be supplied by a fully rectified voltage V_{BB}, as in Fig. 8-19. Firing voltage V_p is shown for an intrinsic standoff ratio (see Chapter 7) of $\eta = 0.7$. Neglecting valley voltage, for simplicity, V_E is shown for three combinations of R_E and C_E. As C_E or R_E is increased, V_E is reduced in magnitude and retarded in phase to delay firing angle Θ_F. As shown, there are three *marker firing angles* at $\Theta_F \cong 20°$, $110°$, and $170°$ for $R_E C_E$ time constants of 0.265, 2.65, and 26.5ms, respectively. For example, if C_E is selected at 0.0265 μF, R_E should equal 10kΩ, 100kΩ, and 1MΩ to give firing angles of approximately 20°, 110°, and 170°, respectively.

Phase angle relationships between V_P and V_E

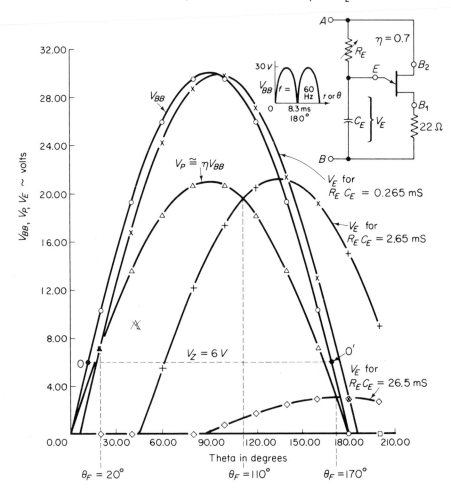

Figure 8-19

Firing angle control with resistance and capacitance.

Full-wave Power Control

As indicated by line OO' on supply voltage curve V_{BB} in Fig. 8-19, a 6V zener can be added across terminals A and B to supply the UJT effectively with a 6V dc supply. Point O is 12° into the V_{BB} sine wave. Use the UJT (or PUT) timing formulas developed in Chapter 7 to find the firing angle, but add a 6° average delay to account for the time necessary for V_{BB} to rise to V_{BB}. For a 60Hz supply, 180° corresponds to 8.3ms, and 6° to about 0.3ms.

In Fig. 8-20, a zener and R_p protect the UJT against excessive voltage in the event it does not fire; C_E and R_E give phase control of the firing angle. When the SCR is triggered on, it controls power to the dc load shown (or to an ac load, if required) and short-circuits the timing circuit. On-off control by temperature or light sensors can be added across terminals AB or BC, as explained in Chapter 7.

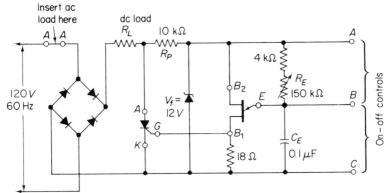

Figure 8-20

Full-wave power control with on-off options.

Battery Charger

A UJT functions as both an oscillator and a voltage level sensor in the battery charger circuit of Fig. 8-21. The battery powers the UJT, resistor R sets the value of V_{BB}, and η in turn sets the value of trigger voltage V_p. The UJT

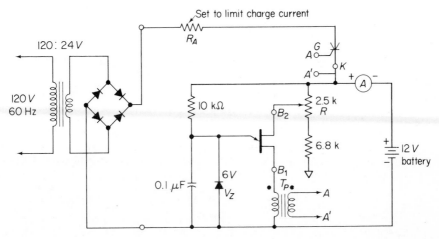

Figure 8-21

Battery charger.

triggers early in each half-cycle to fire a trigger pulse into pulse transformer T_p (such as the Sprague 11Z12 or equivalent). The 1:1 pulse transformer puts a positive-going pulse on the SCR's gate to turn it on and connect charging current to the battery. As battery voltage rises, both V_{BB} and $\eta V_{BB} \cong V_p$ rise. But when V_p equals zener voltage $V_Z = 6V$, V_p can no longer rise and the UJT no longer fires, since ηV_{BB} exceeds V_p. The charging circuit is automatically disconnected from the battery, and charge voltage is adjusted by potentiometer R, and maximum charge current by R_A.

PROBLEMS

1. In Fig. 8-4a, $R_L = 100\Omega$ and $R_C = 10\Omega$. Find C to limit rate of voltage rise to 10V/microsecond. (Reference: Example 8-1.)

2. If $R_1 = 8k\Omega$ and $R_2 = 2k\Omega$ in Fig. 8-8a, find (a) new conduction angle and (b) conduction time. (Reference: Example 8-2).

3. What is the minimum and maximum possible conduction angle in Fig. 8-8a if R_1 is replaced with a 0-25kΩ variable resistor?

4. Solve for currents, voltages, and power in Example 8-3 for a conduction angle of 90° in the half-wave circuit.

5. A half-wave power control circuit has an average dc current of 2A and conduction angle of 90°. What is the peak value of current? Use (a) Eq. 8-3a and (b) Fig. 8-10b. (Reference: Example 8-4).

6. Find the rms current value in problem 5 by (a) Eq. 8-3b and (b) Fig. 8-10b. (Reference: Example 8-5).

7. In Example 8-6 the heat-sinked SCR $\Theta_{CA} = 9°C/W$ conducts an average current of 2A. (a) What power is dissipated by the SCR? (b) What is the SCR's junction temperature if $T_A = 25°C$?

8. How much power can the SCR of Example 8-7 control in an ambient of $T_A = 58°C$?

9. In Example 8-8, for the same conditions but with a conduction angle of 90°, find (a) maximum allowable average current and (b) power we can control.

10. In Example 8-9, minimum R_L is 120V/1.62A $\cong$ 74Ω. If Θ_C is reduced to 90°, voltage; (b) rms load current; (c) load power; and (d) SCR currents.

11. In Example 8-10 assume $\Theta_C = 90°$ for each half-cycle (full-wave). What is the reduced value of (a) load current, (b) load voltage, and (c) load power?

12. In Fig. 8-18, $R_E = 1M\Omega$, $C_E = 10\mu F$ (low leakage). How long will it take to energize the load after switch Sw is thrown to reset?

13. If $R_L = 50\Omega$, $r_{BB} = 8,000\Omega$, and $V = 25V$ in Fig. 8-18, will the threshold voltage fire the SCR? Assume the SCR has a gate characteristic like Fig. 8-7.

14. What firing angle occurs in Fig. 8-19 if $R_E = 265\text{k}\Omega$ and $C_E = 0.01\mu\text{F}$?

15. Neglecting delay time in order for V_{BB} to rise to 12V in Fig. 8-20, what is the firing angle if $R_E = 46\text{k}\Omega$ and $C_E = 0.1\mu\text{F}$?

16. If $R_E = 46\text{k}\Omega$ and $V_p = 8\text{V}$ in Fig. 8-20, what minimum value of resistance, R, must be added across C_E to prevent firing the UJT?

17. If $\eta = 0.6$ and $V_D = 0.4\text{V}$ in Fig. 8-21, the voltage at B_2 with respect to ground should give $V_p \cong 6\text{V}$ when $V_{\text{Battery}} = 12\text{V}$. Find: (a) required voltage at B_2; (b) the setting of resistor R. (Neglect the Thévenin resistance of R and the 6.8kΩ resistor.)

chapter nine
the triac

9-0 INTRODUCTION TO THE TRIAC

The *triac* is a three-terminal, five-layer semiconductor device that acts as a power switch to control the flow of 60Hz alternating current. In Fig. 9-1a the two arrowheads show the direction of conventional current flow. Thus we see current may flow in either direction between *switch terminals* T_1 and T_2. The triac is therefore classified as a *bidirectional* or *bilateral* device. Terminal T_1 is the *reference terminal* from which all voltages are measured. Terminal T_2 is usually the case or header (or metal mounting tab for a plastic encapsulated triac), to which a heat sink is usually attached. The remaining gate terminal exercises limited control over the switching terminals. That is, the switch terminals act as an open switch to block current flow in either direction (off-state). When sufficient positive or negative (with respect to terminal T_1) voltage is applied between gate and terminal T_1, the triac's switch terminals conduct current (on-state) in a direction determined by the external circuit. Once the triac is triggered into conduction by a gate signal, it continues conducting, regardless

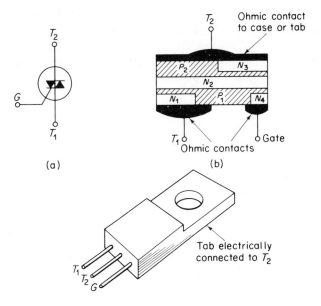

Figure 9-1

Triac symbol, construction, and package (geometry rearranged to simplify instruction).

of gate voltage, until current through T_1 and T_2 is reduced by the external circuit below a specified limit called *holding current*. Holding current, symbolized by I_H, is typically 50mA.

Switching speed of the triac is so fast that it can repeatedly begin conduction at any angle of a half-cycle from a 60Hz line. The angle or time it conducts are determined by when a signal is applied to the gate. Since the triac is intended to conduct in either direction, load current and power dissipations will be measured with rms current. This point is emphasized to differentiate rms measurements with triacs from average or dc measurements with the unidirectional SCR.

9-1 TURNING ON THE TRIAC

Current flow from T_2 to T_1 within the triac is designated as *positive* or *forward* conduction, while current flow from T_1 to T_2 is designated as *negative* or *reverse* conduction. The polarities agree with the potential of T_2 with respect to T_1 when the triac conducts. Figures 9-2a and b illustrate positive conduction through the equivalent positive SCR $P_2N_2P_1N_1$. In Figs. 9-2c and d, the triac has negative conduction through equivalent SCR $P_1N_2P_2N_3$. Plotting the T_2-T_1, I-V characteristic of Figs. 9-2a and b would give a first quadrant curve. Positive conduction is also called *mode I* (read as mode one) operation, to correspond with the quadrant. Current direction and voltage from T_2 to T_1 is reversed for negative conduction, and the negative *I-V* characteristic of T_2-T_1 would fall in the third quadrant. Thus negative conduction operation is represented by the shorthand notation *mode III* (read as mode three).

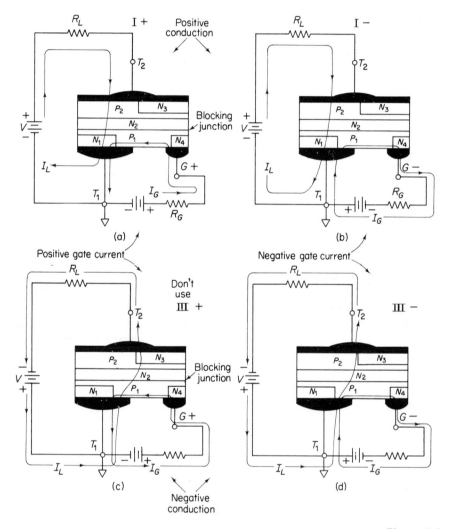

Figure 9-2

The triac is triggered into forward conduction by positive gate current in (a) and negative gate current in (b). Reverse conduction is triggered by positive gate current in (c) and negative gate current in (d).

Conduction in either direction is one principle advantage of the triac over the SCR. *Like* the SCR, the triac can be triggered by making the *gate positive* with respect to reference switch terminal T_1. *Unlike* the SCR, the triac also can be triggered by making the *gate negative* with respect to T_1. In Figs. 9-2a and c, the gate is made positive with respect to T_1 to forward-bias P_1N_1. Regenerative switching action occurs as discussed under SCR turnon. Both positive gate trigger circuits are symbolized by a $+$ sign immediately after the mode symbol. (I $+$ or III $+$). In Figs. 9-2b and d the gate is made negative with respect to T_1 and conduction modes are symbolized by I $-$ and III $-$. Data sheets specify maximum values of gate trigger current I_{GT} or gate trigger voltage V_{GT} required to turn on a particular triac. Values for I_{GT} and V_{GT} are typically 50mA and 3V.

For negative conduction, the positive gate current required in mode III $+$ operation is much higher than for the other three modes. Operation in mode III $+$ is often not recommended, and I_{GT}, V_{GT} may be ommitted for this mode on some data sheets. Triggering sensitivity is best in modes I $+$ and III $-$ and slightly less in mode I $-$. When only one polarity of trigger voltage is available, operation in modes I $-$ and III $-$ (negative gate) is recommended.

9-2 MEASURING ELECTRICAL CHARACTERISTICS OF THE TRIAC

Measurement of I_{GT} and V_{GT}

Maximum gate trigger current I_{GT} and maximum gate trigger voltage V_{GT} are measured with adequate accuracy from the test circuit in Fig. 9-3a. With V_{GG} set to zero, momentarily open the "stop conduction switch" to insure that the triac is nonconducting. Dc voltmeter V_T will indicate a value equal to V_{TT}. Voltage on T_2 is shown positive and gate voltage is positive to measure I_{GT} for mode $I +$. Increase V_{GG} and record the highest values of I_G and V_G that occur just before the triac is triggered into conduction as I_{GT} and V_{GT}. Conduction is evident when V_T drops abruptly to a low *on-state* voltage drop, between T_2 and T_1, typically 1 volt. We will observe that, initially, I_G and V_G rise in step with increasing V_{GG}. However, V_G will rise more slowly as we approach V_{GT}, while I_G continues to rise in step with V_{GG}. This is because bulk resistance of P_1 acts as a resistor between G and T_1 (see Fig. 9-2) and shunts the P_1N_1 junction.

Sweep Measurement of Gate Characteristics

When the P_1N_1 junction sees enough voltage to become forward-biased, the junction tends to hold a relatively constant voltage drop despite increase in junction current. Once enough carriers have been injected into this junction, by increasing I_G to I_{GT}, regenerative switching action triggers the triac into forward

	I_{GT}	V_{GT}
$T_2+, G+$	7 mA	0.7 V
$T_2+, G-$	15 mA	0.8 V
$T_2-, G+$	18 mA	0.9 V
$T_2-, G-$	9 mA	0.8 V

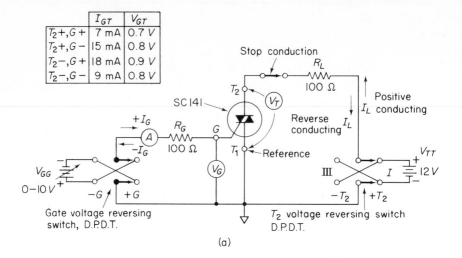

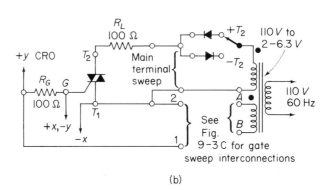

(b)

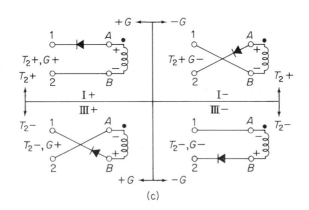

(c)

Figure 9-3

Static gate trigger voltage and current are measured in (a) for each of the four trigger modes. Gate characteristic curves are displayed on a CRO in (d) as measured by the sweep circuits of (b) and (c).

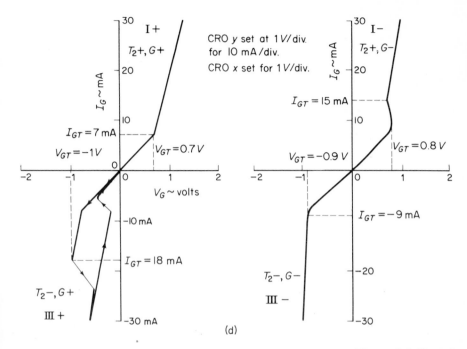

Figure 9-3 (Cont.)

conduction. This action of the gate I_G-V_G characteristics is seen in Fig. 9-3d. The data were obtained from the sweep measuring circuits of Figs. 9-3b and c. One transformer secondary (inexpensive filament transformer) is used to supply switch terminals T_2 and T_1. This ac sweep insures that the triac turns off every half-cycle. An additional secondary varies gate voltage. Details of the gate sweep connections are shown in Fig. 9-3c. Not only does gate voltage polarity have to be reversed, but the windings must be phased properly. For example, when T_2 goes plus, terminal A also goes $+$. For a positive gate signal, A should be wired to the gate through a diode. For a negative gate signal, with T_2 $+$, B should be wired to the gate and a diode should block positive signals from B.

Measuring Maximum Repetitive Peak Off-state Voltage V_{DROM}

Like the SCR, the triac can be triggered on by increasing voltage between T_2 and T_1 (with gate open) until leakage current initiates regenerative switching action. Unlike the SCR, the triac can be triggered on in either direction. Data sheets show this *maximum repetitive peak off-state* voltage as V_{DROM}, and the I-V characteristic between T_2 and T_1, or *principal voltage-current* characteristic, is shown in Fig. 9-4a.

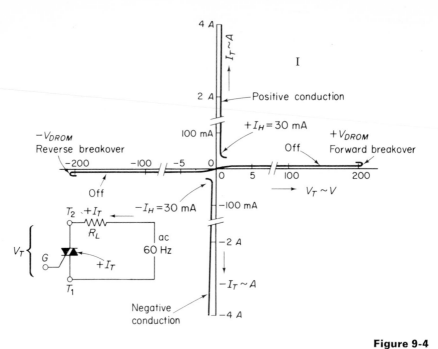

Figure 9-4

Breakover $I_T - V_T$ characteristics of triac switch terminals T_2 and T_1.

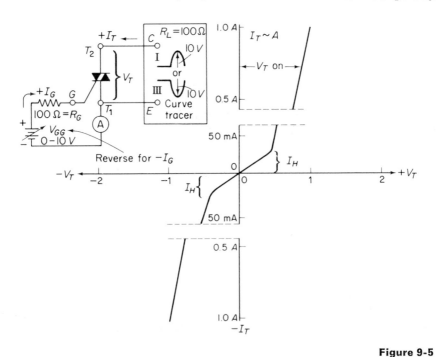

Figure 9-5

Measurement of I_H for a triac.

Measurement of Holding Current

Turnoff occurs in the triac in the same fashion as in the SCR. That is, switch terminal current must be reduced (by reducing line voltage or increasing load resistance) below a specified minimum value called *holding current*, I_H. One way to measure I_H is with a curve tracer, as in Fig. 9-5. Positive or negative gate current, greater than I_{GT}, is injected from an external source or from the curve tracer's step-current generator. (By increasing I_G until the triac fires, we can also measure I_{GT} with this circuit.) Either a positive or negative $I_T V_T$ characteristic is displayed on the curve tracer, depending on whether the collector sweep is $+$ or $-$, respectively. We can also employ Fig. 9-5 to see that $(+)$ or $(-)$ *on-state voltage* V_T is roughly 0.3 to 1 volt for ampere variations in $+$ or $- I_T$.

9-3 TRIAC APPLICATIONS AS A STATIC SWITCH

There are several advantages in replacing mechanical switches with triacs to turn ac power on or off. A resonant reed switch or inexpensive low-power switch interrupts or passes milliamperes of gate current into the triac. In turn,

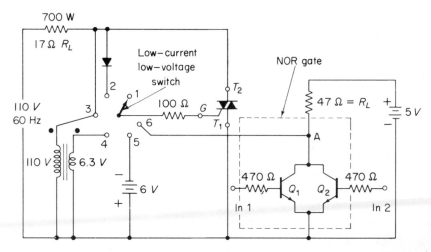

Figure 9-6

Techniques of static switching with a triac. For steady state gate current set $I_G \geqq 2I_{GT}$.

the triac controls amperes of load current. Inexpensive, smaller switches can be used, and switch life extended considerably. The triac always opens the load circuit at essentially zero load current (assuming gate voltage is less than V_{GT}), while a mechanical switch may have to break an inductive load current at its maximum. Thus the triac can eliminate one source of arcing and radio frequency interference. An on-off gate-current control is called *static switching*.

Static Switching Applications for Half-wave Operation

In Fig. 9-6 a six-position, light duty switch illustrates triac applications as a static switch. Point 1 opens the gate and the triac blocks load current flow in both directions, so that T_2 and T_1 act as an open switch. In switch position 2 the gate goes positive when T_2 goes positive, and the triac conducts only during each positive half-cycle. $T_2 +$, $G +$, or mode I + is the operating mode for position 2, and only half power (350W) would be delivered to the 700W load.

Static Switching for Full-wave Operation

In position 4 a remotely-located, low-ac voltage can drive the gate. Transformer polarity is chosen for $T_2 +$, $G +$, and $T_2 -$, $G -$, corresponding to operating modes I + and III −. Load current flows during both half-cycles, and full-wave power of 700W is delivered to the load. Switch position 5 illustrates how a low dc voltage insures flow of negative gate current at all times. The triac has full-wave conduction for $T_2 +$, $G -$, and $T_2 -$, $G -$, corresponding to modes I − and III −, and 700W is delivered to the load.

In switch position 3 gate current is taken directly from the line. Excessive gate current does not flow because as soon as point 3 rises to + or $-V_{GT}$, the triac conducts from T_2 to T_1 and point 3 goes to the low, 1V, on-voltage of T_2. Switch points 2 and 4 are also held at low potentials during T_2-T_1 triac conduction.

Logic Control of Static Switching

Finally, position 6 shows how a two-input NOR gate gives logic control over the triac. If either *in* 1 or *in* 2 go positive enough to saturate Q_1 or Q_2 respectively, point A is essentially at ground potential and disables the triac's gate. Both inputs must be grounded to cut off both Q_1 and Q_2 so that positive gate current can flow into the triac. Mode I + and III + operation results. Circuits similar to the NOR gate-triac combination in Fig. 9-6 are fabricated into one package and called *solid state relays*. Input signals to the NOR gate may be either dc or synchronized with line frequency for phase control.

9-4 LOAD AND TRIAC POWER

We must be able to predict how much power a particular triac can control. Assume we pick an inexpensive plastic triac, rated for maximum continuous *RMS On-state current* of $I_T = 10$A, and want to learn how much full-wave

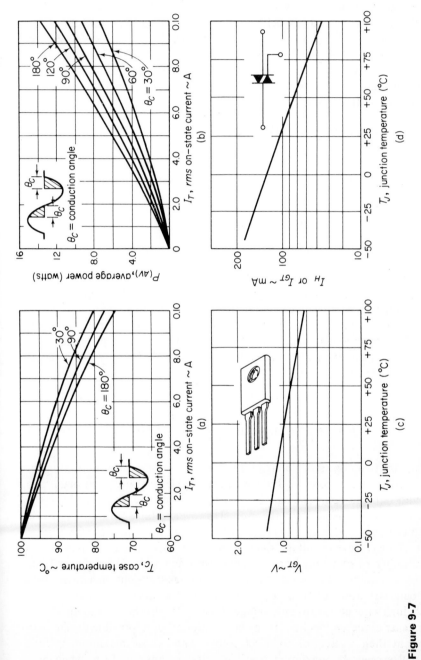

Figure 9-7

Typical maximum case temperature and average power versus *rms* triac current in (a) and (b), respectively. Temperature dependence of gate trigger voltage and current is shown in (c) and (d) for a plastic 10A *rms* triac.

345

power this triac can control. Since heat dissipation is our basic limitation, we must go to a data sheet for electrical specifications that relate rms current to power dissipation in the triac. Triacs intended for lead mounting will have specifications given in terms of ambient temperature, while triacs intended for heat-sink mounting will have specifications in terms of case temperature. From a curve of *Maximum Allowable Case Temperature* T_C versus *RMS On-state Current* I_T, in Fig. 9-7a, at $\Theta_C = 180°$, read $T_{C\,max} \cong 74°C$ at $I_T = 10A$. Also find $T_{J\,max} = 100°C$ from the same curve, where $T_C = 100°C$ at $I_T = 0$. Then, assuming the data sheet gives $\Theta_{JC} = 2°C/W$, we find maximum allowable power dissipation in the triac from

$$T_{J\,max} = T_C + \Theta_{JC}P_{D\,max}$$
$$100 = 74 + 2 \times P_{D\,max}, \text{ and } P_{D\,max} = 13W$$

If the triac must operate in an ambient temperature of 25°C, we find Θ_{CA} by first finding Θ_{JA} from

$$T_{J\,max} = T_A + \Theta_{JA}P_{D\,max}$$
$$100 = 25 + \Theta_{JA}12.5, \text{ and } \Theta_{JA} = 6°C/W$$

Then $\Theta_{CA} = \Theta_{JA} - \Theta_{JC} = 6°C/W - 2°C/W = 4°C/W$. A heat sink is required with $\Theta_{CA} = \Theta_{CS} + \Theta_{SA} = 4°C/W$. A one side area of about 25 square inches of 1/32″ aluminum would suffice, mounted with a silicon greased mica insulating washer. The aluminum piece should measure 5″ × 5″ and be mounted vertically.

Once the triac is heat-sinked to operate at 10A rms for 180° conduction during both half-cycles, smaller firing angles (in the same circuit) will dissipate less heat in the triac.

9-5 SNUBBER NETWORKS AND RADIO INTERFERENCE

There are two effects of voltage transients that have a profound influence on semiconductor components. First, voltage transients can exceed maximum ratings and damage components. Over-voltages on a diac, triac, or SCR (> forward breakover) merely turn them on. However, a false turnon may cause catastrophic injury or damage, such as in a time delay fuse where an SCR turnon ignites a primer. Severe voltage fluctuations may originate from power line faults, lightning, or switching transients, and can reach peak values of over 1,000 volts on your 110 volt line. More common are voltage transients generated from abrupt load variations, or from a power supply's transformer primary or secondary being switched. Transient voltage spikes may also be generated when a diode or zener diode avalanches. It may be impractical or impossible to eliminate transients at their source, or to select components whose maximum voltage ratings will exceed the maximum possible voltage transient. Even when protection devices are installed to limit maximum values of voltage transients, we can still experience turnon problems due to dV/dt (a rate-of-change of voltage with respect to time) effect. For example, suppose a 500V transient is clipped at 200 volts and the 200V spike is applied between T_2 and T_1 of a triac rated at $V_{DRM} = 400V$. If the 200V spike has a rise time of $2\mu s$, rate-of-rise of voltage is $100V/\mu s$. Typical maximum dV/dt is $50V/\mu s$, so the triac will still be turned on by the transient.

Snubber Networks

Snubber networks are arrangements of resistors, capacitors, and occasionally inductors that limit rate of voltage rise in order to eliminate false turnon. Voltage transients occur every time a triac tries to turn off current to an induc-

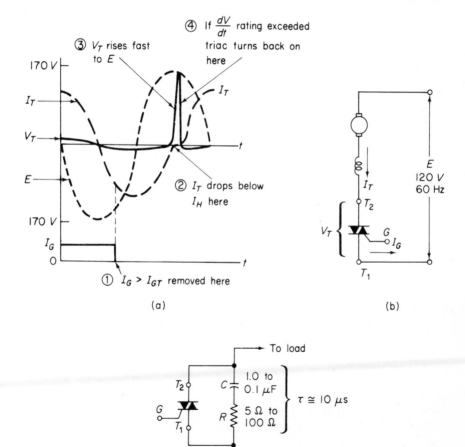

(a)

(b)

(c)

Figure 9-8

False triac turn-on from an inductive load in (a) and (b) is eliminated by added snubber network *RC* in (c)

tive load. In Fig. 9-8a and b, load current I_T lags applied voltage E by 90° because of motor winding inductance. When gate current is removed, the triac continues to conduct between its switch terminals T_2 and T_1 until current drops below I_H. The triac starts to turn off but sees an instantaneous line voltage rise of approximately 150V. This fast voltage rise causes a charging current to flow into the blocking junction, initiating regenerative switching action, and the triac turns back on again when it should not. Scales are distorted for clarity in Fig. 9-8a. Actual peak values are $E_P \cong 170\text{V}$, $I_{TP} \cong 1.4\text{A}$ for 1A rms load, $V_{TP} \cong 1\text{V}$. Voltage rise is reduced in Fig. 9-8c by adding capacitor C across switching terminals of the triac.

However, it is possible for C to be charged to E_P at that instant when a gate signal occurs. During the subsequent triac turnon, the triac or C could be damaged by its large discharge current (through the on-triac) unless series resistor R is added. R and C make up a snubber network in Fig. 12-8c. Time constant $\tau = RC$ is selected for about 10μs and, for $E_P = 170\text{V}$, would restrict dV/dt to $170\text{V}/10μs \cong 17\text{V}/μs$.

Minimizing Radio Interference

Whenever a triac turns on with a *resistive* load, current rises abruptly, causing a transient voltage to propagate out along the supply wires. High frequencies in the voltage pulse can be radiated as radio frequency interference

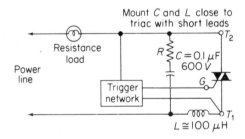

Figure 9-9

LC radio interference filter. Inductor wire for L must be large enough to carry load current.

(RFI) from power wires as if they were a transmitting antenna. An LC network is installed to short-circuit high frequencies around the triac switching terminals. L is chosen for a convenient value of approximately 50 to 100μH, and may be constructed by random winding of about 75 turns of #18 insulated wire around a ferrite core (salvaged from a loop-stick antenna). In Fig. 9-9, capacitor C is placed across the ac line between load and triac with a value designed to re-

sonate at about $f_R = 50\text{kHz}$ from

$$f_R = \frac{1}{2\pi\sqrt{LC}}$$

and

$$C = \frac{1}{(2\pi f_R)^2 L} = \frac{1}{(2 \times 50 \times 10^3)^2 \times 100 \times 10^{-6}} \cong 0.1\mu\text{F}$$

For loads less then 100W, add $R = 10$ to $R = 100\Omega$ in series with C to prevent oscillations that may give false turnons.

9-6 LIGHT CONTROL WITH TRIACS

In Fig. 9-10a, R and C provide phase control for the trigger diode to manu-ally control conduction angle. R can be increased enough for the diac not to fire at all. As R is then decreased, a condition can occur in which the diac fires erratically on alternate half-cycles because capacitor C has a different initial charge on each half-cycle. This condition is known as *hysteresis* and is mani-fested for a lamp load by the lamp suddenly *snapping* on, instead of gradually coming on as R is decreased from its maximum value. Hysteresis can be reduced by adding two resistors and a capacitor in the timing-trigger circuit, as in Fig. 9-10b.

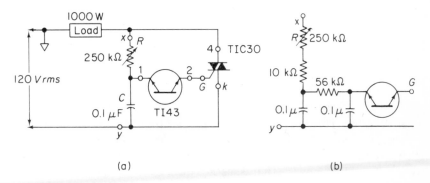

(a)   (b)

Figure 9-10

Diac-triac load power control.

The Asymmetrical Silicon Bilateral Switch-ASBS

An improvement over Fig. 9-10b is Fig. 9-11a. In this circuit only three circuit elements make up an excellent hysteresis-free lamp dimmer. The ST4 is an integrated circuit, semiconductor device with characteristics somewhat

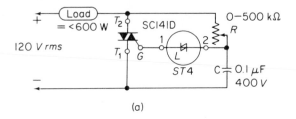

(a)

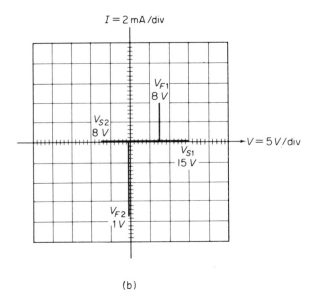

(b)

Figure 9-11

Minimum parts count incandescent lamp dimmer with triac and asymmetrical SBS.

similar to a diac. However, as shown in Fig. 9-11b, its trigger voltages are not symmetrical, and it is classified as an *Asymmetrical Silicon Bilateral Switch* (*ASBS*). Resistor R varies conduction angle Θ_C to give a linear, inverse relation between R and load power. That is, as R is increased by 10 percent, load power is decreased by 10 percent. A light sensitive on-off control can be added to either Figs. 9-10b or 9-11a by connecting a light sensitive resistor across capacitor C.

9-7 TRIAC MOTOR SPEED CONTROL

One advantage of the motor speed control in Fig. 9-12 is that armature and field windings do not have to be separated. A full-wave bridge provides synchronizing rectified ac to the zener and UJT trigger. R_E and C_E give manual

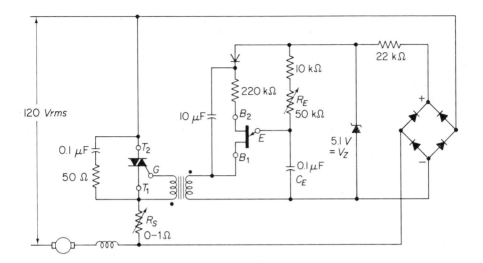

Figure 9-12

Variable-speed motor control for small appliances (up to 5*A* motor rating).

adjustment of conduction angle via UJT trigger pulses that are coupled through a pulse transformer to the triac. Assuming R_E is adjusted to give half-speed at no-load on the motor, the motor would tend to slow down when load was applied. For example, a drill, belt sander, or mixer would normally slow under load unless some way were found to increase motor voltage automatically as mechanical load is increased.

Resistor R_S senses the magnitude of load current. As mechanical load is increased, voltage drop across R_S increases during the conduction time of the triac. This increases input voltage to the bridge and the average value of voltage between the $+$ and $-$ bridge terminals. C_E discharges to this $+$ and $-$ dc voltage during conduction time of the triac. Therefore, as load current increases, the initial voltage across C_E increases at the beginning of the following half-cycle. Thus C_E does not have to charge as much and will fire earlier in the half-cycle (as load current is increased), to provide more voltage to the motor. This automatic advance in the firing angle of C_E offsets the tendency of the motor to slow down under load, and the motor thus tends to maintain a constant speed. Resistor R_S is a variable wire-wound resistor (10-15W). Too small a value of R_S will cause the motor to lose torque; too high a value will cause the motor to plug or surge.

Pulse Transformer

The pulse transformer is connected to give negative gate pulse to the triac for positive pulses from the UJT. A suitable pulse transformer can be wound on a length of ferrite rod salvaged from a loop stick antenna. Wind about 100 turns of #34 to #40 magnet wire in one direction along the full length of a 1″ rod. Identify the start terminal with a dot. Insulate this primary with one layer of plastic electrical tape or coil dope. Wind a secondary in the same direction with 100 turns (on top of the tape), identifying the start with a dot. A better impedance match is provided if the secondary is taped and an additional secondary (100 turns) is wound identical to the first. Connections for this 1:1:1 transformer are shown in Fig. 9-13.

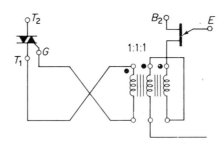

Figure 9-13

Connection of three-winding pulse transformer.

PROBLEMS

1. In Fig. 9-3a, the T_2 switch is on $+T_2$ and the gate switch is on $+G$. (a) What is the mode of operation? (b) If V_{GG} is greater than T_{GT}, is the triac forward- or reverse-conducting?

2. In problem 1, (a) if $V_{T\,on} = 1V$, what is the value of current through R_L? (b) If $V_G = 0.9V$ on the conducting triac, what is the value of gate current?

3. In section 9-4, $P_{D\,max}$ was found to be approximately 13W. Can the same information be obtained from Fig. 9-7b?

4. Six 100 watt, 120V lamps are controlled by a single triac with power dissipation ratings given in Fig. 9-7b. Show that 5W is dissipated in the triac when fully conducting, $\Theta_C = 180°$.

5. What is the maximum allowable case temperature and minimum Θ_{CA}, if $T_A = 25°C$, for problem 4?

6. In problems 4 and 5, let Θ_C be reduced to 90°. From Fig. 8-10a, $I_T = 5A$ at $\Theta_C = 180°$ is reduced to $I_{T\,rms}$ at $\Theta_C = 90° = [I_{peak}$ at $\Theta_C = 180°] \times 0.5; I_{T\,rms}$ at $\Theta_C = 90° = [5/0.70] \times 0.5 = 3.5A$. Find triac power dissipation from Fig. 9-7b.

7. Assuming $\Theta_{CA} = 13°/w$ and $P_{AV} = 2.5W$, find triac case temperature. (Reference: problems 4 to 6.)

8. $C = 0.1\mu F$ in Fig. 9-8c. Find R to form a time constant of $10\mu s$.

9. With $L = 50\mu H$ in Fig. 9-9, find C to resonate at 50kHz.

chapter ten
light sources

10-0 INTRODUCTION

Two technologies—optics and electronics—were at one time discussed, analyzed, and studied separately. In recent years, their merger into optoelectronics has introduced a new field for the creative and imaginative engineer and technician.

Optoelectronic devices fall into one of two categories: (1) devices that convert electric current into light, or (2) devices that convert light into electric current. The first category is the subject of this chapter, and includes the incandescent filament lamp, neon glow lamp and emitting diode. (The expression *emitting diode* is used to encompass both light emitting diodes, LEDs, and infrared emitting diodes. The distinction will become apparent in the following pages.) The second category of devices (more commonly known as photodetectors) is dealt with in Chapter 11. We shall concern ourselves mainly with four types of photodetectors: photoconductive cells, photovoltaic cells, photodiodes, and phototransistors.

In the next two chapters we will be introduced to new terms, expressions, and symbols, each of which will be defined as it is introduced. However, in section 10-17, a glossary of terms for both chapters is given as reference and convenience.

When analyzing light sources we encounter *photometric* and *radiometric* quantities. The reason for two separate quantities is that only a small percentage of total output radiant energy of a lamp is actually perceived by the human eye as light. However, a silicon photodetector is effected by the total radiant energy. Photometric quantities are for visual effects involved with visible light, while radiometric quantities are for *total* output energy of a light source, including invisible energy.

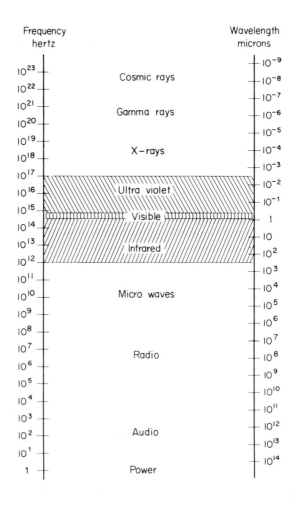

Figure 10-1

Electromagnetic spectrum.

10-1 LIGHT DEFINITION

Light is electromagnetic radiation of such a frequency or wavelength as to be perceived by the human eye. The band of frequencies to which the eye is sensitive is very narrow and would probably not attract much attention if it were not that more information is received by the average human in this band than in the rest of the entire spectrum.

Figure 10-1 shows a plot of the electromagnetic spectrum in both frequency and wavelength.

Wavelength

In the audio or sound range, the reader probably has made all calculations in terms of frequency. At higher frequencies, however, it is conventional to use wavelength (λ) instead of frequency. The relationship between wavelength and frequency is

$$\lambda = \frac{c}{f} \tag{10-1}$$

where λ is wavelength in meters, f is frequency in Hertz or cycles per second, and c is velocity of light at 3×10^8 meters per second (m/s). Wavelength, as the name implies, is the physical length of the wave. As shown in Fig. 10-2, the time for one complete cycle is the period (T) which is the reciprocal of frequency (f). The measured length for one complete cycle is wavelength (λ).

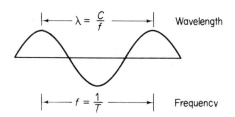

Figure 10-2

Comparison between frequency and wavelength.

The units for wavelengths are microns (μ), millimicrons (mμ) or Angstroms (Å). A micron is one-millionth of a meter, a millimicron is one-thousandth of a micron, and an Angstrom is one-tenth of a millimicron. Thus

$$1 \text{ micron } (1\mu) = 1 \times 10^{-6} \text{ meter}$$

$$1 \text{ millimicron } (1m\mu) = 0.001 \times 10^{-6} \text{ meter} = 1 \times 10^{-9} \text{ meter}$$

$$1 \text{ Angstrom } (1\text{Å}) = 0.1 \times 10^{-9} \text{ meter} = 1 \times 10^{-10} \text{ meter}$$

Example 10-1:

Express the following frequencies in wavelengths: (a) 60kHz; (b) 1×10^{14} Hz; (c) 3×10^{22} Hz.

Solution:

(a) Using Eq. (10-1)

$$\lambda = \frac{c}{f} = \frac{3 \times 10^8 \text{m/s}}{60 \times 10^3 \text{c/s}} = 5 \times 10^3 \text{ meters}$$

In terms of microns

$$5 \times 10^3 \text{m} \times \frac{1 \text{ micron}}{1 \times 10^{-6}\text{m}} = 5 \times 10^9 \text{ microns}$$

In terms of millimicrons

$$5 \times 10^3 \text{m} \times \frac{1 \text{ millimicrons}}{1 \times 10^{-9}\text{m}} = 5 \times 10^{12} \text{ millimicrons}$$

In terms of Angstroms

$$5 \times 10^3 \text{m} \times \frac{1 \text{ Angstrom}}{1 \times 10^{-10}\text{m}} = 5 \times 10^{13} \text{ Angstroms}$$

(b) $\qquad \lambda = \dfrac{c}{f} = \dfrac{3 \times 10^8 \text{m/s}}{1 \times 10^{14}\text{c/s}} = 3 \times 10^{-6} \text{ meters}$

In terms of other units, 3 microns, 3,000 millimicrons, or 3×10^4 Angstroms.

(c) $\qquad \lambda = \dfrac{c}{f} = \dfrac{3 \times 10^8 \text{m/s}}{3 \times 10^{22}\text{c/s}} = 1 \times 10^{-14} \text{ m}$

In terms of other units, 1×10^{-8} microns, 1×10^{-5} millimicrons, or 1×10^{-4} Angstroms.

Note that at low frequencies the wavelengths are very long, and at high frequencies the wavelengths are very short.

10-2 HUMAN RESPONSE TO LIGHT

The human eye is a filter with a response similar to that of a tuned circuit. The response curve of a light adapted eye, more commonly called the standard observer (sort of a standard eyeball), is shown in Fig. 10-3. Note that the eye does not have the same sensitivity for each wavelength. The peak sensitivity is at 0.55 microns and tapers off to zero at 0.4 microns and at 0.76 microns. Figure 10-3 also shows the colors which correspond to a band of wavelengths. The eye is most sensitive to green-yellow and less sensitive to violet (shorter wavelengths) and red (longer wavelengths).

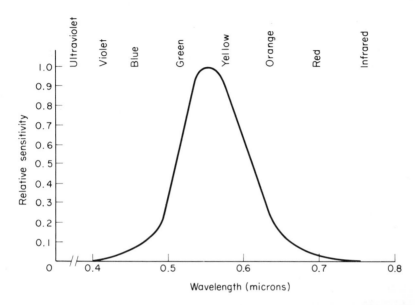

Figure 10-3

Response curve of a standard observer.

The regions at both ends of the visible spectrum are the ultraviolet region and the infrared region (see Fig. 10-3). Although the energies radiated in these regions are often referred to as ultraviolet light and infrared light, technically they are not light because their radiation cannot be perceived by the human eye. The word has been carried over only to indicate that their location on the electromagnetic spectrum is close to the visible region.

10-3 LIGHT GENERATION

Light is generated in one of two ways—*incandescence* and *luminescence*. Incandescence is the generation of light as a result of the temperature of a material. In the incandescent lamp, current flows through a conductor (filament). The resistance of the conductor produces heat, and thus the temperature of the material rises. If the temperature is high enough (approximately 1,800°F), light is emitted.

Luminescence is the generation of light by any method other than temperature of a material. There are four categories of luminescence:

1. *Electroluminescence*—light emission due to hole-electron recombinations; examples are the light emitting diodes and solid state laser.
2. *Chemiluminescence*—light is produced as a result of a chemical reaction; examples are the firefly and rotting wood.
3. *Triboluminescence*—light is produced as a result of friction.
4. *Photoluminescence*—light is generated is a result of absorption of other radiation (either ultraviolet, visible, infrared, x-rays, or γ rays).

Although this text primarily analyzes solid state devices, we could not do justice to light sources without covering the incandescent and gaseous discharge lamps. Both types of lamps have been used for many years, their characteristics are well-known, and they are rather inexpensive. Even more important, they are used to activate solid state light detectors.

10-4 INCANDESCENT LAMPS

Incandescence is the emission of visible light by a material due to its high temperature.

Current flow through any conductor produces heat. In an incandescent lamp, heat is produced by current flowing through the conductor filament. If sufficient electrical energy is applied to the filament to raise its temperature above 875°K (1,800°F), light as well as heat is emitted. Increasing the temperature further causes more light to be emitted. Therefore, an increase in electrical energy into the lamp causes the temperature of the filament to increase, which in turn produces more visible light. For most applications, the operating temperature of incandescent lamps is between 2,500°K and 3,000°K.

Only a few materials are capable of withstanding such high temperatures. The materials noted in Table 10.1 have been used in the manufacture of lamp filaments. In comparison, iron and steel melt at 1,473°K and boil at 3,173°K.

The highest melting point, however, is not the only consideration in choosing a material for a filament. Before a material reaches its melting point, it undergoes a high rate of evaporation. If a filament begins to evaporate, it becomes thinner and eventual failure results. The evaporated material also settles on the inside of the bulb, causing it to darken and reducing light output.

Table 10.1 would seem to indicate that carbon is the best material for filaments. However, carbon begins to evaporate at approximately 2,123°K. The light given off at this low temperature is very little, and thus carbon is not a suitable material for commercial lamps. The next choice is tungsten. Tungsten filaments are capable of operating at temperatures between 2,500°K and 3000°K.

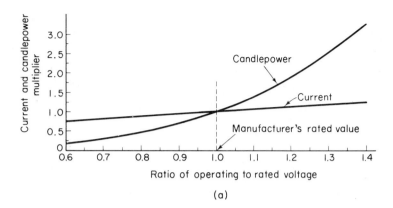

(a)

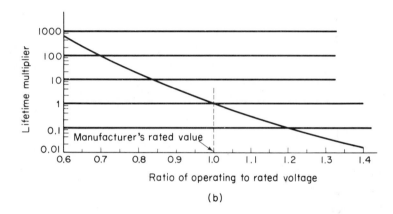

(b)

Figure 10-4

(a) Current and candlepower versus rated value, (b) lifetime versus
multiplier, for incandescent lamps.

The approximate operating temperature of the ordinary household lamp is
2,870°K.

Operating a filament at the highest possible temperature has two major
advantages: (1) Efficiency is higher. Efficiency is the ratio of light output to
input power. (2) Color of the light emitted is whiter, and thus more closely
approaches daylight conditions. To obtain maximum efficiency and long life,
manufacturers specify rated values. Rated values are the best operating conditions (voltage and current) for a particular lamp. If the filament is operated at a
temperature above the manufacturer's recommended value, life of the lamp is
drastically reduced.

Figure 10-4a is a plot of candlepower (light intensity) and current versus
the ratio of applied to rated voltage. Manufacturer's rated voltage is considered
1.0. This figure illustrates that an increase in operating voltage causes an increase

Table 10.1

Materials Used in Lamp Filaments

Material	Melting Point °K
Platinum	2,040
Osmium	2,973
Tantalum	3,173
Tungsten	3,653
Carbon	3,873

in current, which we now know increases the temperature of the filament. A higher filament temperature causes more light output.

10-5 LIGHT OUTPUT

Figure 10-4b shows that an increase in operating voltage drastically reduces lamp lifetime. Note that, for an increase in operating voltage, lifetime decreases much faster than either current or light intensity increases.

Light intensity (photometric quantity) is symbolized by I_L and has units of *candles*. *Total light output* from a lamp is symbolized by F and has units of *lumens*. Household lamps presently manufactured in the United States give total light output on the package. Examples of total lamp output are shown in Table 10.2.

Table 10.2

Total Lamp Output

Wattage Rating	Average Total Light Output (F)	Average Lifetime
15W (soft white)	120 lumens	2,500 hours
25W (soft white)	222	2,500
40W (standard)	440	1,500
60W (soft white)	870	1,000
75W (soft white)	1,180	750

Light intensity is related to total light output by

$$I_L \text{ in candles} = \frac{F \text{ in lumens}}{4\pi \text{ in steradians}} \tag{10-2}$$

Average light intensity is also called *mean-spherical-candlepower* (*MSCP*), but units are still candles. Fig. 10-5 illustrates that the actual light radiation pattern of a tungsten lamp is not uniform in all directions. For example, we might

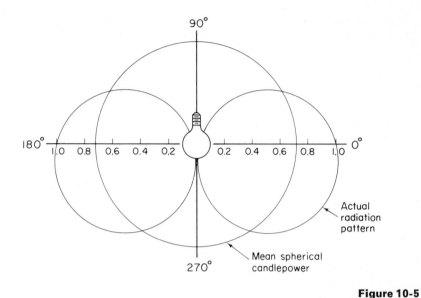

Figure 10-5

Radiation pattern of an incandescent lamp showing typical radiation pattern along with mean spherical candlepower pattern.

conclude that a 60W standard bulb with an average output of 870 lumens would give a uniform light intensity of $870/(4\pi) = 69$ candles symmetrically around the bulb. However, the actual radiation pattern in Fig. 10-5 shows light intensity is maximum in the horizontal directions and minimum in the vertical axis. For miniature lamps, the MSCP is usually given, while for higher wattage lamps it may have to be calculated.

EXAMPLE 10-2:

Calculate the light intensity of a 60W, 120V lamp if the lamp is operated at: (a) 120V; (b) 80V; and (c) 150V.

SOLUTION:

Assuming the 60W lamp is a soft white inside frosted lamp with an average total light output of 870 lumens:

(a) Operating at 120V and using Eq. (10-2)

$$I_L = \frac{F}{4\pi} = \frac{870 \text{ lumens}}{12.56} \cong 69.5 \text{ candles}$$

(b) Since 80V is 0.66 of 120V ($80V/120V = 0.66$), then, from Fig. 10-4a at 0.66, candlepower of the lamp has been reduced to 0.25 of that at 120V, or

$$I_L = (0.25)(69.5\text{cd}) = 17.3\text{cd}$$

(c) Operating at 150V is 150V/120 = 1.25 on the horizontal axis of Fig. 10-4a, thus producing a candlepower multiplier of approximately 2.15. The light intensity has increased to

$$I_L = (2.15)(69.5\text{cd}) \cong 150\text{cd}$$

Note that an increase of 30V (120V to 150V) causes the light intensity to more than double. The next example illustrates the drastic decrease in lifetime for such an increase in intensity.

EXAMPLE 10-3:

Average lifetime of a 60W lamp is 1,000 hours. Calculate the lifetime for each operating voltage of Example 10-2.

SOLUTION:

(a) 120V: Since manufacturer's rated voltage is 120V, then operating the lamp at 120V results in an average lifetime of 1000 hours.
(b) Operating at 80V yields 80V/120V = 0.66 as the horizontal intercept or rated value multiplier in Fig. 10-4b. At 0.66, lifetime multiplier is 200 and lifetime is increased to (200) (1,000 hours) = 200,000 hours.
(c) Horizontal intercept operating at 150V is 150V/120V = 1.25, and the corresponding lifetime multiplier is approximately 0.07. The lifetime is thus decreased to (0.07) (1,000 hours) = 70 hours.

Therefore, to increase light output of a lamp we must be willing to accept a greatly reduced lifetime. Conversely, accepting a low light output level significantly increases the life of a lamp. Remember, lumens is the unit for total light output, F, while candles is the unit for light intensity, I_L (lumens/4π).

10-6 ILLUMINATION

Illumination (E) is a measure of light intensity (I_L) falling on a surface of a distance d from the lamp. For a perpendicular distance between the lamp and surface (this surface may be a window of a photodetector, as illustrated in Chapter 11), the expression for illumination E is

$$E \text{ in footcandles} = \frac{I_L \text{ in candles}}{d^2} \qquad (10\text{-}3)$$

If I_L is measured in candles and d in feet, then E is in candles/foot^2, more commonly referred to as *footcandles*.

EXAMPLE 10-4:

For the lamp of Example 10-2 operating at rated value, calculate the intensity at distances of (a) 2 feet, (b) 5 feet, and (c) 10 feet.

SOLUTION:

At rated value 120V, $I_L = 69.5$cd.

(a) $$E = \frac{I_L}{d^2} = \frac{69.5\text{cd}}{(2\text{ft})^2} = 17.6\text{ft-cd}$$

(b) $$E = \frac{69.5\text{cd}}{(4\text{ft})^2} = 4.35\text{ft-cd}$$

(c) $$E = \frac{69.5\text{cd}}{(10\text{ft})^2} = 0.695\text{ft-cd}$$

Note that a lamp's intensity decreases according to the square of the distance. Figure 10-6 is a plot of intensity versus distance for a standard 60W, 870 lumen lamp.

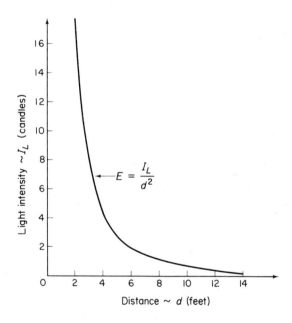

Figure 10-6

Plot of light intensity versus distance for Example 10-4.

10-7 COLOR TEMPERATURE

It is customary to give the temperature of a lamp in terms of color temperature (CT), which is *not* the actual temperature of the lamp's filament. CT is the temperature of an ideal "blackbody" radiator that produces the same visual color sensation as the lamp. CT is a photometric term because it is a visual effect. Although color temperature and filament temperature are not equal, they are usually within 100°K of one another. Because we will encounter the term color temperature in data sheets, in this book all lamp temperatures will be color temperatures.

For most standard lamps operating at rated value, color temperature is approximately 2,870°K. This is convenient, because most manufacturers of photodetectors give the device's output characteristics with a reference of 2,870°K. However, output current of a photodetector is extremely dependent on the color temperature of the light source. If the rated value of voltage is decreased to extend the life of a lamp, its color temperature decreases. This, in turn, will change a photodetector's characteristics.

Luminous Efficiency

Color temperature may be determined from Fig. 10-7, which is a plot of color temperature versus *luminous efficiency*, *p*. Luminous efficiency is the ratio of mean-spherical-candlepower (*MSCP*) to input power. *MSCP* is given by the

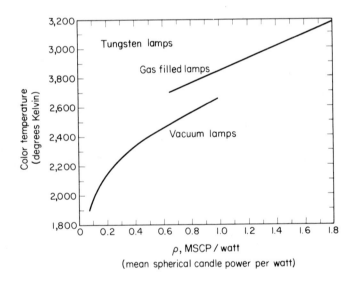

Figure 10-7

Plot of color temperature versus luminous efficiency. (Courtesy of General Electric Semiconductor Products Department, Syracuse, N. Y.)

manufacturer for miniature lamps, and may be calculated for standard lamps ($I_L = F/4\pi$). Figure 10-7 has two curves, a plot for vacuum lamps and another for gas-filled lamps. Lamps designed to operate at 5V or less, and less then 10W, may be considered vacuum lamps. Manufacturer's data will specify whether a lamp is vacuum or gas-filled.

EXAMPLE 10-5:

A No. 47 lamp is a general purpose miniature vacuum lamp with rated values of 6.3V, 0.15A, an average lifetime of 3000 hours, and a light intensity of 0.52cd. If the lamp is operated at 5V, determine: (a) light intensity; (b) lifetime; (c) illumination 3 inches from the lamp; and (d) color temperature.

SOLUTION:

(a) Operating at 5V results in the point along the horizontal axis in Fig. 10-4a of 5V/6.3V = 0.795, and a candlepower multiplier of 0.49. Therefore

$$I_L = (0.49)(0.52\text{cd}) = 0.253\text{cd}$$

(b) Fig. 10-4b at 0.795 yields a lifetime multiplier of 20, and the life of the lamp is increased to (20)(3,000 hours) = 60,000 hours.
(c) Converting distance to feet, $d = 3$ in $= 0.25$ft. Using Eq. (10-3)

$$E = \frac{I_L}{d^2} = \frac{0.253\text{cd}}{(0.25\text{ft})^2} \cong 4.04\text{ft-cd}$$

(d) From Fig. 10-4a at a rated value of 0.795, the current multiplier is (0.87) and the current through the lamp operating at 5V is (0.87)(0.15A) $\cong$ 0.13A. Input power $P_{in} = (5\text{V})(0.13\text{A}) = 0.65$W. Luminous efficiency is

$$\rho = \frac{MSCP}{P_{in}} = \frac{0.253\text{cd}}{0.65\text{W}} = 0.39\text{cd/W}$$

From Fig. 10-7, color temperature at 0.39cd/W is approximately 2,320°K.

10-8 PHOTOMETRIC AND RADIOMETRIC TERMINOLOGY

Terms of total light output, light intensity, and illumination, along with their units of lumens, candles, and footcandles, are all concerned with visual effects. Thus they are grouped under a general listing of *photometric terminology.* However, most light sources and all incandescent filament lamps also emit radiation outside the visible spectrum. Since silicon photodetectors are sensitive to radiation in both the visible and near infrared region, we must be able to determine the *total radiation* emitted from a source. In order to accomplish this, we must expand from the visible spectrum and photometric terminology to *radiometric* terms and units. Figure 10-8 is a plot of a response of the human eye and a silicon phototransistor (to be studied in Chapter 11) along with the radiation pattern of an incandescent lamp operating at 2,870°K. We can clearly see from Fig. 10-8 that if we limit ourselves to the visible spectrum, there is no accurate way of determining the effect on a phototransistor or any photodetector.

In photometric units, we started with total *light* output in units of lumens. In radiometric units we start with total *radiant* power or energy in watts. A tungsten filament is very efficient in converting electrical power to radiant power. For gas-filled lamps it is about 80 percent, while for vacuum lamps it is about 90 percent. This efficiency must *not* be confused with luminous efficiency. Although a tungsten filament lamp is very efficient in converting electrical power to radiant power, only between 5 to 10 percent of the radiant power falls in the visible spectrum. This point is illustrated in Fig. 10-8. The area under the tungsten-filament curve is the total radiant energy from a lamp operating at 2,870°K, but only the shaded portion is actually perceived by the human eye.

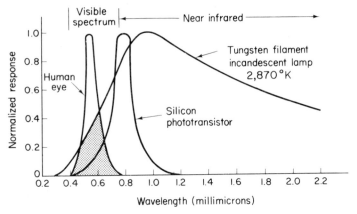

Figure 10-8

Comparison of response of a human eye, silicon phototransistor, and a tungsten filament lamp.

10-9 IRRADIANCE

Irradiance (H) is radiant energy striking a surface which is a fixed distance from the source. H is the radiometric equivalent to E_L, and may be approximated by

$$H = \frac{P_{out}}{4\pi d^2} \tag{10-4}$$

where P_{out} is radiant output power in milliwatts, d is the distance between source and detector in centimeters. H is then expressed in milliwatts per centimeter squared (mW/cm²).

EXAMPLE 10-5:

Calculate the irradiance 10cm from a No. 47 lamp operating at rated value.

SOLUTION:

From Example 10-4, rated values are 6.3V and 0.15A. Since a No. 47 lamp is a vacuum lamp, efficiency is approximately 90 percent. Then P_{out} = (0.9)(6.3V)(0.15A) $\cong$ 0.85W = 850mW. From Eq. (10-4)

$$H = \frac{P_{out}}{4\pi d^2} = \frac{850mW}{4(3.14)(10cm)^2} = 0.676mW/cm^2$$

Note P_{out} in watts is first converted to milliwatts before using Eq. (10-4).

In this book we will consider only *point sources*. That is, if the distance d between lamp and detector is approximately 10 times the diameter of the lamp, then the lamp is essentially a point source. Table 10.3 lists the relationships for a point source for both radiometric and photometric quantities.

Table 10.3

Radiometric and Photometric Terminology

Location	*Characteristic*	*Photometric*	*Radiometric*
At source	Total output	Total light output F in lumens	Total energy output P_{out} in watts
	Intensity	Lumination I_L in lumens/ steradian or candles $I_L = \dfrac{F}{4\pi}$	Radiance I_r in watts/steradian $I_r = \dfrac{P_{out}}{4\pi}$
At detector	Intensity	Illumination E in lumens/foot² or footcandles $E = \dfrac{I_L}{d^2} = \dfrac{F}{4\pi d^2}$	Irradiance H in mW/cm² $H = \dfrac{I_r}{d^2} = \dfrac{P_{out}}{4\pi d^2}$

10-10 GASEOUS DISCHARGE LAMPS

Incandescent lamps produce light as a result of heating a filament. *Discharge lamps* produce light as a result of recombinations between free electrons and ionized gas atoms. Figure 10-9 illustrates that the principle components of gas discharge lamps are a glass tube, rare gases, and a pair of electrodes.

In the normal state gases are insulators and do not contain a large number of free electrons. Gases may be made to conduct and emit light through positive ionization. A positive ion is an atom that has lost one or more electrons. The ionization process is accomplished by an electron striking a gas atom. If one electron with enough energy collides with a neutral gas atom, a second electron is freed, thus giving us two free electrons where before there was only one. Since each ionization frees an electron that can ionize another atom, free electrons and positive ions, and consequently current, increase very rapidly. To limit current so that the lamp is not destroyed, an external resistor must be connected in series with the lamp. This resistor is called a *ballasting* resistor, and the value depends on applied voltage and current.

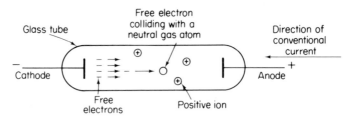

Figure 10-9

Principle components and operation of gaseous discharge lamps.

Note that current in gas discharge lamps is a result of both free electrons and ionized gas atoms. Free electrons move away from the cathode toward the anode. Since the mass of an atom is much greater than that of an electron, the ionized gas atoms move slower under the same potential difference. Therefore, current is practically a result only of the movement of free electrons.

Light is produced in gas discharge lamps when a free electron recombines with an ionized gas atom. Energy lost by the free electron is released in the form of light.

Incandescent lamps produce a continuous spectrum of radiation, from the ultraviolet to the infrared region, as illustrated in Fig. 10-8. Discharge lamps, however, emit radiation in discrete wavelengths, as shown in Fig. 10-10. The radiation spectrum of discharge lamps is dependent on the type of gas and pressure within the glass tube. Figure 10-10a is the radiation response for a low pressure (approximately 1 atmosphere), while Fig. 10-10b is the response for higher pressure lamps (2 to 10 atmospheres). As pressure increases the spectrum becomes more continuous, but still does not approach the spectrum of the incandescent lamp.

Some of the more commonly used discharge lamps are: (1) sodium lamps, (2) mercury lamps, (3) xenon lamps, (4) flashbulbs, (5) neon lamps, and (6) fluorescent lamps. The neon glow lamps are used in conjunction with electronic circuits, because neon lamps may be used as either a circuit element or an indicator, and are low in cost.

Neon glow lamps consist of a pair of electrodes placed close together, with the glass envelope filled with neon. This lamp exhibits a breakdown characteristic when sufficient voltage is applied between the electrodes. When break-

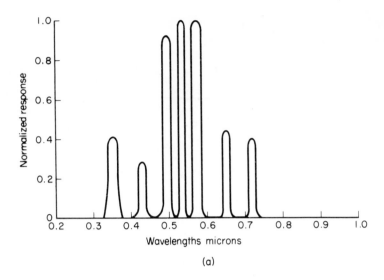

(a)

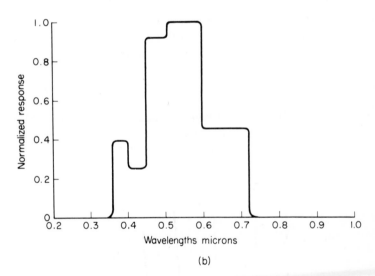

(b)

Figure 10-10

Spectral output for a gaseous discharge lamp (a) low pressure—1 atmosphere, (b) high pressure—2 to 10 atmospheres.

down occurs, the voltage across the neon lamp drops very quickly to a voltage called *maintaining voltage* (see Fig. 10-11). At this time, a glow surrounds the negative electrode. Thus the neon glow lamp has a breakdown characteristic which makes it useful as a triggering device (see section 7-15) and a glow characteristic which makes it useful as an indicator lamp.

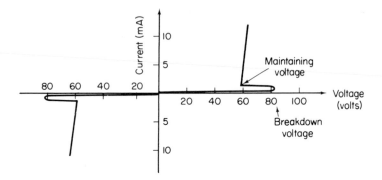

Figure 10-11

Current-voltage characteristic of a neon glow lamp.

10-11 EMITTING DIODES

An emitting diode is a solid state diode and, depending on the materials used in its construction, emits either visible or infrared light. A diode that emits radiation in the visible spectrum is classified as a *light emitting diode* or LED. A diode, emitting radiation in the infrared region is classified as an infrared emitting diode. Sometimes both types of diodes are referred to as LEDs, but we will maintain the fundamental definition of light as that radiation capable of being perceived by the human eye.

Emitting diodes have current-voltage characteristics similar to that of ordinary germanium or silicon diodes, except that the threshold voltage is higher ($\cong$ 1.0 to 1.6V—see Fig. 10-12). Materials used in the manufacture of emitting

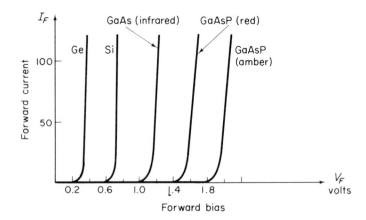

Figure 10-12

Forward characteristics of both ordinary diodes and emitting diodes.

diodes determine the peak wavelength of emitted light. For example, a gallium arsenide (GaAs) diode emits radiant energy in the infrared region ($\cong$ 9000Å). A gallium arsenide phosphide (GaAsP) LED emits light in the visible red or amber region, depending on the amount of phosphide.

The generation of light in emitting diodes is classified as *electroluminescence*. Remember that luminescence is the production of light by means other than high temperature. Emitting diodes generate light (visible or infrared) as a result of recombination of holes and electrons.

Filaments of incandescent lamps are encapsuled in an evacuated bulb; otherwise, the filament would be quickly consumed if exposed to air. Emitting diodes are encased only to protect their delicate wires, since the presence of air has no effect on such diodes.

Physics

The two outermost energy levels of a semiconductor are the valence band and the conduction band, between which is a region called the forbidden gap. Figure 10-13a illustrates a simplified two-dimensional model.

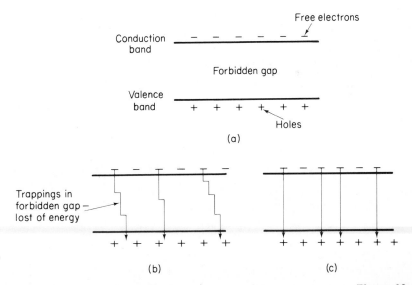

Figure 10-13

(a) Two-dimensional model illustrating energy levels of a diode. (b) Indirect gap diodes. (c) Direct gap diodes.

Forward-biasing a *pn* junction causes electrons from the *n* side to be injected into the *p* side. The injected electrons (which are in the conduction band) re-combine with holes in the *p* side (which are in the valence band). Recombination means that an electron falling from the conduction band fills a hole in the valence band. An electron that falls from a higher to a lower energy level gives up its excess energy in the form of heat and/or light. The degree to which either heat or light is emitted depends on whether the falling electron is temporarily trapped in the forbidden region.

In ordinary germanium or silicon, an electron will most probably be trapped one or more times in the forbidden gap. These trappings cause the electron's energy to be given up mostly in the form of heat, and only slightly in the form of light. Note that when a germanium or silicon diode is forward-biased, a small quantity of light in the infrared region is emitted. However, the efficiency of producing light in an ordinary diode is very small. Because of the many trappings, these diodes are called *indirect* gap semiconductor diodes.

In contrast with the *indirect* gap are the *direct* gap diodes, to which emitting diodes belong. In a direct gap diode an electron is not trapped in the forbidden gap but falls directly from the conduction band to the valence band. The excess energy of the electron is converted into light energy upon recombination. Production of light in this type of semiconductor is so efficient that it is the basis for solid state *lasers* as well as for emitting diodes.

Although the generation of light (visible and infrared) at the *pn* junction of an emitting diode is very efficient, only a small fraction actually reaches the surface, primarily because of *internal absorption* and *reabsorption*. Internal absorption is a result of the path of light and the transparency of the materials used in the diode. The longer the path, the greater the internal absorption, and thus less light reaches the surface. As light strikes the surface some of it is reflected back into the diode; this is reabsorption. To lessen the reabsorption problem, diodes are now manufactured with dome lenses.

10-12 SWEEP MEASUREMENT OF EMITTING DIODES

Connect a gallium arsenide phosphide LED (such as the MLED 50) in the sweep circuit of Fig. 10-14. Increasing the variac voltage increases diode voltage V_F. As the threshold voltage ($\cong$ 1.4V for GaAsP red LED) is reached, diode current I_F increases. At the threshold point, a reddish glow is emitted. As I_F is further increased the light becomes brighter. The brightness of the light becomes uncomfortable to view at close ranges if I_F is over 100mA. If I_F is over 200mA, the diode may be destroyed from heat. The 1kΩ load resistor and the 20:1 step-down transformer in Fig. 10-14 limit maximum current flow through the LED.

The circuit of Fig. 10-14 may also be used to obtain the forward characteristics of an infrared (GaAs) emitting diode. Of course a GaAs diode emits radiation in the infrared region, and therefore no visible light is seen.

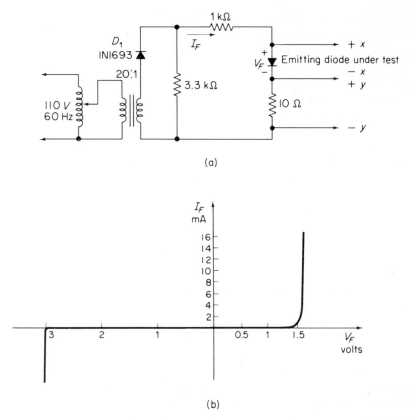

(a)

(b)

Figure 10-14

(a) Sweep circuit to obtain characteristics. (b) Forward and reverse
characteristics for a GaAsP red *LED*.

Diode D_1 may be reversed to obtain the reverse characteristics of the diode.
The reverse breakdown voltage of emitting diodes is approximately 3V. At the
breakdown voltage for visible LEDs, light is emitted. Operating emitting diodes
in the reverse condition will quickly destroy them.

10-13 DATA SHEETS AND THE LED

To choose the proper emitting diodes for a particular application, one or
more of the following must be considered: peak wavelength, input power requirements, output power, frequency response, efficiency, emitting region,
spectral response, turnon and turnoff time, mounting arrangement, and, for

visible LEDs, light intensity and brightness. On manufacturers' specification sheets the above information is given in one of four ways:

1. Maximum ratings—power and temperature.
2. Thermal characteristics.
3. Electro-optical characteristics.
4. Characteristic curves.

Not all manufacturers give the same information, nor does the same manufacturer give the same information for different diodes. Therefore, it often becomes confusing and discouraging when comparing manufacturers' data.

Maximum Ratings

Maximum ratings include reverse voltage, power dissipation, operating temperature range, continuous forward current, and peak forward current.

Reverse voltage, V_R, is the breakdown voltage and is approximately 3 to 4 volts for emitting diodes. This breakdown voltage is much lower than that of a germanium or silicon diode. Although light is emitted at the breakdown point, an emitting diode should *not* be operated in the reverse region because excess heat will quickly destroy it.

Power dissipation at 25°C ambient varies but is usually over 100mW. This rating is the maximum power that the device is capable of dissipating in an environment of 25°C (77°F). If the device is dissipating maximum power and the ambient temperature is 25°C, then the junction temperature is also at its maximum rating. The relationship between power dissipation and junction temperature is the same as that for other solid state devices.

$$T_J = T_A + \theta_T P_D \qquad\qquad (10\text{-}5)$$

where T_J is the junction temperature in °C, T_A is the ambient temperature in °C, θ_T is the thermal resistance, junction to ambient, in °C/W, and P_D is the power dissipated in watts.

EXAMPLE 10-6:
Motorola's MLED50 is a visible red light emitting diode. Its maximum ratings are:

$$V_R = 3V \qquad \text{reverse voltage}$$

$$I_F = 40\text{mA} \qquad \text{forward current—continuous}$$

$$P_D = 120\text{mW} \qquad \text{total dissipation at } T_A = 25°C$$

$$T_J = 85°C \qquad \text{maximum junction temperature}$$

$$\theta_{JA} = 500°C/W \qquad \text{thermal resistance, junction to ambient}$$

Calculate $P_{D\,max}$ from Eq. (10-5).

SOLUTION:

From Eq. (10-5)

$$P_{D\,max} = \frac{T_J - T_A}{\theta_T}$$

$$P_{D\,max} = \frac{85°C - 25°C}{500°C/W} = 120mW$$

This checks with the manufacturer's data, given in the above ratings and in Appendix 4.

Under operating conditions, power dissipated by the device is given by

$$P_D = V_F I_F \qquad (10\text{-}6)$$

where V_F is the forward voltage across the diode and I_F is the forward current through the diode. Maximum forward continuous current for the MLED50 is 40mA. Forward voltage at this value of current may be given or can be found from a diode curve similar to Fig. 10-14b. From the diode curve supplied by the manufacturer at $I_F = 40mA$, $V_F \cong 1.65V$

$$P_D = V_F I_F \cong (1.7V)(40mA) = 68mW$$

Therefore, operating the diode with the forward continuous current will not exceed the maximum junction temperature. With 68mW having to be dissipated, the junction temperature will rise to 59°C.

$$T_J = T_A + \theta_T P_D$$
$$T_J = 25°C + (500°C/W)(68mW) = 59°C$$

Theoretically, with only 68mW having to be dissipated, the ambient temperature could rise to

$$T_{A\,max} = T_{J\,max} - \theta_T P_D$$
$$T_{A\,max} = 85°C - (500°C/W)(68mW) = 51°C$$

before exceeding $T_{J\,max}$. Practically, however, it is undesirable to operate the device near $T_{J\,max}$ because:

1. The higher the junction temperature, the lower the output brightness.
2. The life of the diode decreases with an increase in temperature.
3. The peak wavelength increases as the junction temperature increases.
4. The forward voltage (V_F) was taken from the typical curves given by the manufacturer.

It could be possible, for the particular diode being used, that V_F is higher (by as much as 25 percent), and thus the junction temperature would be exceeded and the diode destroyed. If the device had to be operated in an environment of 51°C, a solution would be to lower the thermal resistance (θ_T) by heat-sinking the diode.

10-14 PULSED OPERATION OF EMITTING DIODES

At times it is desirable to obtain more light output than can be obtained from operating the LED under continuous conditions. This can be accomplished by applying high current pulses. Although these pulses do produce more light output, they must be of short duration (less than 10μs) or they will quickly destroy the diode. Not only do we not want to destroy the diode, we do not want to increase the junction temperature any more than is necessary. As stated before, an increase in junction temperature increases peak wavelength, decreases the light output, and shortens the life of the diode.

Figure 3 in Appendix 4 is a plot of brightness versus instantaneous current for a junction temperature of 25°C. Thus T_J is held constant, and either the ambient temperature (T_A) must be lowered, or the pulses be applied for such a short period of time that there is no noticeable increase in T_J (less than 5°C increase). To account for these pulses, Eq. (10-5) is modified to

$$T_J = T_A + \theta_T V_F I_F D \qquad (10\text{-}7)$$

where T_J is the junction temperature in °C, T_A is the ambient temperature in °C, θ_T is the total thermal resistance in °C/W, V_F is instantaneous forward voltage in volts, I_F is instantaneous forward current in amperes, and D is the duty cycle in which

$$\text{duty cycle} = \text{pulse width} \times \text{pulses per second, or} \qquad (10\text{-}8a)$$

$$\text{duty cycle} = \frac{\text{pulse width}}{\text{period}} \qquad (10\text{-}8b)$$

EXAMPLE 10-7:

The current waveform of Fig. 10-15 is applied to the MLED 50. Calculate T_J if $T_A = 25$°C.

SOLUTION:

From the forward characteristics of the MLED 50 in Fig. 2 of Appendix 4, when $I_F = 1$A, $V_F = 3.0$V. Using Eq. (10-8b)

$$D = \frac{5\mu s}{5ms} = 1 \times 10^{-3}$$

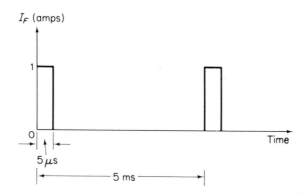

Figure 10-15

Waveform for Example 10-7.

Since $T_J = T_A + \theta_T V_F I_F D$, then

$$T_J = 25°C + (500°C/W)(3.0V)(1A)(1 \times 10^{-3})$$
$$T_J = 25°C + 1.5°C$$
$$T_J = 26.5°C$$

Note that, even in an ambient temperature of 25°C, it is possible with pulsed operation *not* to have a large increase in junction temperature. Therefore, with $T_J \cong 25°C$, we may use manufacturer's typical data with negligible error. Since frequency is the reciprocal of the period, the pulsed waveform has a frequency of

$$f = \frac{1}{5ms} = 200pps \text{ (pulses per second)}$$

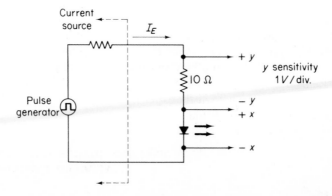

Figure 10-16

Sweep circuit to measure V_F for pulsed operation and to determine the minimum pulses needed per second for constant I_F brightness.

The reader should be aware that enough pulses per second must be applied so that the LED does not appear to blink. The best method for obtaining this lower frequency limit can be demonstrated experimentally. In the test circuit of Fig. 10-16, decrease the frequency of the pulse generator to find the lower pulse limit. At the same time a CRO can be used to obtain the forward I-V characteristics—up to 1A where V_F may be measured. Typical lower pulse limits are 8 to 30 pps.

10-15 BRIGHTNESS OF LEDS

Some manufacturers of visible LEDs give data in terms of *footlamberts*, which are the units for brightness. However, this data may easily be misunderstood, because we must distinguish between *physical* brightness and *apparent* brightness. Brightness is *not* light intensity (candles), discussed in section 10-5. Physical brightness is that which is measured, while apparent brightness is that which is seen. For example, a lit streetlight appears very bright at night, but is hardly visible during the day. Physical brightness (that which is measured) is the same in both cases because it depends only on the emission of light. Apparent brightness, on the other hand, depends on surrounding conditions. For this reason, apparent brightness is very subjective.

Those manufacturers which give data in units of footlamberts do so from a measured value. The measurement may be taken at the brightest spot on the LED chip and could easily yield readings of over 1,000 ft-L (footlamberts). Since the eye is inclined to see areas and surrounding conditions rather than a spot, these values become meaningless when the LED is viewed from a reasonable distance (1 foot or farther). Other manufacturers may give the brightness of the surface of the LED chip (approximately 200 ft-L). Without knowing anything else, we cannot conclude that one manufacturer's data of 1,000 ft-L is more informative than another's of 200 ft-L. Therefore, if brightness is the sole criterion for choosing an LED, do *not* rely on the manufacturer's measured data. The best test is to view different diodes with the same conditions under which the LED will be used—background conditions, room lighting, and power consumption. In short, buy it and try it. To obtain the same effect, one LED may require more current than another, and therefore more of a load on an entire system. Remember, however, that if more light is desired from any visible LED, pulsed operation may be used at the expense of having to supply a large value of current for a short period of time.

10-16 IRRADIANCE CALCULATION FOR INFRARED EMITTING DIODES

When using infrared emitting diodes, one primary concern is to determine irradiance at a known distance from the source. To make this calculation we will first have to solve for operating junction temperature T_J and duty cycle (if any). In addition, the manufacturer's data sheet (see Appendix 5) is needed to obtain: (a) forward characteristics; (2) radiant power output versus forward current; (3) radiant power output versus junction temperature; and (4) divergence angle at one-half power output. The following examples show how irradiance is calculated and will, be used in Chapter 11 as infrared light sources in examples for photodetectors.

The expression for irradiance, H (in mW/cm²), striking a detector is

$$H = \frac{4P_o}{\pi d^2 \theta^2} \qquad (10\text{-}9)$$

where P_o is *total radiant output*, d is distance in cm between emitting diode and detector, and θ in radians is the *divergence angle* at one-half radiant output power. θ is either given as a number in degrees or found from the spatial radiation pattern on data sheets.

EXAMPLE 10-8:

What is the irradiance striking a surface 1 cm, away from an MLED 930? The forward diode current I_F is 10mA and $T_A = 25°C$.

SOLUTION:

From Fig. 2 in Appendix 5 at $I_F = 10$mA, $V_F \cong 1.1$V. Therefore, power *dissipated* (not radiant output power) is

$$P_D = V_F I_F = (1.1\text{V})(10\text{mA}) = 11\text{mW}$$

From Eq. (10-5)

$$T_J = T_A + \theta_T P_D$$
$$T_J = 25°C + (400°C/W)(11\text{mW}) = 29.4°C$$

Radiant output power is found from Fig. 4 of Appendix 5 at $I_F = 10$mA, $P_o = 0.07$mW, but this is with $T_J = 25°C$. Radiant output power versus junction temperature is shown in Fig. 3 of Appendix 5 at $T_J = 29.4°C$; P_o is approximately 0.95 of that at $T_J = 25°C$. Then radiant output power for this example is

$$P_o = (0.95)(0.07\text{mW}) \cong 0.067\text{mW}$$

The divergence angle, θ, is found from Fig. 5 of Appendix 5 to be 30°. Converting to radians, 30° = 0.535rad. Using Eq. (10-9)

$$H = \frac{4(0.067\text{mW})}{(3.14)(1\text{cm})^2(0.535)^2} = 0.296\text{mW/cm}^2$$

EXAMPLE 10-9:

An MLED 930 is used under pulsed operation with a peak current of $I_F =$ 1A, pulses of $10\mu S$ for every 1ms, as in Fig. 11-26. Find irradiance H at a distance of 3 feet.

SOLUTION:

From Fig. 2 of Appendix 5, at $I_F = 1A$, $V_F \cong 1.6V$ and

$$P_D = V_F I_F = (1.6V)(1A) = 1.6W$$

Duty cycle is

$$D = \frac{\text{pulse width}}{\text{period}} = \frac{10\mu s}{1ms} = 0.01$$

Applying Eq. (10-7)

$$T_J = T_A + \theta_T P_D D$$
$$T_J = 25°C + (400°C/W)(1.6W)(0.01) = 31.4°C$$

Note that duty cycle has kept junction temperature down. At 1A, $P_o =$ 5.5mW from Fig. 4 of Appendix 5. Operating at 31.4°C, P_o is reduced to

$$P_o = (0.92)(5.5mw) \cong 5.1mW$$

where Fig. 3 of Appendix 5 was used to find 0.92. Since the same emitting diode is used as in Example 10-8, divergence angle is the same, $30° = 0.535$ rad. Converting feet to centimeters, 3ft = 91.5cm. From Eq. (10-9)

$$H = \frac{4(5.1mW)}{(3.14)(91.5cm)^2(0.535)^2} = 2.7\mu W/cm^2$$

10-17 GLOSSARY OF TERMS

Angstrom (Å)—a unit of measurement of a wavelength, equal to one ten-thousandth of a micron. Visible spectrum is from 4,000 to approximately 7,600 angstroms.

Brightness—Physical brightness is a measured quantity, commonly expressed in footlamberts. Apparent brightness is what the eye perceives and depends on surroundings.

Candle—unit of luminous intensity. A point source of one candlepower emits one lumen into a solid angle of one steradian. To put it another way, a point source of one candle power radiating uniformly in all directions emits 4π lumens.

Direct gap—electrons in a semiconductor material recombine with holes by falling directly from the conduction band to the valence band. Released energy is mostly in the form of light and not heat (basic principle of emitting diodes).

Electroluminescence—direct conversion of electrical energy into light.

Flux (F)—energy per unit time. For optics, number of photons passing through a unit area per unit time. Photometric unit—lumens; radiometric unit—watts.

Focal length—distance between a lens and the point where light rays passing through the lens converge.

Footcandle—unit of illumination. It is the illumination at a surface one foot from a one-candle power source.

Footlambert—unit of brightness. It is the uniform brightness of surface emitting or reflecting at the rate of one lumen per square foot. Footlambert equals $1/\pi$ candles per square foot.

Illumination—amount of light (visible radiation) striking a surface, expressed in footcandles. Illumination is a photometric quantity.

Incandescence—generation of light as a result of the temperature of a material.

Indirect gap—electrons in the conduction band of a semiconductor material do not recombine directly with holes in the valence band, but rather fall in steps from the conduction band to the valence band because of trapping levels between the bands.

Infrared light (IR)—radiation just above the visible spectrum (range from approximately 7,800Å to 30,000Å).

Irradiance—radiometric quantity, expressed in either watts or milliwatts per distance squared (W/cm^2, mW/cm^2). The distance is measured between light source and photo detector.

LASCR—light-activated-silicon-controlled-rectifier, a *pnpn* device in which light (actually radiant energy) is used to trigger the device.

Light—electromagnetic radiation of such a frequency or wavelength so as to be perceived by the human eye. Visible spectrum is approximately 4,000Å to 7,600Å.

Light emitting diode—a solid state device that produces visible light.

Lumen—unit of measurement for total light output of a source.

Luminescence—generation of light by any means other than incandescence.

Luminous intensity—total light output of a source per unit solid angle, lumens/4π = candles.

Micron (μ)—unit of length equal to one-millionth of a meter, $1\mu = 10^{-6}$m.

Millimicron (mμ)—unit of length equal to one-thousandth of a micron or 10^{-9} meters.

Optoelectronics—study of optical and electronic circuitry.

Opto-isolator—a single package containing both a light source and photo-detector.

Photoconductive cell—a two-terminal device whose resistance varies inversely to light intensity striking the photoconductive sensitive material.

Photodetector—any device that converts radiant energy into electric current.

Photodiode—a *pn* solid state device in which radiant energy striking the diode causes an increase in leakage current. Normal operation is for the diode to be reverse-biased.

Photon—a distinct quantity of energy.

Phototransistor—construction similar to a bipolar transistor, except a window permits radiant energy to strike the reverse-biased collector-base junction, thereby increasing leakage current. Unlike a photodiode, leakage current in a phototransistor is multiplied by the transistor's current gain.

Photovoltaic cell—generates an output voltage because of light striking the cell's surface. Unlike other photodetectors, this device needs no external power supply.

Radiant energy—Energy transmitted in the form of electromagnetic waves.

Solar cell—a photovoltaic cell that converts sunlight striking the cell directly into electrical energy.

Spatial response—radiation pattern of a light (visible or infrared) source.

Steradian—solid angle at the center of a sphere subtending an area on the surface. The area equals the square of the radius.

Tungsten filament lamp—emits light as a result of the temperature of the filament. Filament material is tungsten.

Visible spectrum—that portion of the electromagnetic spectrum that is visible to the human eye.

Wavelength (λ)—distance between two successive points in a wave which are of the same phase.

PROBLEMS

1. Express the following frequencies in wavelengths (in units of meters): (a) 100kHz; (b) 1GHz; (c) 3.3×10^{14}Hz. (Reference: Example 10-1.)
2. Express the frequencies of problem 1 in units of: (a) microns; (b) millimicrons; (c) Angstroms.
3. Convert the following wavelengths to frequency: (a) $0.8m\mu$; (b) 5500Å; (c) 0.6μ.

4. From data given in section 10-5, determine average light intensity for a 40W standard lamp.

5. Calculate the light intensity of a 40W standard lamp at (a) 120V (rated value), (b) 105V, and (c) 135V.

6. What is the average lifetime, for the 40W lamp, for each operating voltage of problem 5? (Reference: Example 10-3.)

7. Calculate the illumination of a 75W (soft white) lamp operating at (a) rated value, (b) 80 percent of rated value. The distance is 3 feet.

8. A No. 756 indicator lamp has rated values of 14V and 0.08A, a light intensity of 0.31 cd, and an average lifetime of 15,000 hours. If the lamp is operated at 10V determine: (a) light intensity; (b) lifetime; (c) illumination 2 feet from the lamp; and (d) color temperature. The 756 is a vacuum filled lamp. (Reference: Example 10-5.)

9. A general purpose miniature lamp must deliver 3 ft-cd at a distance of 4 inches and be able to last at least 4,000 hours. Will a No. 47 lamp meet these specifications? Characteristics for a No. 47 lamp are given in Example 10-5.

10. Calculate the irradiance for the conditions of problem 9.

11. Using the No. 756 lamp of problem 8, calculate the irradiance 2 feet from the lamp. Note that the lamp is operated at 10V.

12. For an MLED 50 calculate: (a) power dissipated by the device at an $I_F = 20$mA; (b) junction temperature for the operating point $I_F = 20$mA, $T_A = 25°C$. Typical data for the MLED 50 are given in Appendix 4.

13. An MLED 50 is to be used for pulsed operation $I_F = 0.5$A. What duty cycle is necessary to keep $T_J = 25°C$ if $T_A = 23°C$? Typical data for the MLED 50 are given in Appendix 4.

14. An MLED 50 is used with pulsed operation $I_F = 1.2$A and $D = 5 \times 10^{-3}$. Determine: (a) power dissipated, P_D; (b) junction temperature if $T_A = 25°C$; (c) decrease in brightness.

15. Repeat Example 10-8 for a distance of 2cm.

16. What is the irradiance striking a surface 3 inches away from an MLED 930, if $I_F = 400$mA, $T_A = 25°C$, and $D = 0.025$?

17. Repeat Example 10-9 with pulses of $1\mu s$ every $50\mu s$.

18. The irradiance 1 foot from an MLED 930 is $50\mu W/cm^2$. Calculate P_o. If $T_J \cong 35°C$, determine: (a) I_F; (b) V_F; (c) D, $T_A = 25°C$.

chapter eleven
photodetectors

11-0 INTRODUCTION

To complete an introduction to the study of optoelectronics, and to complement Chapter 10 on light sources, this chapter deals with devices that convert radiant energy into an electrical output. We will concern ourselves mainly with five devices: (1) photoconductive cells, (2) photovoltaic cells, (3) photodiodes, (4) phototransistors, and (5) light-activated-silicon-controlled-rectifiers (LASCR). Photodetectors may be classified either by their differences in construction or by their differences in mode of operation. In construction, photodetectors are grouped as either *bulk-* or *junction*-type devices. Bulk photodetectors have *no pn junction*. They are made from a single layer of photosensitive material, the resistance of which changes as light level changes. This device displays properties similar to that of a thermistor, studied in Chapter 1, except that heat is replaced by light. Photoconductive cells are bulk-type devices.

As the name implies, junction-type devices are constructed to have a *pn* junction. Photovoltaic cells, photodiodes, phototransistors, and LASCRs are examples of junction-type devices.

Classifying photodetectors according to their mode of operation, we have *photogeneration* (or photovoltaic) and *photoconduction* types. Photogeneration devices do not require any external bias supply because they generate their own voltage across a *pn* junction. Photovoltaic cells are such devices. Photoconductive devices require an external bias for operation and include photoconductive cells, photodiodes, phototransistors, and LASCRs. Basic operation of *all* photodetectors involves radiant energy of the proper wavelength striking the device and creating hole-electron pairs. For junction-type photodetectors with an external bias, these hole-electron pairs are created in the vicinity of a *reverse*-biased *pn* junction. This causes an increase in *minority* carriers, resulting in an increase in minority current. In phototransistors this current is multiplied by the transistor's current gain, β, and because of the gain, phototransistors are more sensitive than photodiodes. For LASCRs this increase in minority current is the triggering mechanism referred to in Chapter 8. A summary of photodetector classification is shown in Fig. 11-1.

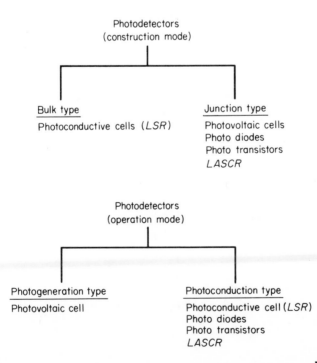

Figure 11-1

Two methods of classifying photodetectors.

11-1 PHOTOCONDUCTIVE CELLS

When enough energy is applied to a semiconductor material, valence electrons can escape from their parent atoms, thereby creating hole-electron pairs. For thermistors, energy in the form of heat is required, while photoconductive cells need light energy. The amount of energy needed depends on the photosensitive material. Most photoconductive cells are either Cadmium Sulfide (CdS)

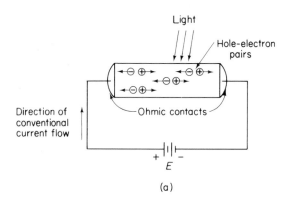

(a)

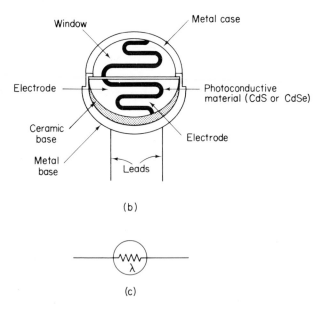

(b)

(c)

Figure 11-2

(a) Two-dimensional model, (b) structure, (c) circuit symbol for a photoconductive cell.

or Cadmium Selenide (CdSe), whose spectral response occurs between 4,000 and 10,000Å. Since light sources (sunlight, incandescent lamps, fluorescent lamps, neon glow lamps) all emit radiation in this band, hole-electron pairs will be created within the cell. The more intense the light, the more energy strikes the cell, and the more hole-electron pairs are created, thus decreasing the cell's resistance. The simplified two-dimensional model in Fig. 11-2a illustrates that an external power supply is necessary to generate a direction and provide a path for current to flow. The value of E varies from a few volts to several hundred volts, depending on the photocell and application. CdS cells are used in light measuring applications, while CdSe cells are used in counting applications. Both may be used as on-off switches.

Figure 11-2b illustrates the structure of a typical photoconductive cell. These devices are manufactured by depositing the photoconductive material (either CdS or CdSe) and electrodes on a ceramic base. Modern technology has allowed electrodes to be bonded very close to the photosensitive material, thereby increasing the cell's overall sensitivity. At the surface of the cell the photosensitive material separates the two electrodes. A lead is brought out to each electrode and then, with a window placed over the sensitive area, the entire cell is enclosed in a metal case.

Advantages of photoconductive cells over other photodetectors are: (1) high sensitivity, (2) low cost, (3) ease of handling, and (4) high dark-to-light resistance ratios exceeding 100:1. Disadvantages include: (1) slow response, (2) narrow spectral response (possible advantage in some applications), and (3) light history effects.

Spectral Response

Spectral response of a photoconductive cell, as with any photodetector, depends on the material used. Figure 11-3 illustrates a typical response of two Clairex cells—CL905L and CL5M3. From Fig. 11-3a we see that a CL905L CdS cell has a peak wavelength at 5,500 Angstroms, thereby closely matching the spectral response of the human eye. This cell may be used in light measuring applications or as a photodetector for incandescent, fluorescent, or neon light sources. Figure 11-3b is the spectral response of a CL5M3 CdSe cell whose peak wavelength is at 7,350 Angstroms. This cell is most sensitive to near infrared and is a suitable photodetector for incandescent and neon light sources.

The vertical axis of Fig. 11-3 is plotted as *relative sensitivity*. This indicates the cell's relative resistance, at any particular wavelength, to the cell's maximum resistance at the peak wavelength. For example, from Fig. 11-3b, the sensitivity at 8,000Å is 0.25 of that at 7,350Å, and therefore R (at 8,000Å) $= 0.25R$ (at 7,350Å).

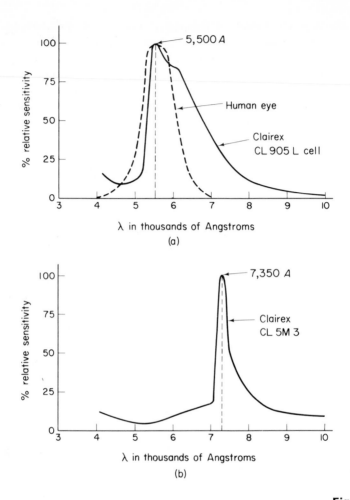

Figure 11-3

Spectral response of a cadmium sulfide in (a) and a cadmium selenide in (b).

Light History Effect

Light history effect (also known as light memory, hysteresis, or light history phenomena) is the expression that describes what happens when a photoconductive cell is suddenly subjected to a different light level. The resistance of a photoconductive cell is also a function of the light level to which the cell was previously subjected. For example, if a cell is kept at some particular light level (designated reference level in Fig. 11-4), the resistance of the cell will stabilize. Now let us keep the cell at a higher light level and then suddenly subject it to the reference light level. As shown by curve *A* in Fig. 11-4, resistance will aim for the previous reference light level but will overshoot and then approach the reference value asymptotically. Conversely, keeping a cell at a lower light level and then suddenly subjecting it to reference light level results in the resistance undershooting the reference light level to rise asymptotically, as shown by curve *B* in Fig. 11-4. The greater the difference between two light levels, the greater the overshoot or undershoot error and the longer the time needed for resistance to stabilize.

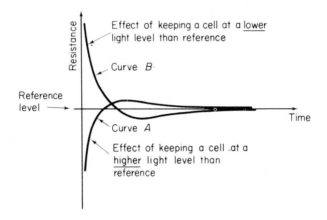

Figure 11-4

Light history effect for a typical photoconductive cell.

Light history effect may be minimized by not subjecting photoconductive cells to large variations in light levels. CdSe cells stabilize faster, although the error at first is greater than with CdS cells.

11-2 RESISTANCE MEASUREMENT OF PHOTOCONDUCTIVE CELLS

The conventional method of depicting a photoconductive cell's characteristics is a plot of resistance versus illumination (units are footcandles), and *not* the radiometric quantity of irradiance. The characteristics of other photodetectors introduced in following sections will be plotted with respect to irradiance (mW/cm^2).

To obtain a resistance curve, an ordinary household incandescent lamp may be used. For our applications and for most normal work, an ordinary incandescent lamp used at rated voltage operates at approximately $2,870°K$. This is the color temperature commonly employed by manufacturers for their measurements. For the curve plotted in Fig. 11-5, a standard 75W inside frosted lamp was used, with an *average* light output of 1,180 lumens. (Average light output is marked on the package of all lamps presently manufactured in the United States.) Applying Eq. (10-2), light intensity (or candlepower) is

$$I_L = \frac{F}{4\pi} = \frac{1180 \text{ lumens}}{12.56} = 94\text{cd}$$

and from Eq. (10-3), illumination E is

$$E \text{ in footcandles} = \frac{\text{candlepower}}{d^2} = \frac{94\text{cd}}{d^2}$$

where d is the distance in feet between the lamp and the photocell. These formulas are derived for a point source, so we should try to keep the distance d approximately 10 times the diameter of the bulb, although even 6 times the diameter does not produce significant error. For a 75W lamp, the minimum distance for d is approximately 1 foot to produce $94\text{cd}/(1\text{ft})^2 = 94$ footcandles of illumination at the surface of the photocell.

Voltage E in Fig. 11-5 should be set equal to the voltage in the circuit in which the photocell is to be used or, in general, at a test voltage used by the manufacturer. For photocell CL905L, $E = 10$V, and for a CL5M3, $E = 12$V. Ratio of E and current reading I read from the ammeter is the resistance of the

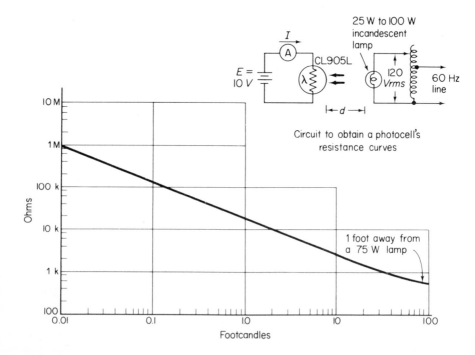

Circuit to obtain a photocell's resistance curves

Figure 11-5

Plot of resistance versus illumination for a CL905L photocell.

cell at a specific value of illumination. By varying d we vary illumination and thus resistance. Since a CL905L photocell is also sensitive to fluorescent light (normally used to light laboratories), a laboratory should be dark in order to minimize error. However, if a photocell is to be used in a well-lighted environment, it should be tested under such conditions. Otherwise, more current will flow in the system, which could either produce erroneous results or damage other components.

The auto transformer in Fig. 11-5 insures that the lamp is operated at rated voltage, near a color temperature of 2,870°K. Log-log graph paper is necessary to plot a photocell's resistance curve because of large variations in both resistance and illumination. Consider 0.01 footcandles as the cell's dark resistance (no light striking the cell's surface). This condition may be established by placing a *black* covering over the cell. For comparison, the sun gives an illumination of about 10,000 footcandles at noon on a clear day.

The plot in Fig. 11-5 shows that a CL905L photocell decreases almost linearly with an increase in illumination. Thus the photoconductive cell is also called a *light-sensitive-resistor* (LSR). After each large variation in illumination we must allow enough time (5 to 10 seconds) for the cell's resistance to stabilize, in order to eliminate light history effects.

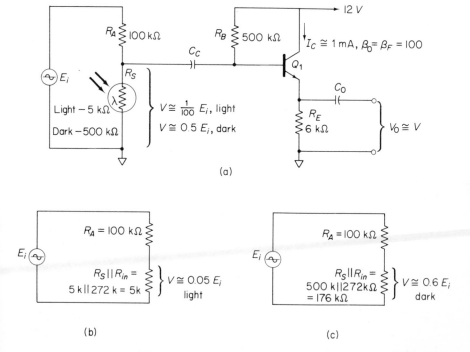

(a)

(b)

(c)

Figure 11-6

Light control of audio level in (a) is shown for light and dark conditions in (b) and (c), respectively.

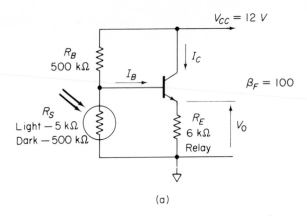

(a)

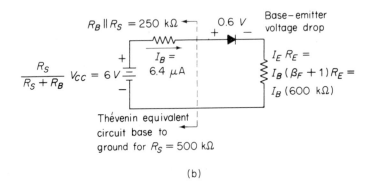

(b)

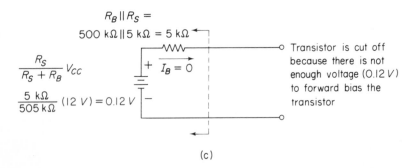

(c)

Figure 11-7

(a) Oil burner control; (b) dc biasing circuit for dark conditions; (c) dc biasing circuit for light conditions.

11-3 APPLICATIONS OF PHOTOCONDUCTIVE CELLS

Volume Control

In Fig. 11-6 resistor R_A and LSR R_S form a volume control that is controlled by light level. With full illumination R_S assumes a value of about 5kΩ and is in parallel with $R_B = 500$kΩ and $R_i = (\beta_o + 1)R_E = 600$kΩ, or $R_{in} = R_B \,//\, R_L = 272$kΩ. E_i divides between $R_A = 100$kΩ and this parallel combination of about 5kΩ in Fig. 11-6b. Input voltage V to the emitter follower is

$$V = \frac{(R_s \,//\, R_{in})E_i}{R_A + R_S \,//\, R_{in}} = \frac{5\text{k}\Omega}{105\text{k}\Omega} \times E_i = 0.05E_i$$

With no illumination, as in Fig. 11-6c, E_i divides between $R_A = 100$kΩ and the parallel combination of $R_{in} = 272$kΩ and dark resistance of $R_S = 500$kΩ. Since $R_S \,//\, R_{in} = 176$kΩ, $V = 0.6E_i$. The emitter follower allows V_o to equal V, so there is a 12 to 1 change in volume for an illumination change from light to dark.

On-Off Control

The light actuated on-off control in Fig. 11-7a has the photoconductive cell connected in the bias network between base and ground. This circuit functions as an on-off control for an oil burner. If the flame goes out, a dark condition exists and cell resistance is 500kΩ. To determine base current we need the Thévenin equivalent circuit, base-to-ground, in Fig. 11-7b, from which

$$I_B = \frac{6\text{V} - 0.6\text{V}}{250\text{k}\Omega + 600\text{k}\Omega} = \frac{5.4\text{V}}{850\text{k}\Omega} \simeq 6.4\mu\text{A}$$

$$I_C = \beta_F I_B = (100)(6.4\mu\text{A}) = 0.64\text{mA}$$

The relay must be chosen so that 0.64mA is sufficient to close it, and it must be connected so that, with its contacts closed, the oil pump is shut down and a signal is activated to indicate ignition failure. As long as the flame is burning a light condition exists and cell resistance is approximately 5kΩ. Figure 11-7c shows that the Thévenin equivalent voltage is only 0.12V, which is *not* sufficient to forward-bias the transistor. This causes the transistor to be cut off, and since $I_C = 0$, the relay is open and the system is operating normally.

Light Meter

The basic light meter of Fig. 11-8a uses a 1.5V battery, 0-1 dc milliamme-ter, and a photocell whose characteristics are shown in Fig. 11-5. If we consider the photocell's characteristics as linear from 0.01 to 20 footcandles, we can calibrate the milliameter in footcandles. For example, if the meter has a full-scale deflection, then 1mA flows and the resistance in the circuit is 1.5V/1mA = 1.5kΩ. From Fig. 11-5 at 1.5kΩ, light level is approximately 20 footcandles. If the meter indicated 0.1mA, then the resistance is 1.5V/0.1mA = 15kΩ, and light level is approximately 1.5 footcandles. With more calculations a face plate similar to that of Fig. 11-8b is obtained. By removing the milliammeter and in its place inserting a 0-100 microammeter, we will have a light meter sensitive to the range between total darkness and 1.5 footcandles. The face plate of the microammeter may be calibrated similar to that of Fig. 11-8b.

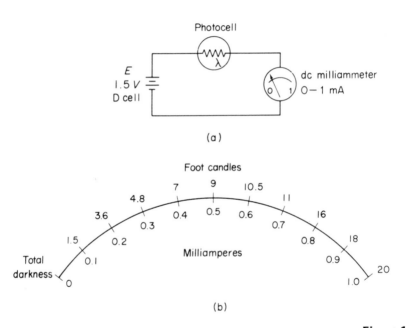

(a)

(b)

Figure 11-8

(a) Basic light meter. (b) Calibrated face plate for the milliammeter.

11-4 INTRODUCTION TO PHOTOVOLTAIC CELLS

Generation of voltage due to radiant energy striking a device is known as a *photovoltaic effect*, and such devices are referred to as photovoltaic cells. These devices have a *pn* junction, but, unlike photoconductive cells, require no external power supply. Two of the more common materials found in photovoltaic devices are selenium and silicon. A selenium cell closely approximates the spectral response of the human eye (see Fig. 11-10a) and is used in such applications as automatic exposure devices in cameras, roadside flashers, and lighting controls. Silicon cells have a peak spectral response in the near (or short) in-

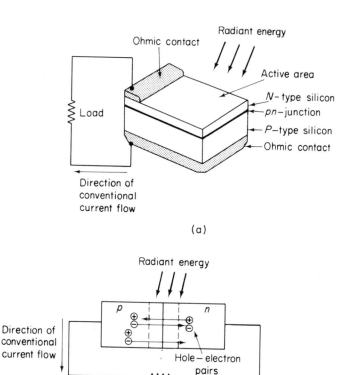

Figure 11-9

(a) Structure of photovoltaic cell; (b) simplified two-dimensional model; (c) schematic symbol.

frared region, with applications as power supplies for space vehicles, transistor radios, and for detection of infrared rays.

In Fig. 1-8b and sections 1-5 and 1-6 we saw that, at a *pn* junction, a space charge region is developed. If no external bias is connected, the locked ions that cause the space charge region affect the flow of both majority and minority carriers. This region prevents majority carriers (holes for *p* material and electrons for *n* material) from crossing the junction. Minority carriers (electrons for *p* material and holes for *n* material) that diffuse into the space charge region are accelerated and swept across the junction. Therefore, a minority current will flow if we are able to create minority carriers near the space charge region. Photovoltaic cells are designed so that minority carriers will be created by radiant energy striking the junction.

Figure 11-9 illustrates a typical photovoltaic cell composed of a thin layer of *n*-type silicon diffused onto a *p*-type silicon substrate. Radiant energy of the proper wavelength striking a valence electron creates a hole-electron pair. If a hole-electron pair is created on the *n*-side, the electron remains as a majority carrier while the hole may diffuse into the space charge region and be swept across the junction. Similarly, if a hole-electron pair is created on the *p*-side, the hole remains as a majority carrier while the electron may diffuse into the space charge region and be swept across the junction. This movement of minority carriers constitutes an electric current. The value of current depends on the number of hole-electron pairs created, which in turn depends on the amount of radiant energy striking the cell.

Spectral Response

Figure 11-10a is a plot of the spectral response of silicon and selenium cells, along with the response of the human eye. The response range of selenium cells is from 2,500 to 7,500 Angstroms and peaks at 5,500Å. Since selenium cells peak at the same wavelength as that of the human eye, these cells are used primarily in photographic applications.

The approximate response range of silicon cells is from 3,500 to 11,500Å, with a peak response at 8,300Å. Therefore, silicon cells are used primarily with light sources that peak in the near infrared range. Figures 11-10b, c, d, and e compare the response of silicon cells with radiation patterns of other common light sources. Note that either a tungsten lamp or an infrared emitting diode paired with a silicon cell is an efficient, reliable light source and detector. An emitting diode and silicon cell has the advantage of being an all solid state system with inherent long life and ruggedness.

Figure 11-10b shows that peak wavelengths of sunlight and silicon cells do not coincide. If the response of a selenium cell is superimposed on this graph, its peak wavelength would be close to the peak wavelength of sunlight. However, silicon and not selenium cells are most often used to convert sunlight into electrical emergy in power supplies for space vehicles. The reason is that silicon cells are approximately 20 times more efficient than selenium cells. Photovoltaic cells that are used primarily to convert sunlight into electrical energy are called *solar cells.*

11-5 MEASURING CHARACTERISTICS OF PHOTOVOLTAIC CELLS

Characteristic Curves

Figure 11-11 illustrates a test circuit to measure voltage-current characteristics of a silicon photovoltaic cell for different values of irradiance (mW/cm²). With no radiant energy striking a cell, the dark curve is similar to that of any *pn*

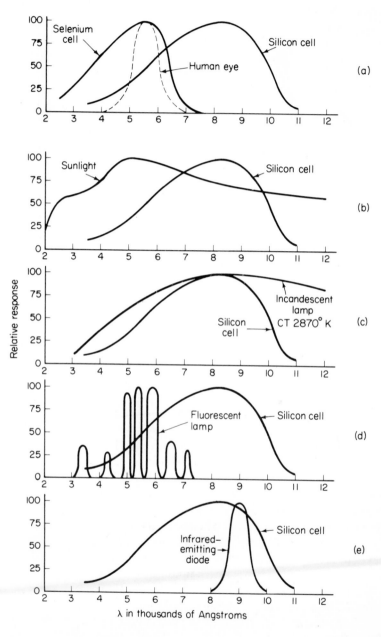

Figure 11-10

Spectral response of silicon photovoltaic cell and different light sources: (b) sunlight, (c) tungsten lamp 2850 K, (d) fluorescent lamp, (e) infrared-emitting diode.

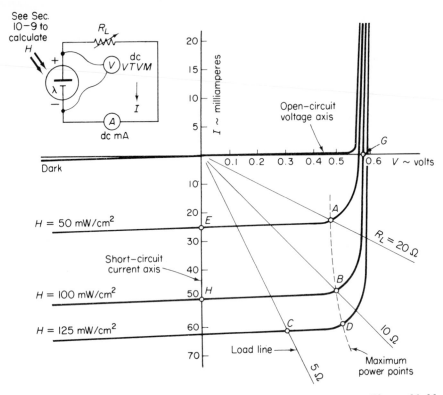

Figure 11-11

Typical current-voltage characteristics for a silicon photovoltaic cell.

junction. That is, current flows easily when forward-biased (low forward resistance) and blocks current flow when reverse-biased (high reverse-resistance). When radiant energy strikes a cell, the self-generated current produces curves in the fourth quadrant. The fourth quadrant is created by positive voltage $(+V)$ and current exiting from the positive terminal, or negative current $(-I)$. Since power is the product of voltage and current, then in the fourth quadrant power is $(+V)(-I) = -P$. The negative sign indicates that device does *not* dissipate energy, but instead *delivers* energy. Thus, when radiant energy strikes a photovoltaic cell, electrical energy is the resultant output.

The characteristic curves in Fig. 11-11 are expressed in terms of radiant intensity H, with radiometric unit mW/cm^2 (unit for irradiance). Sometimes we may find these curves expressed in terms of the photometric unit footcandle (unit for illumination).

Open-Circuit Voltage

The circuit of Fig. 11-12 measures open-circuit voltage of a photovoltaic cell. Under open-circuit conditions, total current I from the cell must be zero.

To maintain a total current equal to zero, the minority current generated as a result of radiant energy striking the cell must be offset by a current produced by power supply E. The power supply may be used to obtain zero current. A null meter allows quick indication of zero current, and if it is not zero, needle direc-

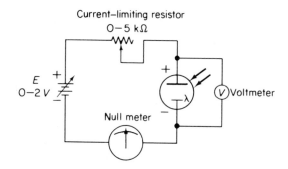

Current–limiting resistor
0–5 kΩ

E
0–2 V

Null meter

Voltmeter

Figure 11-12

Circuit to measure open-circuit voltage.

tion will indicate whether the cell's generated current is greater than that produced by the power supply, or vice versa. When zero current occurs, open-circuit voltage is read from voltmeter V. If radiant energy level is changed, the circuit of Fig. 11-12 allows us to find the new open-circuit voltage. A plot similar to that of Fig. 11-13a may be obtained. Open-circuit voltage for a photovoltaic cell is also refered to as *photovoltaic potential.*

Short-Circuit Current

When both external leads of a photovoltaic cell are connected together, a current flows. This short-circuit current is directly proportional to the hole-electron pairs, which, in turn, are directly proportional to radiant energy striking the cell. This linear relationship is illustrated in Fig. 11-13b, short-circuit versus irradiance.

Output Power

If a load resistor, R_L, is connected between the cells' external leads, as in Fig. 11-11, output voltage and current can be read from the volt-ampere characteristics. The operating point is at the intersection of a load line and I-V curve at a particular illumination. The load line is drawn from the origin through the I-V curve with a slope of

$$\frac{V}{I} = -R_L \qquad (11\text{-}1)$$

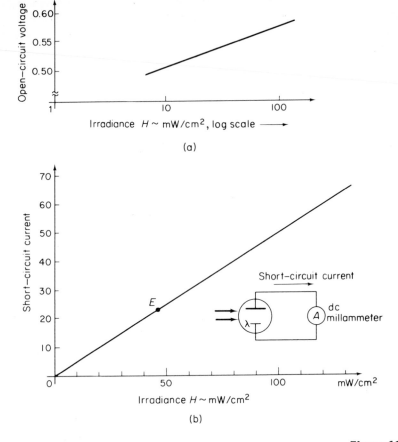

Figure 11-13

(a) Plot of open-circuit voltage versus irradiance; (b) plot of short-circuit current versus irradiance.

The minus sign in Eq. (11-1) only indicates a negative slope. For the three different loads in Fig. 11-11

(a) $$\frac{-V}{I} = R_L = \frac{-(0 - 0.46)V}{(23 - 0)mA} = 20\Omega$$

(b) $$\frac{-V}{I} = R_L = \frac{-(0 - 0.47)V}{(47 - 0)mA} = 10\Omega$$

(c) $$\frac{-V}{I} = R_L = \frac{-(0 - 0.3)V}{(62 - 0)mA} = 5\Omega$$

Output power is the product of operating point voltage and current, $P_o = VI$. For a 10Ω value of load resistance, output power is a maximum at point B

in Fig. 11-11. Note that at the two operating points of open-circuit voltage (G) and short-circuit current (H) output power is zero.

EXAMPLE 11-1:

The photovoltaic cell of Fig. 11-11 is exposed to irradiance of 125mW/cm². What power is delivered to loads of (a) 20Ω and (b) 5Ω?

SOLUTION:

(a) At the intersection of $R_L = 20\Omega$ and curve $H = 125\text{mW/cm}^2$, read $V = 0.58\text{V}$, $I = 30\text{mA}$ for a power of $P_o = 0.58\text{V}(30\text{mA}) = 17.4\text{mW}$. (b) At point C, $P_o = (0.3\text{V})(62\text{mA}) = 18.6\text{mW}$.

EXAMPLE 11-2:

Evaluate (a) maximum power output available at an irradiance of $H = 125\text{mW/cm}^2$, and (b) required load R_L.

SOLUTION:

(a) At point D in Fig. 11-11, read $V = 0.5\text{V}$ and $I = 62\text{mA}$ so that $P_o = 0.5\text{V}(62\text{mA}) = 31\text{mW}$. (b) $R_L = V/I = 0.5/0.062 \cong 8\Omega$.

Efficiency

Efficiency of a photovoltaic cell is the ratio of electrical output power to radiant input power. In percent

$$\eta = \frac{P_o}{H \cdot A} \times 100 \tag{11-2}$$

where η is the symbol for efficiency, P_o is electrical output power, H is irradiance (radiant energy striking the cell $\approx$ mW/cm²), and A is the active area of the cell in cm². For the same value of irradiance, maximum efficiency occurs at the maximum power point.

EXAMPLE 11-3:

Calculate cell efficiency at points A and C on Fig. 11-11. Dimensions of the cell are 1cm by 2cm.

SOLUTION:

Area of the cell is $A = 1\text{cm} \times 2\text{cm} = 2\text{cm}^2$. Point A: From Fig. 11-11, $P_o = (23\text{mA})(0.46\text{V}) = 10.6\text{mW}$ and $H = 50\text{mW/cm}^2$. Applying Eq. (11-2)

$$\eta = \frac{10.6\text{mW} \times 100}{(50\text{mW/cm}^2)(2\text{cm}^2)} = 10.6\%$$

Point C: From Fig. 11-11, $P_o = (61mA)(0.3V) = 18.3mW$ and $H = 125mW/cm^2$. Applying Eq. (11-2)

$$\eta = \frac{(18.3mW)(100)}{(125mW/cm^2)(2cm^2)} \cong 7.3\%$$

11-6 APPLICATIONS OF PHOTOVOLTAIC CELLS

Irradiance Meter

A solar cell, with characteristics similar to those in Fig. 11-11, drives a low-resistance (less than 1Ω meter movement) 0-100mA milliameter to make a sunlight (radiant) intensity (H) meter. Since the meter is essentially a short circuit, the relation between meter current and irradiance H is given by Fig. 11-13b.

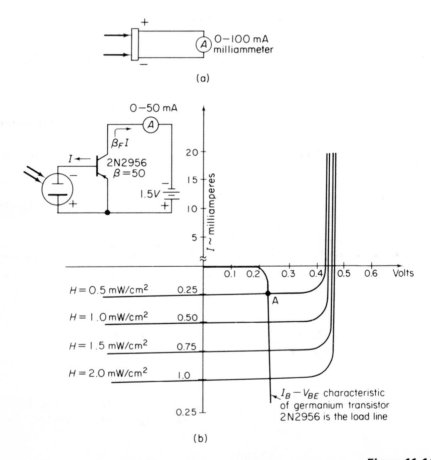

(a)

(b)

Figure 11-14

Sensitivity of the basic sun intensity lightmeter in (a) is increased by the transistor in (b).

EXAMPLE 11-4:

If the meter of Fig. 11-14a reads 25mA, what value of irradiance H should be calibrated at this point?

SOLUTION:

Refer to point E on either Figs. 11-13b or 11-11 to read $H = 50\text{mW/cm}^2$.

At low light intensity, characteristics of a typical photovoltaic cell are shown in Fig. 11-14b. To increase sensitivity, a germanium transistor is added to multiply the lower cell current that enters its base by β.

EXAMPLE 11-5:

The photovoltaic cell in Fig. 11-14b is exposed to an irradiance of 0.5mW/cm^2. What is the meter reading?

SOLUTION:

The circuit operating point is located by A at the intersection of load line and $H = 0.5\text{mW/cm}^2$ characteristic in Fig. 11-14b. Read base current $I = 0.25\text{mA}$ and meter current is $0.25\text{mA} \times 50 = 12.5\text{mA}$.

Frequency Doubler

A lamp-photovoltaic cell combination in Fig. 11-15a gives a frequency doubler circuit that eliminates tuning elements. Since light output is propor-

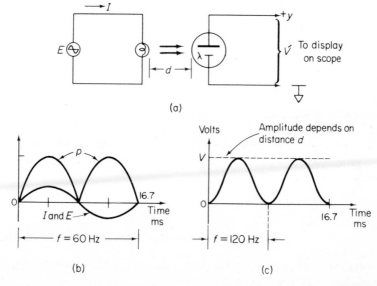

(a)

(b)   (c)

Figure 11-15

(a) Frequency doubler circuit, (b) lamp current, voltage, and power, (c) output voltage waveform of photovoltaic cell.

tional to *instantaneous power p* in Fig. 11-15b, instantaneous power has twice the frequency of applied voltage E, and photovoltaic cell output voltage V is at twice the frequency of E, as in Fig. 11-15c.

Solar Cell Power Supply

To obtain more output power, cells may be connected in series and/or in parallel. Series connection increases output voltage, but current is set by the cell with the lowest output current. Parallel connection increases available output current, but since output voltage is determined by the lowest voltage cell, cells with matched characteristics should be used.

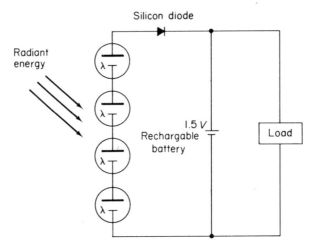

Figure 11-16

A simplified circuit to illustrate the principle of a solar battery charger.

A satellite power supply is illustrated in Fig. 11-16. Four series cells are selected to generate about 2.2V in sunlight. The voltage forward-biases the diode, connecting energy from the cells to recharge the battery. When dark, cell voltage drops below battery voltage, reverse-biasing the diode, and the load is fed only by the battery.

11-7 INTRODUCTION TO THE PHOTODIODE

Photodiodes, like photovoltaic cells, are a *junction-type* photodetector. Unlike photovoltaic cells, photodiodes require an external bias voltage. These devices provide fast response, but they are limited in sensitivity due to the small area of the junction. We will consider two types of photodiodes—*pn* junction photodiode and *pin* photodiode.

Theory and Structure of *pn*-Junction Photodiodes

A *pn* junction photodiode is operated by first reverse-biasing the junction and then illuminating the junction. As was studied in Chapter 1, reverse-biasing a *pn* junction increases the space charge region and creates a greater electric field at the junction. When radiant energy of the proper wavelength (ultraviolet, visible, or infrared light) strikes a diode, hole-electron pairs are generated, thereby increasing the minority carriers (holes in *n*-type material and electrons in *p*-type material). These minority carriers, with the aid of the space charge region, are swept across the junction, contributing to the flow of minority current (see Fig. 11-17). As the radiant energy becomes more intense, more hole-electron pairs are created and the minority current increases.

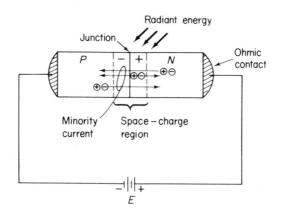

Figure 11-17

Two-dimensional model of a *pn*-junction photodiode.

Note that if the *pn* junction were forward-biased, there would only be a very small increase in majority carriers and a relatively insignificant increase in current. Therefore, normal operation for a photodiode is to reverse-bias the *pn* junction.

Under dark conditions (no radiant energy striking the diode) a photodiode is like any ordinary *pn* junction diode. That is, a tiny reverse-saturation current flows because of the reverse bias. To minimize this effect, almost all photodiodes presently manufactured use silicon as the base material (reverse saturation current is in the order of nanoamperes). When one of these photodiodes is exposed to radiant energy, we can safely assume that the output current is a result of the *newly* created hole-electron pairs. Applications for photodiodes are detection (visible and infrared light), demodulation, switching, and logic circuits that require stability and high speed.

11-8 THE *PIN* PHOTODIODE

Construction

Pin photodiodes differ from *pn* junction photodiodes in that an *intrinsic* or *I* layer is sandwiched between the *p*- and *n*-type material (see Fig. 11-18). Intrinsic means that a semiconductor is pure—no external doping has been added. Therefore, the intrinsic material is neither a *p*-type nor an *n*-type material. Undoped semiconductor materials have very high resistance, and it is this high resistance region that gives a *pin* photodiode several advantages over a *pn* junction photodiode. Two of these advantages are: (1) a decrease in capacitance between the *p* and *n* regions because capacitance is inversely proportional to distance, and (2) the possibility of a greater electric field between the *p* and *n* regions. A decrease in junction capacitance allows a faster response time, and thus the frequency of input optical signals can be greatly increased. In fact, *pin* photodiodes are used in *RF* applications. Increasing the electric field greatly enhances hole-electron pair generation, and thus *pin* photodiodes can process weaker optical signals than do *pn* junction photodiodes. Other advantages of *pin* photodiodes are a broad spectral response and generation of very little noise.

Advantages

Pin photodiodes have also made inroads into the market of the multiplier phototubes that are ordinarily used for low-level optical signals. Advantages of *pin* photodiodes over multiplier phototubes are:

1. Reduced size and weight.
2. Lower cost.
3. Lower power supply voltage.
4. Wider spectral response.
5. Ruggedness.
6. Similar characteristics among units.
7. Stability.

Operation

Operation of *pin* photodiodes is not that much different from *pn* junction photodiodes, except that most of the hole-electron pairs are created in the intrinsic layer. When a hole-electron pair is generated in the *I layer*, because of the polarity of the electric field, the hole moves toward the p material while the freed electron moves toward the n material. Thus radiant energy has created hole-electron pairs contributing to an increase in minority carriers, and subsequently to a large increase in minority current flow.

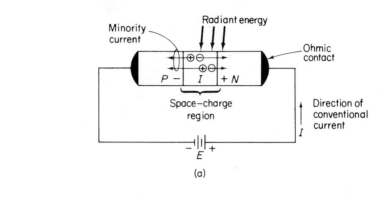

(a)

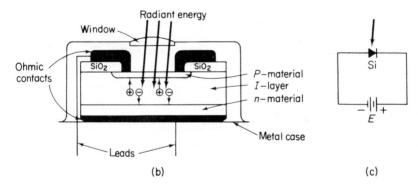

(b)   (c)

Figure 11-18

(a) Simplified two-dimensional model, (b) cross-sectional view, (c) circuit symbol of a *pin* photodiode.

Figure 11-18a is included only to show the polarity of the electric field and direction of minority carriers and conventional current, while Fig. 11-18b is actually more representative of a cross-sectional view (although not drawn to scale) of a *pin* photodiode. Note in Fig. 11-18b that the p material is very much thinner than either the I layer or the n material. A thin p region allows more radiant energy through to the I layer, thus creating more hole-electron pairs. However, the thinner the p region, the higher is its resistance. P material's resistance is a factor that limits high-frequency performance. Therefore, thickness of the p layer is a manufacturer's compromise between a high-frequency limit and quantum efficiency (electrons generated for the amount of input radiant energy). For most applications the external bias voltage E need not be greater than 20V, and is usually between 10 and 20 volts.

11-9 PHOTODIODE APPLICATION

The circuits of Fig. 11-19 form an optical link which electrically isolates input E and output I_E. The input part of the link is an infrared emitting diode driven by a constant current source. The output part of the isolator is a high-gain amplifier using a *pin* photodiode and a silicon transistor. Photo-induced current in a diode is given by

$$I_\lambda = \delta S_R H \tag{11-3}$$

where I_λ is photo-induced current, δ is relative response which is a function of wavelength of the radiant energy, S_R is sensitivity at peak wavelength, and H is irradiance.

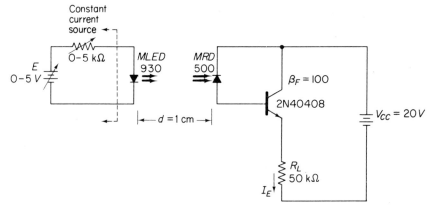

Figure 11-19

Optical link using an infrared-emitting diode and a pin photodiode.

From Appendix 6, peak wavelength of the MRD 500 is 0.8 millimicrons, and at this wavelength S_R equals 6.6mA/mW/cm². However, the peak wavelength of the MLED 930 is 0.9 millimicrons. Relative response δ at 0.9μm is 0.8, found from Fig. 7 in Appendix 6. The irradiance H at 1cm from an MLED 930 was calculated in Example 10-8 as $H = 0.296$mW/cm². From Eq. (11-3)

$$I_\lambda = (0.8)(6.6\text{A/mW/cm}^2)(0.296\text{mW/cm}^2) = 1.56\mu\text{A}$$

which is the base current of the transistor. Current through R_L is

$$I_E \approx I_C = \beta_F I_\lambda = (100)(1.56A) = 0.156mA$$

If more load current is needed, either irradiance must be increased and/or another transistor gain stage be used.

11-10 INTRODUCTION TO PHOTOTRANSISTORS

In Chapter 2 we saw that, in order to operate a bipolar transistor in the active region of its collector characteristics, the emitter-base junction must be forward-biased while the base-collector junction is reverse-biased. The reverse-bias develops a leakage current (as in any reverse-biased pn junction). For a silicon transistor the magnitude of this leakage current is negligible. Phototransistors, however, operate on the principle of increasing the leakage current to develop more collector current. In a phototransistor the base-collector junction is exposed via a window or lens to light (actually radiant energy, visible and/or infrared). This energy creates hole-electron pairs in the vicinity of the base-collector junction, thereby increasing the leakage current. The intensity of the energy source will determine the number of hole-electron pairs and thus the magnitude of output current.

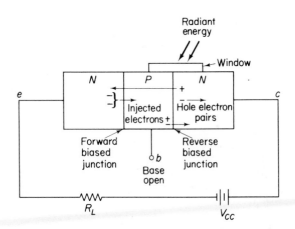

Figure 11-20

Two-dimensional model of a phototransistor.

Figure 11-20 shows hole-electron pairs created in both base and collector. Holes generated in the base remain in the base to replenish those lost due to recombination. Holes created in the collector are drawn into the base because of the reverse-bias at the junction. These holes diffuse toward the emitter-base junction and are then swept into the emitter. Because of the high efficiency of the emitter in injecting electrons into the base, a hole crossing from base to emitter causes several electrons to be injected into the base. At this time we have normal transistor action. The injected electrons diffuse across the base and are swept into the collector. These electrons, along with those created from hole-electron pairs, result in the total collector current.

In the above discussion the base terminal has been left open, yet transistor action has still occured. Therefore, *unlike* bipolar transistors, phototransistors can be *either* a two- or a three-terminal device. In the two-terminal device the base lead is not brought out. Although these devices look like a photodiode, they are not, because they have current gain. In the three-lead version the base lead is brought out, and these phototransistors may be operated as a bipolar transistor with or without light as an input. Since many applications require only light as the input, two-terminal phototransistors are very common.

11-11 PHOTOTRANSISTOR CHARACTERISTICS AND CIRCUIT MODEL

One of the most useful ways of depicting information about a standard bipolar transistor is by its collector characteristics. These curves are a plot of

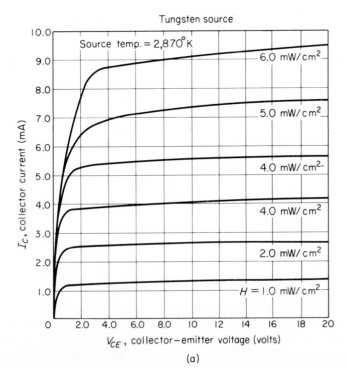

(a)

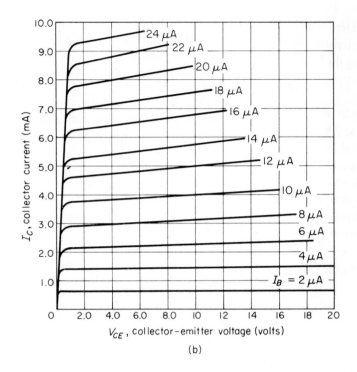

(b)

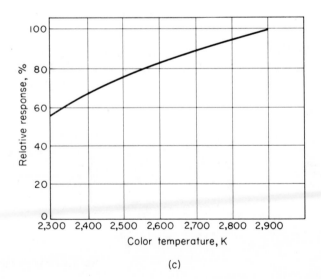

(c)

Figure 11-21

Collector characteristics may be generated with input steps of irradiance
in (a) or base current in (b). Relative response for different color
temperature is given in (c).

collector current, I_C, versus collector-emitter voltage, V_{CE}, for steps of base current, I_B, as shown in Fig. 11-21b. A similar set of curves are supplied by manufacturers of phototransistors, but, instead of base current steps, collector characteristic curves are a function of irradiance, H (that is, radiant energy striking the lens of the phototransistor). Figure 11-21a is a set of typical collector characteristics for Motorola's MRD 300. Note that the units for irradiance are mW/cm²—energy-per-distance squared. The distance is measured from the light source to the phototransistor. The light source is a tungsten filament lamp operating at 2,870°K. As we shall see in section 11-12, a correction will have to be included in any application if the light source is operating at a different temperature.

A set of curves may be obtained for a three-terminal device as a function of base current by using either commercially available curve tracers or sweep techniques (see section 2-3). When obtaining the electrical characteristic curves, the lens of the phototransistor should be covered; otherwise, collector current will be a function of *both* base current and photo-induced current. The reason for obtaining the curves as a function of base current is to bias the phototransistor (if so desired) at an operating point, as was done for a standard bipolar transistor. Note of caution: Although it is possible to bias a phototransistor at maximum current gain, the sensitivity of the device will not be increased, because the external bias resistor (connected between base and ground in the hybrid π model) shunts some current around the base-emitter junction, and this shunted current never gets multiplied by β_F.

Hybrid π Model

The hybrid π model used for bipolar transistors is applicable to phototransistors, as illustrated in Fig. 11-22. Current generator I_λ between base and collector represents the current induced by radiant energy striking the photo-

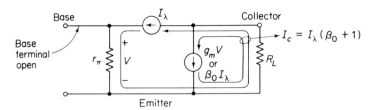

Figure 11-22

Hybrid-π model of a phototransistor.

transistor's lens. The hybrid π model may be used to represent either two- or three-terminal phototransistors. For two-terminal phototransistors, the base lead is permanently open. As with standard bipolar transistors, r_π represents the ratio of ac base-emitter voltage to ac base current, and $g_m V$ is a dependent current generator (its value depends on the voltage-across r_π). In Fig. 11-22 the base terminal is open, therefore $V = I_\lambda r_\pi$ and $g_m V = \beta_o I_\lambda$. The output or ac collector current is

$$I_c = \beta_o I_\lambda + I_\lambda = I_\lambda(\beta_o + 1) \tag{11-4}$$

If a phototransistor is biased similarly to a regular bipolar transistor at a particular operating point, then the external bias resistor (or resistors) must be included in the hybrid π model. Figure 11-23a illustrates external biasing, while 11-23b is the equivalent hybrid π model. Note that R_B shunts the photo-induced current I_λ around r_π and, therefore, the less current through r_π, the less current is multiplied by β_o. From Fig. 11-23b we see that R_B should be large so that almost all of I_λ flows through r_π.

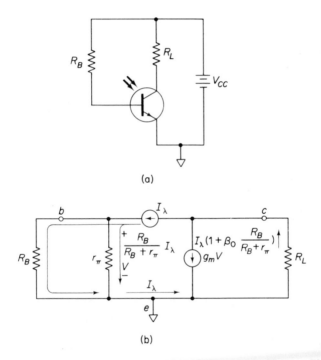

(a)

(b)

Figure 11-23

The phototransistor circuit in (a) is modeled in (b).

11-12 DESIGN EXAMPLE—PHOTOTRANSISTOR WITH AN INCANDESCENT LAMP

Principles introduced in Chapters 10 and 11 will be reviewed and applied in the following example using a photodetector and an incandescent lamp. An application of the LED will be studied in the next section.

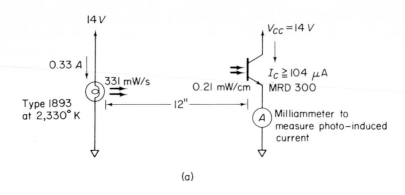

(a)

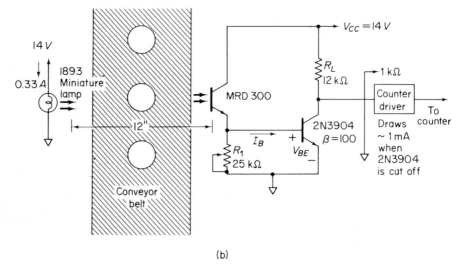

(b)

Figure 11-24

Circuit to measure photo induced collector current in (a) for the conveyor
belt counter system in (b).

EXAMPLE 11-6:

Design an optical counter for a conveyor belt using an MRD 300 as the
photodetector and an 1893 miniature incandescent lamp as the light source.
Distance between lamp and detector is 12 inches, as in Fig. 11-24b.

SOLUTION:

Light source: Analysis of the light source is the same as that developed in
section 10-9. From manufacturer's data for an 1893 lamp: Design voltage
14V, design current 0.33A, and a mean-spherical-candlepower (*MSCP*)
of 2 candles. As was stated previously, efficiency of a tungsten filament lamp

in converting electrical energy into *radiant* energy is high, approximately 90%. (Efficiency of a tungsten filament lamp in converting electrical energy into *light* energy is low—5 to 20%). Since an MRD 300 phototransistor detects a large amount of radiant energy, then the total radiated output of an 1893 lamp is

$$P_{out} = (0.9)P_{in} = (0.9)(14V)(0.33A) = 4.16W$$

A light source is considered as a point source if the diameter of the lamp is less than one-tenth of the distance between the light source and photodetector. For a point source, the source radiant intensity is:

$$I_r = \frac{P_{out}}{4\pi} = \frac{4.16W}{12.56} = 331mW/steradian$$

Converting the distance between source and detector to centimeters (1in = 2.54cm), then

$$12in \times 2.54cm/in = 30.5cm$$

Therefore, the radiant energy striking the phototransistor 12 inches from the lamp is given by

$$H = \frac{I_r}{d^2} = \frac{331mW/st}{(30.5cm)^2} = 0.36mW/cm^2$$

Now we must consider the phototransistor in order to determine the effect an 1893 lamp produces in an MRD 300 which is 12 inches away.
Photodetector: Collector characteristics for an MRD 300 are generated using a tungsten filament lamp operating at a color temperature of 2,870°K. We wish to determine the color temperature of an 1893 lamp operating at its design voltage and current. In order to find color temperature of a lamp, the ratio of mean spherical candlepower to input power must first be calculated.

$$\rho = \frac{2cd}{(14V)(0.33A)} = 0.433cd/W$$

Therefore, from Fig. 10-7, which is a plot of color temperature versus ρ, the color temperature of an 1893 at design voltage and current is 2,330°K. Note that this temperature is 540°K (2870° − 2330°) below the color temperature at which the phototransistor characteristics were obtained. Figure 11-21c is a plot of the relative response of Motorola's MRD phototransistors versus color temperature. From this curve we see that at 2330°K the sensitivity is 60% of the sensitivity at 2870°K.

Therefore, the effective radiant energy striking the phototransistor is not $H = 0.36\text{mW/cm}^2$, but rather

$$H' = (0.60)(0.36\text{mW/cm}^2) = 0.21\text{mW/cm}^2$$

We may wish to design this system so that *any* MRD 300 phototransistor can be used. Therefore, the minimum value of collector-emitter radiation sensitivity, S_{RCEO}, is obtained from the manufacturer's data sheet, Fig. 1 of Appendix 7. At 0.21mW/cm^2, minimum sensitivity is 0.5mA/mW/cm^2. The minimum collector current that can flow as a result of radiant energy from an 1893 lamp striking the phototransistor is

$$I_C = S_{RCEO} \times H' = [0.5\text{mA/(mW/cm}^2)] \times (0.21\text{mW/cm}^2) = 106\mu\text{A}$$

Remember that $106\mu\text{A}$ is the minimum collector current for any MRD 300 phototransistor. Typical characteristics would generate a current of at least $210\mu\text{A}$. But by actually setting up the circuit of Fig. 11-24, we could measure collector current for the particular transistor.

Gain stage: If the counter driver in Fig. 11-24 needs 1mA, we must add gain transistor 2N3904 to boost I_C to 1mA. With the 2N3904 off, R_L must be $14\text{V}/1\text{mA} = 14\text{k}\Omega$, or a standard $12\text{k}\Omega$ resistor. When the 2N3904 is driven into saturation (neglecting V_{CE}), it must divert 1mA away from the counter, and needs a minimum base current of $1\text{mA}/\beta = 1\text{mA}/100 = 10\mu\text{A}$. Since $I_C = I_E \cong 100\mu\text{A}$ of the MRD 300 (at full illumination), R_1 should carry less than $90\mu\text{A}$ to allow $10\mu\text{A}$ base current. Assuming $V_{BE} \cong 0.6\text{V}$

$$R_1 \cong \frac{0.6\text{V}}{90\mu\text{A}} = 6.7\text{k}\Omega$$

Select R_1 as a $25\text{k}\Omega$ pot. to allow compensation for differences between transistors. To set R_1: (1) allow full illumination; (2) connect a voltmeter from collector of the 2N3904 to ground; (3) adjust R_1 unitil the 2N3904 just begins to come out of saturation, as indicated by a slight rise in the voltmeter reading from about 0.2V. Then, when an object interrupts illumination, the transistor cuts off and actuates the counter.

11-13 PHOTOTRANSISTORS WITH EMITTING DIODES

In Chapter 10 we saw that one advantage of emitting diodes is that they may be considered a monochromatic light source. That is, all of their radiant energy is at the peak wavelength, so color temperature that describes a filament lamp is meaningless when applied to emitting diodes. However, phototransistor characteristics are given in relationship to a filament lamp at a specific color temperature. Therefore, to determine a phototransistor's response to an emitting diode, we must be able to relate the diode's output to the phototransistor's characteristics. In Chapter 10 we found how to determine the amount of radiant energy at a known distance from an emitting diode. The procedure will be reviewed in the following example to illustrate analysis of an emitting diode-phototransistor circuit.

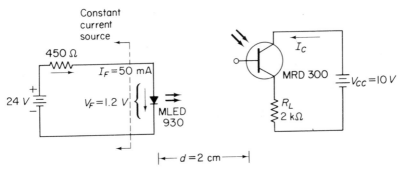

Figure 11-25

Circuit diagram for Example 11-7.

EXAMPLE 11-7:

Calculate the output current for an MRD 300 if an MLED 930 is placed 2cm away from it, as in Fig. 11-25. Assume $I_F = 50$mA (continuous) for the diode, and, for the transistor, $R_L = 2$kΩ and $V_{CC} = 10$V.

SOLUTION:

From Fig. 11-25 at $I_F = 50$mA, $V_F = 1.2$V, and $P_D = V_F I_F = (1.2$V$)$ $(50$mA$) = 60$mW. Junction temperature of the diode is found from Eq. (10-7). Assume $T_A = 25°$C, and from manufacturer's data, Appendix 5, $\Theta_{JA} = 400°$C/W.

$$T_J = T_A + \Theta_{JA} P_D = 25°C + (400°C/w)(60mW) = 49°C$$

From Appendix 5, Fig. 4, at $I_F = 50$mA the instantaneous power output is 0.325mW. However, from Fig. 3 of the same Appendix note that, at a junction temperature of 49°C, output power is 0.8 of that at $T_J = 25°$C, therefore

$$P_o = (0.8)(0.325mW) = 0.26mW$$

Applying Eq. (10-9), we determine the irradiance at a known distance from the source. The divergence angle for an MLED 930 is 30° (or 0.535 rad).

$$H = \frac{4P_o}{\pi d^2 \theta^2} = \frac{4(0.26mW)}{(3.14)(2cm)^2(0.535)^2} = 0.289mW/cm^2$$

Figure 1—Collector-Emitter Sensitivity—of Appendix 7 specifies the reponse of an MRD 300 phototransistor in relationship to a tungsten source operating at 2,870°K. The radiation from such a source is only about 25% effective on a phototransistor. The radiation from an infrared emitting-diode is about 90% effective on a phototransistor. Therefore, the photo-transistor's collector-emitter sensitivity, S_{RCEO}, is increased to

$$S'_{RCEO} = \frac{0.9}{0.25} S_{RCEO} = 3.6 S_{RCEO}$$

where S'_{RCEO} is the phototransistor collector-emitter sensitivity to an infrared emitting diode. From Fig. 1 of Appendix 7 at 0.289mW/cm² the typical sensitivity from a tungsten source is approximately 1.25mA/mW/cm². For the MLED 930 the sensitivity is

$$S'_{RCEO} = 3.6(1.25\text{mW/cm}^2) = 4.5\text{mA/mW/cm}^2$$

The collector current for the phototransistor is

$$I_C = S'_{RCEO}H = (4.5\text{mA/mW/cm}^2)(0.289\text{mW/cm}^2) = 1.3\text{mA}$$

In this example we used typical characteristics. If you were using a photo-transistor with minimum specification, the collector current would be approximately 55% of typical values, or $I_C = (1.3\text{mA})(0.55) = 0.715\text{mA}$.

Infrared Alarm System

Figure 11-26 is an infrared alarm system for a window or door three feet wide. The transmitter includes a power supply designed in Chapter 6 and a UJT oscillator designed in Chapter 7. The receiving unit is an MRD 300 driving an SCR. Once the SCR latches ON the relay is closed, activating an audible alarm. Frequency of oscillation is determined by R and C, $f = 1/RC = 1/(10\text{k}\Omega)(1\mu\text{F})$ = 100Hz. Irradiance three feet from an MLED 930 infrared emitting diode was calculated in Example 10-9 to be $H \cong 0.003\text{mW/cm}^2$. The effective sensitivity of a phototransistor is increased when the light source is an emitting diode and not a tungsten filament source. The increase (as shown in the previous example) is

$$S'_{RCEO} = 3.6 S_{RCEO}$$

Minimum value of S_{RCEO} for an MRD 300 is 0.8mA/mW/cm², thus

$$S'_{RCEO} = 3.6(0.8\text{mA/mW/cm}^2) = 2.88\text{mA/mW/cm}^2$$

and the photo-induced collector current is $I_C = S'_{RCEO} H = (2.88\text{mA/mW/cm}^2)(0.003\text{mW/cm}^2) \cong 8.6\mu\text{A}$.

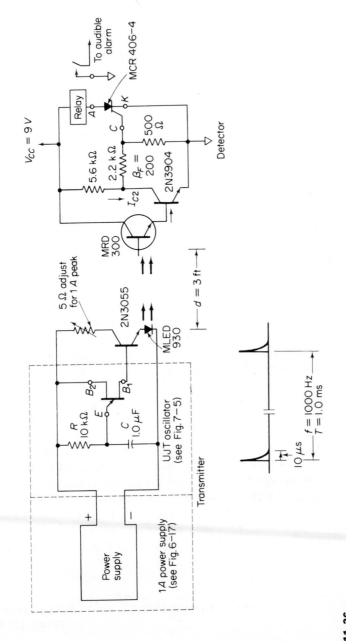

Figure 11-26

Infrared detection alarm system.

This current is now the base current of the 2N3904, with a $\beta_F = 200$. Its collector current is

$$I_{C2} = (200)(8.6\mu A) = 1.73mA$$

Minimum value of collector current needed to saturate the 2N3904 is $9V/5.6k\Omega$ = 1.6mA. Therefore, we guarantee that the transistor is saturated, but not so deep into saturation that the switching time is drastically reduced. The $2.2k\Omega$ and 500Ω gate resistors divide $V_{CE\,sat}$ of the 2N3904 to prevent false turnons.

11-14 INTRODUCTION TO THE LASCR

Figure 11-27 shows that construction of an LASCR is similar to that of an ordinary SCR (Fig. 8-1)—a four-layer *p-n-p-n* device with two forward-biased anode and cathode junctions and one reverse-biased control junction. Normal operation of light-activated SCRs is: (1) radiant energy of the proper wavelength (2) creates hole-electron pairs in the vicinity of the control junction, (3) which increase the minority current flow across the junction, and (4) SCR regenerative switching action triggers the device from the stable *off*-state to the stable *on*-state.

Of course the radiant energy falling on the device must be greater than a *threshold value* below which the LASCR will not turn on. For a GE L8 LASCR the *effective* irradiation at $T_J = 25°C$ is 10mW/cm², which corresponds for this device to 40mW/cm² at the window from a tungsten filament lamp operating at 2,800°K. In the *on* or *conducting* state the voltage drop from anode to cathode is the same as that of any SCR—approximately 1 volt.

Characteristics of a GE L8 LASCR are obtained with an external gate resistor of 56kΩ connected between gate and cathode. This resistor shunts current around the cathode junction, thereby reducing the gain of the *npn* part of the LASCR. Recall from Chapter 8 that SCRs and LASCRs turn on if the leakage current through the control junction is increased, regardless of what causes the increase. Since LASCRs are quite sensitive, the external gate resistor will require more leakage current for turn on, but causes a sacrifice in the decrease of light sensitivity. Removing the resistor allows maximum light sensitivity. Applications of the LASCR are similar to those in Chapter 8, but with light as the trigger.

11-15 INTRODUCTION AND MEASUREMENTS OF OPTICAL ISOLATORS

In the past, electrical isolation between two circuits was accomplished by using relays, isolation transformers, or some other arrangements whose bulk,

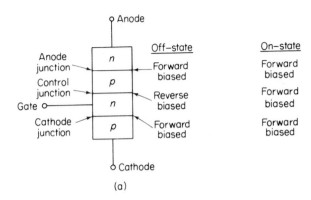

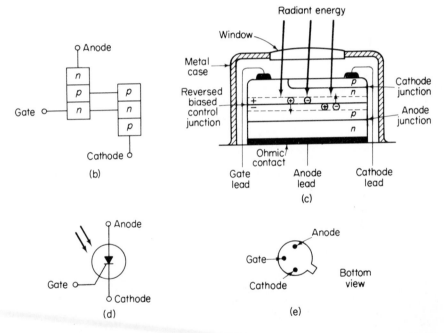

Figure 11-27

(a) Simplified model and off-on conditions of junctions; (b) two-transistor model; (c) structure of LASCRs; (d) circuit symbol; (e) pin connections.

weight, frequency response, or enviromental limits were often distinct disadvantages in some systems. For many applications, circuit designers have found better solutions in *optical isolators*. Optical isolators are also known as *optical couplers* or *photon couplers*. As shown in Fig. 11-28, the optical isolator is made in a single package containing an infrared emitting diode and photodetector. The photodetector is either a photoconductive cell, photodiode, phototransistor,

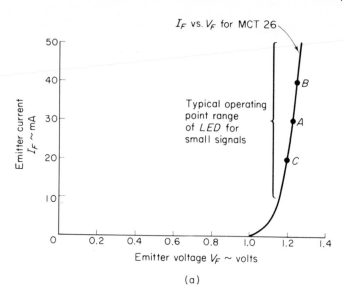

I_F vs. V_F for MCT 26

Typical operating point range of *LED* for small signals

Emitter current $I_F \sim$ mA

Emitter voltage $V_F \sim$ volts

(a)

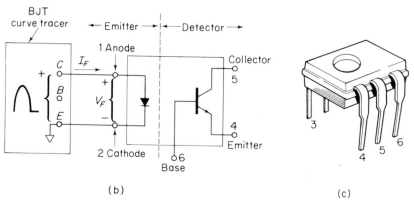

(b)

(c)

Figure 11-28

$I_F - V_F$ characteristic of the optical insulator's *LED* in (a) is measured by the sweep circuit in (b). Packaging is shown in (c).

or LASCR. Advantages are an all solid state unit with its inherently long life, ruggedness, and reliability.

Optical isolators actually may be considered more an electrical system than an optoelectronic device, since we deal only with the unit's electrical parameters—current and voltage—and *not* its optical properties. Input electrical signals are applied to the emitting diode's terminals, turning it on and producing an electrical output at the photodetector's terminals. Input power is low because an infrared emitting diode (depending on the diode) only requires 1.2 to 1.8V and a current of 10 to 100mA. The input forward characteristics I_F-V_F in Fig. 11-28a are obtained from a sweep circuit, as in Fig. 11-28b.

Separation between the light source and photodetector provides electrical isolation comparable to electromechanical relays. Typical resistance between source and detector is 10^{11} ohms.

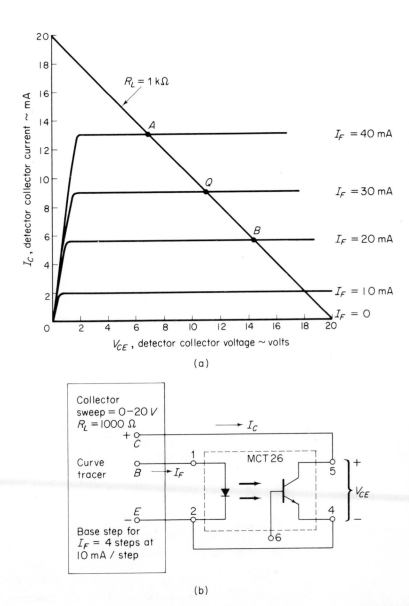

(a)

(b)

Figure 11-29

Detector output characteristics of the optical isolator in (a) are obtained from the sweep circuit in (b).

Detector output characteristics in Fig. 11-29a are obtained by sweeping the photodetector's terminals 4 and 5 while maintaining input current I_F constant. A sweep circuit or standard BJT curve tracer can be used, as in the measuring curcuit of Fig. 11-29b.

Dc current gain of the optical isolator is the ratio of output current I_C to input current I_F, and is given in percent:

$$\text{dc current gain (in percent)} = \frac{I_C}{I_F} \times 100 \qquad (11\text{-}5)$$

Dial a vertical display of collector current and a horizontal display of base current on a BJT curve tracer (or sweep circuit). Using the same test connections and control settings as in Fig. 11-29 ($V_{CE} = 2V/\text{div}$, $I_C = 2mA/\text{div}$, $I_B = 10mA/\text{step}$), we obtain the plot of output current I_C versus input current I_F in Fig. 11-30. (This plot is analogous to an I_C–I_B plot of a BJT.) Evaluating dc current gain at operating point Q, we get $I_C/I_F = (9mA/30mA) \times 100 = 30\%$.

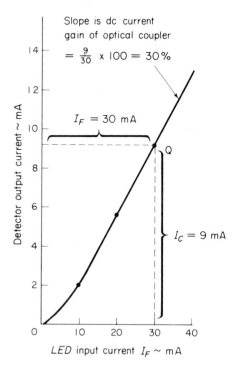

Figure 11-30

Output current versus input current for an MCT26 phototransistor opto-isolator.

11-16 APPLICATIONS OF OPTICAL ISOLATORS

Isolation of Gain Stages

In Fig. 11-31, E_i could represent output of a brain wave monitor. An optical isolator is employed to isolate the patient from high-power control circuitry driven by V_o. Analysis of the ac signal processing will be undertaken in the following examples.

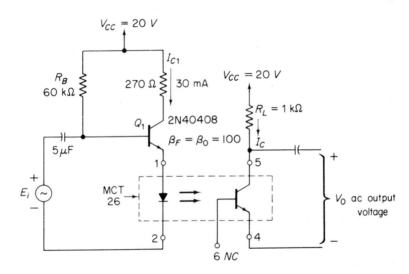

Figure 11-31

Optical isolation of gain stages.

EXAMPLE 11-8:

The emitting diode in the optical coupler of Fig. 11-31 is biased to operate at 30mA. Calculate E_i to superimpose a 20mA peak-to-peak ac signal on the bias current.

SOLUTION:

Locate 30mA on Fig. 11-28a (point A). Superimposing a 20mA peak-to-peak ac signal about point A is a swing from B to C. Emitter diode ac resistance r_d is found either graphically of from Eq. (1-5) as

$$r_d = \frac{\Delta V}{\Delta I} = \frac{(1.26 - 1.2)\text{V}}{(40 - 20)\text{mA}} = \frac{0.06\text{V}}{20\text{mA}} = 3\Omega$$

Evaluate r_π of Q_1 from

$$g_m = \frac{I_{C1}}{25\text{mV}} = \frac{30\text{mA}}{25\text{mV}} = 1.2\mho$$

and rearranging Eq. (2-26) yields

$$r_\pi = \frac{\beta_o}{g_m} = \frac{100}{1.2\mho} = 83\Omega$$

Input resistance to the transistor is

$$R_i = r_\pi + (\beta_o + 1)r_d$$
$$R_i = 83\Omega + (101)(3\Omega) = 386\Omega$$

Since $R_B \gg R_i$, then $R_{in} \simeq R_i = 386\Omega$. For a 20mA ac peak-to-peak diode input current of $\Delta I_F = 20\text{mA}$, input peak-to-peak base current to Q_1 must be $I_F/\beta_o = 20\text{mA}/100 = 0.2\text{mA}$. Therefore

$$E_i = (386\Omega)(0.2\text{mA}) = 77\text{mV}$$

EXAMPLE 11-9:

What change in detector current I_C, or ΔI_C, results from the change in emitting diode current ΔI_F of Example 11-8?

SOLUTION:

Assuming ac and dc current gain of the optical isolator are equal (Fig. 11-30) at 30%, then from Eq. (11-5)

$$\Delta I_C = \Delta I_F(\text{current gain}) = (20\text{mA})(0.30) = 6\text{mA}$$

A graphical solution is shown from points A and B in Fig. 11-29, where $\Delta I_C = (13 - 5.8)\text{mA} \simeq 7\text{mA}$ and is in reasonable agreement.

EXAMPLE 11-10:

What output voltage V_o results from the circuit in Examples 11-8 and 11-9?

SOLUTION:

The *change* in detector output current of $\Delta I_C = 6\text{mA}$ developes a *change* in voltage across R_L of $\Delta I_C R_L = V_o = (6\text{mA})(1\text{k}\Omega) = 6\text{V}$. From the voltage change at $V_{CE} = 14\text{V}$ at point B, to $V_{CE} = 7\text{V}$ at point A in Fig. 11-29a, we obtain reasonable agreement from a graphical solution of $14 - 7 = 7\text{V} = V_o$. As a matter of interest, system voltage gain is $V_o/E_i = 6/0.077 \simeq 80$.

Solid State Latch

An optical isolator replaces a pulse transformer in Fig. 11-32 and isolates a low-voltage logic circuit from a higher voltage power circuit. When both inputs to the NOR gate are high, both Q_1 and Q_2 are in saturation and no current flows into the emitter diode. When a negative-going pulse drives "in 1" to ground during a time when "in 2" is grounded, Q_1 and Q_2 are cut off. An input pulse of approximately 10mA enters the LED and (from Fig. 11-30) generates a detector output pulse of 2mA to trigger the SCR into its conduction state.

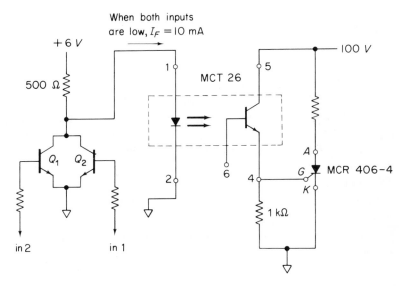

Figure 11-32

Solid state latch using an optical coupler.

PROBLEMS

1. Would the CdS cell of Fig. 11-3a or the CdSe cell of Fig. 11-3b offer the best spectral match for: (a) flourescent lamp? (b) MLED 50 light emitting diode (data given in Appendix 4)? (c) incandescent lamp ($CT \cong 2870°K$)?

2. In the circuit of Fig. 11-6a, if $V_o = 0.1E_i$, what is the light level? Data for the photocell are shown in Fig. 11-5.

3. If a 60W lamp with an average total light output of 870 lumens is placed two feet from the light control circuit of Fig. 11-6a, calculate V_o in terms of E_i. Data for the photocell are given in Fig. 11-5.

4. If the relay in Fig. 11-7a requires 0.5mA to close, and the amount of light striking the photocell is two footcandles, determine whether the relay is open or closed. Data for the photocell are given in Fig. 11-5.

5. A 0-100 dc microammeter is inserted in place of the 0-1 dc milliammeter in Fig. 11-8. (a) What is the range in footcandles? (b) Calibrate a meter face plate in steps of $10\mu A$ and corresponding values of footcandles.

6. Using Fig. 11-11, determine the maximum output power for the 10Ω load for irradiance values of: (a) 50mW/cm²; (b) 100mW/cm²; (c) 125mW/cm².

7. If four photovoltaic cells, with typical characteristics of Fig. 11-11, are connected in parallel, what power is delivered to a 20Ω load?

8. Repeat problem 7 with the four photovoltaic cells connected in series.

9. Calculate cell efficiency at point B on Fig. 11-11. Dimensions of the cell are 2cm × 2cm.

10. If the photovoltaic cell of Fig. 11-14b is exposed to an irradiance of 2mW/cm², calculate the meter reading. Compare result with Example 11-5.

11. The distance in Fig. 11-19 is increased to 2cm. Calculate load current I_E.

12. The 1893 lamp used in the optical counter of Fig. 11-24b is mistakenly replaced with an 1896 lamp whose rated values are 18V, 0.15A, and $MSCP$ = 1.8cd. (a) Calculate the irradiance striking the phototransistor. (b) Will the emitter current of the MRD 300 develop 0.7V across R_1 to saturate the 2N3904 transistor?

13. Calculate collector current I_C in Fig. 11-25 if the distance is increased to 4cm.

14. Determine the dc current gain of the optical coupler in Fig. 11-29 at I_F = 10mA.

appendices

APPENDIX 1

Internally-Compensated Operational Amplifier 741

Courtesy of Fairchild Semiconductor, a Division of Fairchild Camera and Instrument Corporation.

GENERAL DESCRIPTION — The μA741 is a high performance monolithic operational amplifier constructed on a single silicon chip, using the Fairchild Planar* epitaxial process. It is intended for a wide range of analog applications. High common mode voltage range and absence of "latch-up" tendencies make the μA741 ideal for use as a voltage follower. The high gain and wide range of operating voltage provides superior performance in integrator, summing amplifier, and general feedback applications.

- **NO FREQUENCY COMPENSATION REQUIRED**
- **SHORT-CIRCUIT PROTECTION**
- **OFFSET VOLTAGE NULL CAPABILITY**
- **LARGE COMMON-MODE AND DIFFERENTIAL VOLTAGE RANGES**
- **LOW POWER CONSUMPTION**
- **NO LATCH UP**

ABSOLUTE MAXIMUM RATINGS

Supply Voltage	
Military (312 Grade)	±22 V
Commercial (393 Grade)	±18 V
Internal Power Dissipation (Note 1)	
Metal Can	500 mW
Ceramic DIP	670 mW
Silicone DIP	340 mW
Mini DIP	310 mW
Flatpak	570 mW
Differential Input Voltage	±30 V
Input Voltage (Note 2)	±15 V
Storage Temperature Range	
Metal Can, Ceramic DIP, and Flatpak	−65° C to +150° C
Mini DIP and Silicon DIP	−55° C to +125° C
Operating Temperature Range	
Military (312 Grade)	−55° C to +125° C
Commercial (393 Grade)	0° C to + 70° C

CONNECTION DIAGRAMS
(TOP VIEW)

8 LEAD METAL CAN

NOTE PIN 4 CONNECTED TO CASE

ORDER PART NOS.

U5B7741312
U5B7741393

14 LEAD DIP

FOR CERAMIC DIP ORDER PART NOS.

U6A7741312
U6A7741393

FOR SILICONE DIP ORDER PART NO.

Metal Can, Ceramic DIP and Flatpak (60 seconds) 300° C
Mini DIP and Silicone DIP (10 seconds) 260° C
Output Short Circuit Duration (Note 3) Indefinite

FLATPACK

N C
OFFSET NULL
INVERTING INPUT
NON INVERTING INPUT
V-

10 N C
9 N C
8 V+
7 OUTPUT
6 OFFSET NULL

ORDER PART NO.

U3F7741312

MINIDIP

N C
V+
OUTPUT
OFFSET NULL

8
7
6
5

OFFSET NULL
INVERT INPUT
NON INVERT INPUT
V-

1
2
3
4

ORDER PART NO.

U9T7741393

*Planar is a patented Fairchild process.

EQUIVALENT CIRCUIT

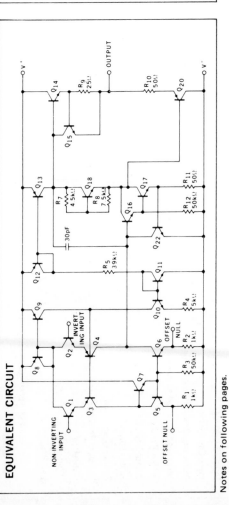

Notes on following pages.

312 GRADE

ELECTRICAL CHARACTERISTICS ($V_S = \pm 15$ V, $T_A = 25°C$ unless otherwise specified)

PARAMETERS (see definitions)	CONDITIONS	MIN.	TYP.	MAX.	UNITS
Input Offset Voltage	$R_S \leq 10$ kΩ		1.0	5.0	mV
Input Offset Current			20	200	nA
Input Bias Current			80	500	nA
Input Resistance		0.3	2.0		MΩ
Input Capacitance			1.4		pF
Offset Voltage Adjustment Range			± 15		mV
Large-Signal Voltage Gain	$R_L \geq 2$ kΩ, $V_{out} = \pm 10$ V	50,000	200,000		
Output Resistance			75		Ω
Output Short-Circuit Current			25		mA
Supply Current			1.7	2.8	mA
Power Consumption			50	85	mW
Transient Response (unity gain)	$V_{in} = 20$ mV, $R_L = 2$ kΩ, $C_L \leq 100$ pF				
Risetime			0.3		μs
Overshoot			5.0		%
Slew Rate	$R_L \geq 2$ kΩ		0.5		V/μs

The following specifications apply for $-55°C \leq T_A \leq +125°C$:

Input Offset Voltage	$R_S \leq 10$ kΩ		1.0	6.0	mV
Input Offset Current	$T_A = +125°C$		7.0	200	nA
	$T_A = -55°C$		85	500	nA
Input Bias Current	$T_A = +125°C$		0.03	0.5	μA
	$T_A = -55°C$		0.3	1.5	A

432

Input Voltage Range					
Common Mode Rejection Ratio	$R_c \leq 10\ K\Omega$	70	90		dB
Supply Voltage Rejection Ratio	$R_S \leq 10\ K\Omega$		30	150	$\mu V/V$
Large-Signal Voltage Gain	$R_L \geq 2\ K\Omega,\ V_{out} = \pm 10\ V$	25,000			
Output Voltage Swing	$R_L \geq 10\ K\Omega$	± 12	± 14		V
	$R_L \geq 2\ K\Omega$	± 10	± 13		V
Supply Current	$T_A = +125°C$		1.5	2.5	mA
	$T_A = -55°C$		2.0	3.3	mA
Power Consumption	$T_A = +125°C$		45	75	mW
	$T_A = -55°C$		60	100	mW

TYPICAL PERFORMANCE CURVES
312 GRADE

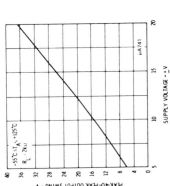

OPEN LOOP VOLTAGE GAIN AS A FUNCTION OF SUPPLY VOLTAGE

OUTPUT VOLTAGE SWING AS A FUNCTION OF SUPPLY VOLTAGE

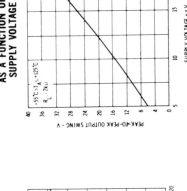

INPUT COMMON MODE VOLTAGE RANGE AS A FUNCTION OF SUPPLY VOLTAGE

393 GRADE

ELECTRICAL CHARACTERISTICS ($V_S = \pm 15$ V, $T_A = 25°C$ unless otherwise specified)

PARAMETERS (see definitions)	CONDITIONS	MIN.	TYP.	MAX.	UNITS
Input Offset Voltage	$R_S \leq 10$ kΩ		2.0	6.0	mV
Input Offset Current			20	200	nA
Input Bias Current			80	500	nA
Input Resistance		0.3	2.0		MΩ
Input Capacitance			1.4		pF
Offset Voltage Adjustment Range			±15		mV
Input Voltage Range		±12	±13		V
Common Mode Rejection Ratio	$R_S \leq 10$ kΩ	70	90		dB
Supply Voltage Rejection Ratio	$R_S \leq 10$ kΩ		30	150	μV/V
Large-Signal Voltage Gain	$R_L \geq 2$ kΩ, $V_{out} = \pm 10$ V	20,000	200,000		V/V
Output Voltage Swing	$R_L \geq 10$ kΩ	±12	±14		V
	$R_L \geq 2$ kΩ	±10	±13		V
Output Resistance			75		Ω
Output Short-Circuit Current			25		mA
Supply Current			1.7	2.8	mA
Power Consumption			50	85	mW
Transient Response (unity gain)	$V_{in} = 20$ mV, $R_L = 2$ kΩ, $C_L \leq 100$ pF				
Risetime			0.3		μs
Overshoot			5.0		%
Slew Rate	$R_L \geq 2$ kΩ		0.5		V/μs

The following specifications apply for $0°C \leq T_A \leq +70°C$:

Input Offset Voltage				7.5	mV
Input Offset Current				300	nA

TYPICAL PERFORMANCE CURVES
393 GRADE

OPEN LOOP VOLTAGE GAIN AS A FUNCTION OF SUPPLY VOLTAGE

OUTPUT VOLTAGE SWING AS A FUNCTION OF SUPPLY VOLTAGE

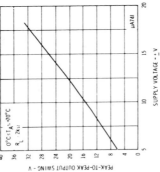

INPUT COMMON MODE VOLTAGE RANGE AS A FUNCTION OF SUPPLY VOLTAGE

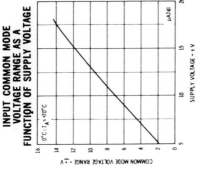

NOTES

1. Rating applies to ambient temperatures up to 70°C. Above 70°C ambient derate linearly at 6.3 mW/°C for the Metal Can, 8.3 mW/°C for the Ceramic DIP, 6.3 mW/°C for the Silicone DIP, 5.6 mW/°C for the Mini DIP and 7.1 mW/°C for the Flatpak.
2. For supply voltages less than ±15 V, the absolute maximum input voltage is equal to the supply voltage.
3. Short circuit may be to ground or either supply. Rating applies to +125°C case temperature or 75°C ambient temperature.

TYPICAL PERFORMANCE CURVES (312 GRADE)

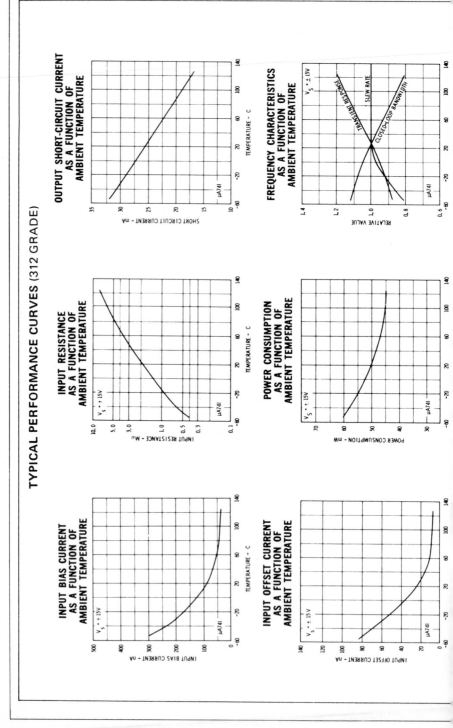

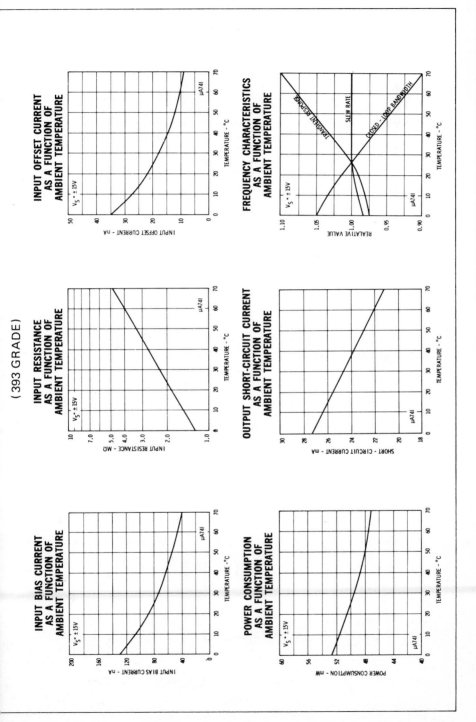

(393 GRADE)

INPUT BIAS CURRENT AS A FUNCTION OF AMBIENT TEMPERATURE

INPUT RESISTANCE AS A FUNCTION OF AMBIENT TEMPERATURE

INPUT OFFSET CURRENT AS A FUNCTION OF AMBIENT TEMPERATURE

POWER CONSUMPTION AS A FUNCTION OF AMBIENT TEMPERATURE

OUTPUT SHORT-CIRCUIT CURRENT AS A FUNCTION OF AMBIENT TEMPERATURE

FREQUENCY CHARACTERISTICS AS A FUNCTION OF AMBIENT TEMPERATURE

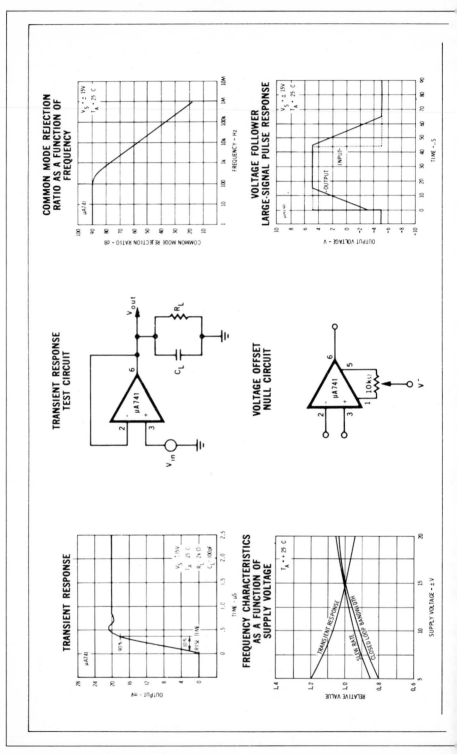

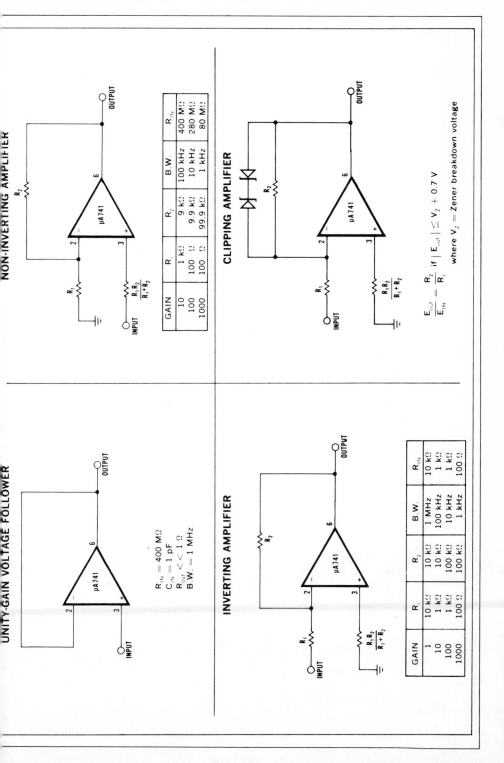

NON-INVERTING AMPLIFIER

GAIN	R_1	R_2	B.W.	R_{IN}
10	1 kΩ	9 kΩ	100 kHz	400 MΩ
100	100 Ω	9.9 kΩ	10 kHz	280 MΩ
1000	100 Ω	99.9 kΩ	1 kHz	80 MΩ

UNITY-GAIN VOLTAGE FOLLOWER

$R_{IN} = 400$ MΩ
$C_{IN} = 1$ pF
$R_{out} << 1$ Ω
B.W. $= 1$ MHz

CLIPPING AMPLIFIER

$$\frac{E_{out}}{E_{IN}} = \frac{R_2}{R_1} \text{ if } |E_{out}| \leq V_Z + 0.7 \text{ V}$$

where V_Z = Zener breakdown voltage

INVERTING AMPLIFIER

GAIN	R_1	R_2	B.W.	R_{IN}
1	10 kΩ	10 kΩ	1 MHz	10 kΩ
10	1 kΩ	10 kΩ	100 kHz	1 kΩ
100	1 kΩ	100 kΩ	10 kHz	1 kΩ
1000	100 Ω	100 kΩ	1 kHz	100 Ω

439

TYPICAL PERFORMANCE CURVES (312 AND 393 GRADES)

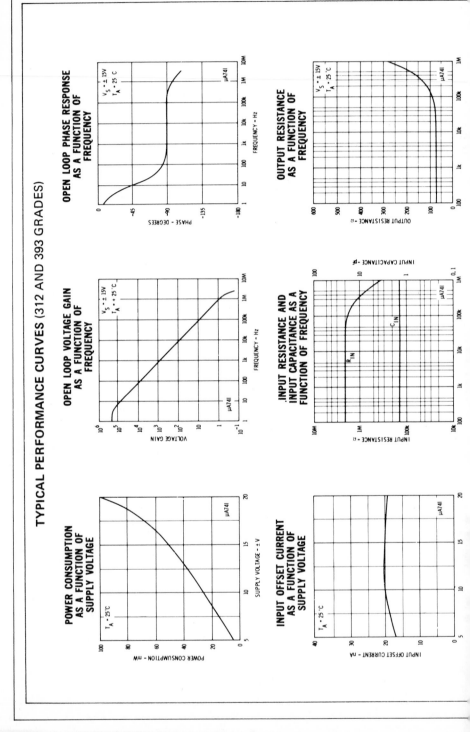

OPEN LOOP PHASE RESPONSE AS A FUNCTION OF FREQUENCY

OPEN LOOP VOLTAGE GAIN AS A FUNCTION OF FREQUENCY

POWER CONSUMPTION AS A FUNCTION OF SUPPLY VOLTAGE

OUTPUT RESISTANCE AS A FUNCTION OF FREQUENCY

INPUT RESISTANCE AND INPUT CAPACITANCE AS A FUNCTION OF FREQUENCY

INPUT OFFSET CURRENT AS A FUNCTION OF SUPPLY VOLTAGE

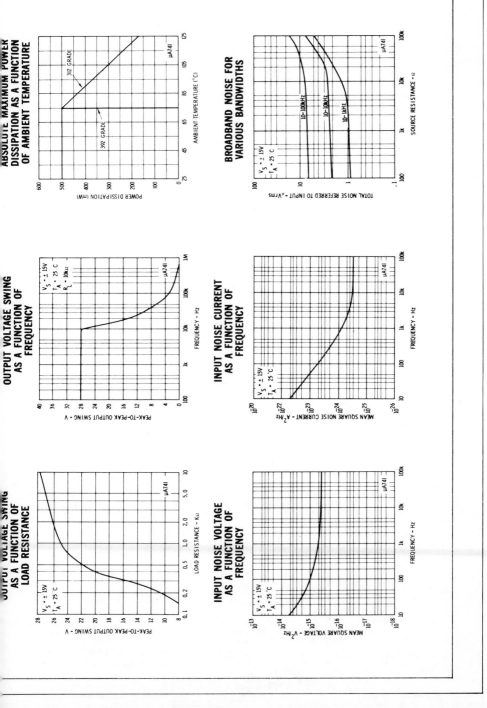

441

Programmable Unijunction Transistor

Courtesy of General Electric Semiconductor Department, Syracuse, N.Y.

Silicon

Programmable Unijunction Transistor
(PUT)

60.20 1/71
Supersedes 60.20 11/67

D13T SERIES
2N6027
2N6028

The General Electric PUT is a three-terminal planar passivated PNPN device in the standard plastic low cost TO-98 package. The terminals are designated as anode, anode gate and cathode.

The 2N6027 and 2N6028 have been characterized as Programmable Unijunction Transistors (PUT), offering many advantages over conventional unijunction transistors. The designer can select R_1 and R_2 to program unijunction characteristics such as η, R_{BB}, I_P and I_V to meet his particular needs.

The 2N6028 is specifically characterized for long interval timers and other applications requiring low leakage and low peak point current. The 2N6027 has been characterized for general use where the low peak point current of the 2N6028 is not essential. Applications of the 2N6027 include timers, high gain phase control circuits and relaxation oscillators.

10 Outstanding Features of the PUT:	Applications:
1. Planar Passivated Structure	• SCR Trigger
2. Low Leakage Current	• Pulse and
3. Low Peak Point Current	Timing Circuits
4. Low Forward Voltage	• Oscillators
5. Fast, High Energy Trigger Pulse	• Sensing Circuits
6. Programmable η	• Sweep Circuits
7. Programmable R_{BB}	
8. Programmable I_P	
9. Programmable I_V	
10. Low Cost	

SYMBOL	INCHES		MILLIMETERS	
	MIN.	MAX.	MIN.	MAX.
A	.170	.265	4.32	6.73
ϕb_2	.016	.019	.406	.483
ϕD	.165	.205	4.19	5.21
E	.110	.155	2.79	3.94
e	.095	.105	2.41	2.67
e_1	.045	.055	1.14	1.40
L	.500		12.70	
Q_2		.075		1.90
s	.080	.115	2.03	2.92

NOTE I: LEAD DIAMETER IS CONTROLLED IN THE ZONE BETWEEN .070 AND .250 FROM THE SEATING PLANE BETWEEN .250 AND END OF LEAD A MAX OF .021 IS HELD.

Operation of the PUT as a unijunction is easily understood. Figure 1(a) shows a basic unijunction circuit. Figure 2(a) shows identically the same circuit except that the unijunction transistor is replaced by the PUT plus resistors R_1 and R_2. Comparing the equivalent circuits of Figure 1(b) and 2(b), it is seen that both circuits have a diode connected to a voltage divider. When this diode becomes forward biased in the unijunction transistor, R_1 becomes strongly modulated to a lower resistance value. This generates a negative resistance characteristic between the emitter E and base one (B_1). For the PUT, the resistors R_1 and R_2 control the voltage at which the diode (anode to gate) becomes forward biased. After the diode conducts, the regeneration inherent in a PNPN device causes the PUT to switch on. This generates a negative resistance characteristic from anode to cathode (Figure 2(b)) simulating the modulation of R_1 for a conventional unijunction.

Resistors R_{B2} and R_{B1} (Figure 1(a)) are generally unnecessary when the PUT replaces a conventional UJT. This is illustrated in Figure 2(c). Resistor R_{B1} is often used to bypass the interbase current of the unijunction which would otherwise trigger the SCR. Since R_1 in the case of the PUT, can be returned directly to ground there is not current to bypass at the SCR gate. Resistor R_{B2} is used for temperature compensation and for limiting the dissipation in the UJT during capacitor discharge. Since R_2 (Figure 2) is *not* modulated, R_{B2} can be absorbed into it.

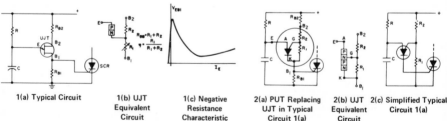

| 1(a) Typical Circuit | 1(b) UJT Equivalent Circuit | 1(c) Negative Resistance Characteristic | 2(a) PUT Replacing UJT in Typical Circuit 1(a) | 2(b) UJT Equivalent Circuit Using PUT | 2(c) Simplified Typical Circuit 1(a) |

Figure 1 Unijunction Transistor   **Figure 2 PUT Equivalent of UJT**

GENERAL ELECTRIC

absolute maximum ratings: (25°C)

Voltage
*Gate-Cathode Forward Voltage   +40 V
*Gate-Cathode Reverse Voltage   –5 V
*Gate-Anode Reverse Voltage   +40 V
*Anode-Cathode Voltage   ±40 V

Current
*DC Anode Current†   150 mA
 Peak Anode, Recurrent Forward
  (100 μsec pulse width, 1% duty cycle)   1 A
 *(20 μsec pulse width, 1% duty cycle)   2 A
 Peak Anode, Non-recurrent Forward
  (10 μsec)   5 A
*Gate Current   ±20 mA

Capacitive Discharge Energy††   250 μJ

Power
*Total Average Power†   300 mW

Temperature
*Operating Ambient†
 Temperature Range   –50°C to +100°C

†Derate currents and powers 1%/°C above 25°C
††E = ½ CV² capacitor discharge energy with no current limiting

Figure 3

electrical characteristics: (25°C) (unless otherwise specified)

	Fig. No.	2N6027 (D13T1) Min.	2N6027 (D13T1) Max.	2N6028 (D13T2) Min.	2N6028 (D13T2) Max.	
*Peak Current (V_s = 10 Volts)	I_P 3					
(R_G = 1 Meg)			2		.15 μA	
(R_G = 10 k)			5		1.0 μA	
*Offset Voltage (V_s = 10 Volts)	V_T 3					
(R_G = 1 Meg)		.2	1.6	.2	.6 Volts	
(R_G = 10 k)		.2	.6	.2	.6 Volts	
*Valley Current (V_s = 10 Volts)	I_V 3					
(R_G = 1 Meg)			50		25 μA	
(R_G = 10 k)		70		25	μA	
(R_G = 200 Ω)		1.5		1.0	mA	
Anode Gate-Anode Leakage Current						
*(V_s = 40 Volts, T = 25°C)	I_{GAO} 4		10		10 nA	
(T = 75°C)			100		100 nA	
Gate to Cathode Leakage Current						
(V_s = 40 Volts, Anode-cathode short)	I_{GKS} 5		100		100 nA	
*Forward Voltage (I_F = 50 mA)	V_F		1.5		1.5 Volts	
*Pulse Output Voltage	V_O 6	6		6	Volts	
Pulse Voltage Rate of Rise	t_r 6		80		80 nsecs.	

*JEDEC registered data

Figure 4

Figure 5

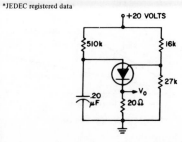

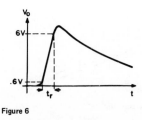

Figure 6

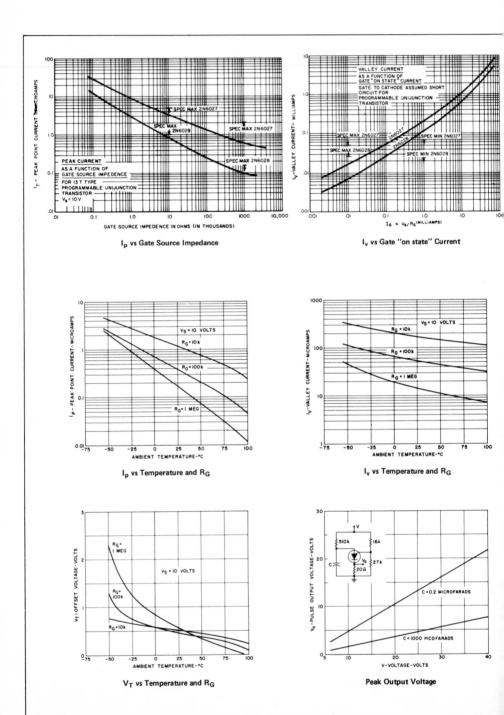

I_p vs Gate Source Impedance

I_v vs Gate "on state" Current

I_p vs Temperature and R_G

I_v vs Temperature and R_G

V_T vs Temperature and R_G

Peak Output Voltage

APPLICATIONS

TYPICAL UNIJUNCTION CIRCUIT CONFIGURATIONS

Here are four ways to use the PUT as a unijunction. Note the flexibility due to "programmability." Applications from long time interval latching timers to wide range relaxation oscillators are possible.

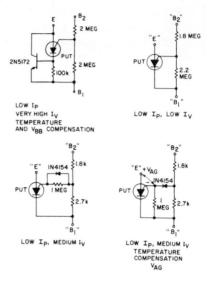

LOW I_P
VERY HIGH I_V
TEMPERATURE
AND V_{BB} COMPENSATION

LOW I_P, LOW I_V

LOW I_P, MEDIUM I_V

LOW I_P, MEDIUM I_V
TEMPERATURE
COMPENSATION
V_{AG}

HOUR TIME DELAY SAMPLING CIRCUIT

This sampling circuit lowers the effective peak current of the output PUT, Q2. By allowing the capacitor to charge with high gate voltage and periodically lowering gate voltage, when Q1 fires, the timing resistor can be a value which supplies a much lower current than I_P. The triggering requirement here is that minimum charge to trigger flow through the timing resistor during the period of the Q1 oscillator. This is not capacitor size dependent, only capacitor leakage and stability dependent.

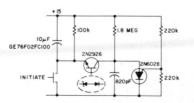

1 SECOND, 1kHz OSCILLATOR

Here is a handy circuit which operates as an oscillator and a timer. The 2N6028 is normally on due to excess holding current through the 100 kohm resistor. When the switch is momentarily closed, the 10 μF capacitor is charged to a full 15 volts and 2N6028 starts oscillating (1.8 Meg and 820 pF). The circuit latches when 2N2926 zener breaks down again.

PRINTED IN U.S.A.

SEMICONDUCTOR PRODUCTS DEPARTMENT
ELECTRONICS PARK, SYRACUSE, N. Y. 13201

GENERAL ⊛ ELECTRIC

In Canada, Canadian General Electric Company Ltd, Toronto, Ont
Outside the U.S.A. and Canada, by Electronic Component Sales
I.G.E. Export Division, 159 Madison Avenue, New York, N. Y. 10016

445

Courtesy of Motorola Semiconductor Products, Inc.

MCR406-1
thru
MCR406-4

PLASTIC THYRISTORS

. . . Annular† PNPN devices designed for high volume consumer applications, such as temperature, light, and speed control, process and remote control, and warning systems where reliability of operation is important. Sensitive gate trigger permits operation as a switch directly from low level sensors.

- Annular† Passivated Surface for Reliability and Uniformity
- True Power Rated — 4.0 Amp @ T_C = 97°C
- Low Level Gate Characteristics — I_{GT} = 200 μA @ T_A = 25°C
- Higher Surge Current Rating — I_{TSM} = 30 Amp
- Flat, Rugged, Thermopad†† Construction — for Low Thermal Resistance, High Heat Dissipation, and Durability

PLASTIC SILICON CONTROLLED RECTIFIERS

4.0 AMPERES RMS
30 thru 200 VOLTS

FEBRUARY 1969 — DS 8529

MAXIMUM RATINGS

Rating	Symbol	Value	Unit
Peak Reverse Blocking Voltage (Note 1)	V_{RRM}		Volts
MCR406-1		30	
MCR406-2		60	
MCR406-3		100	
MCR406-4		200	
Forward Current RMS (All Conduction Angles)	$I_{T(RMS)}$	4.0	Amp
Peak Forward Surge Current (1/2 cycle, 60 Hz, T_J = -40 to +110°C)	I_{TSM}	30	Amp
Circuit Fusing Considerations (T_J = -40 to 110°C, t = 1.0 to 8.3 ms)	I^2t	3.6	A^2s
Peak Gate Power — Forward	P_{GFM}	0.5	Watt
Average Gate Power — Forward	$P_{GF(AV)}$	0.1	Watt
Peak Gate Current — Forward	I_{GFM}	0.2	Amp
Peak Gate Voltage — Reverse	V_{GRM}	6.0	Volts
Operating Junction Temperature Range	T_J	-40 to +110	°C
Storage Temperature Range	T_{stg}	-40 to +150	°C
Mounting Torque (6-32 screw) (Note 2)	—	12	in. lb.

†Annular Semiconductor Patented by Motorola Inc.   ††Trademark of Motorola Inc.

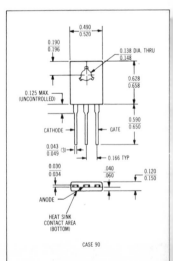

0.490
0.520
0.190
0.196
0.138 DIA. THRU
0.148
0.628
0.658
0.125 MAX. (UNCONTROLLED)
CATHODE
GATE
0.590
0.650
0.043
0.049 (3)
0.166 TYP
0.030
0.034
.040
.060
0.120
0.150
ANODE
HEAT SINK CONTACT AREA (BOTTOM)
CASE 90

MCR406-1 thru MCR406-4

ELECTRICAL CHARACTERISTICS (T_C = 25°C unless otherwise noted, R_{GK} = 1000 Ohms)

Characteristic	Symbol	Min	Typ	Max	Unit
Peak Forward Blocking Voltage	V_{DRM}				Volts
(T_J = 110°C) Note 1 MCR406-1		30	–	–	
MCR406-2		60	–	–	
MCR406-3		100	–	–	
MCR406-4		200	–	–	
Peak Forward Blocking Current (Rated V_{DRM} @ T_J = 110°C)	I_{DRM}	–	–	100	μA
Peak Reverse Blocking Current (Rated V_{RRM} @ T_J = 110°C)	I_{RRM}	–	–	100	μA
Forward "On" Voltage (I_{TM} = 4.0 A peak)	V_{TM}	–	–	2.2	Volts
Gate Trigger Current (Continuous dc) (Anode Voltage = 7.0 Vdc, R_L = 100 Ohms)	I_{GT}	–	–	200	μA
Gate Trigger Voltage (Continuous dc) (Anode Voltage = 7.0 Vdc, R_L = 100 Ohms)	V_{GT}	–	–	0.8	Volts
(Anode Voltage = Rated V_{DRM}, R_L = 100 Ohms, T_J = 110°C)	V_{GD}	0.2	–	–	
Holding Current (Anode Voltage = 7.0 Vdc)	I_H	–	–	3.0	mA
Turn-On Time	t_{on}		Circuit Dependent. Consult Manufacturer.		
Turn-Off Time	t_{off}				
Forward Voltage Application Rate (T_J = 110°C)	dv/dt	–	10	–	V/μs
Thermal Resistance, Junction to Case	θ_{JC}	–	–	2.0	°C/W
Thermal Resistance, Junction to Ambient	θ_{JA}	–	–	50	°C/W

NOTES:
1. V_{DRM} and V_{RRM} for all types can be applied on a continuous dc basis without incurring damage. Ratings apply for zero or negative gate voltage but positive gate voltage shall not be applied concurrently with a negative potential on the anode. When checking forward or reverse blocking capability, thyristor devices should not be tested with a constant current source in a manner that the voltage applied exceeds the rated blocking voltage.

2. Torque rating applies with use of torque washer (Shakeproof WD19522 #6 or equivalent). Mounting torque in excess of 8 in. lbs. does not appreciably lower case-to-sink thermal resistance. Anode lead and heatsink contact pad are common.

For soldering purposes (either terminal connection or device mounting), soldering temperatures shall not exceed +225°C. For optimum results, an activated flux (oxide removing) is recommended.

FIGURE 1 – CASE TEMPERATURE versus CURRENT

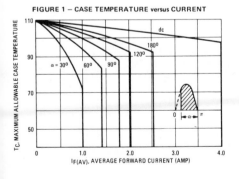

FIGURE 2 – AMBIENT TEMPERATURE versus CURRENT

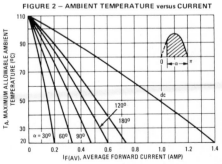

MOTOROLA *Semiconductor Products Inc.* • • • • • • • • • • • • •

FIGURE 3 – FORWARD CONDUCTION CHARACTERISTICS

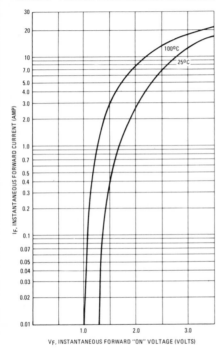

I_F, INSTANTANEOUS FORWARD CURRENT (AMP)

V_F, INSTANTANEOUS FORWARD "ON" VOLTAGE (VOLTS)

FIGURE 4 – P_D, POWER DISSIPATION

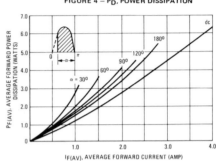

$P_{F(AV)}$, AVERAGE FORWARD POWER DISSIPATION (WATTS)

$I_{F(AV)}$, AVERAGE FORWARD CURRENT (AMP)

FIGURE 5 – 60 Hz SURGES

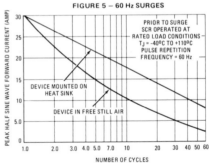

PEAK HALF SINE WAVE FORWARD CURRENT (AMP)

PRIOR TO SURGE
SCR OPERATED AT
RATED LOAD CONDITIONS
T_J = -40°C TO +110°C
PULSE REPETITION
FREQUENCY = 60 Hz

DEVICE MOUNTED ON HEAT SINK

DEVICE IN FREE STILL AIR

NUMBER OF CYCLES

FIGURE 6 – THERMAL RESPONSE

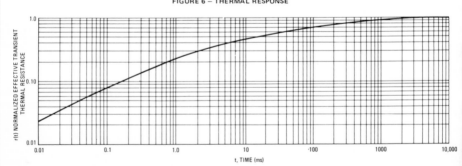

r(t) NORMALIZED EFFECTIVE TRANSIENT THERMAL RESISTANCE

t, TIME (ms)

MOTOROLA *Semiconductor Products Inc.*

MCR406-1 thru MCR406-4

FIGURE 7 – TYPICAL GATE TRIGGER CURRENT

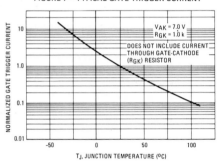

FIGURE 8 – TYPICAL GATE TRIGGER VOLTAGE

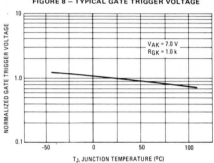

FIGURE 9 – TYPICAL HOLDING CURRENT

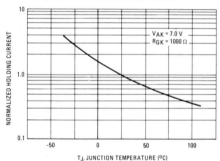

SELECTED THYRISTOR-TRIGGER APPLICATION NOTES

AN-227A – Thyristor Trigger Circuits for Power-Control Applications
AN-240 – SCR Power Control Fundamentals
AN-290A – Mounting Procedure for, and Thermal Aspects of, Thermopad†† Plastic Power Devices
AN-295 – Suppressing RFI in Thyristor Circuits
AN-422 – Testers for Thyristors and Trigger Diodes
AN-453 – Zero Point Switching Techniques

To obtain copies of these notes list the AN number(s) on your company letterhead and send your request to:

Technical Information Center
Motorola Semiconductor Products, Inc.
P.O. Box 20924
Phoenix, Arizona 85036

MOTOROLA Semiconductor Products Inc.

BOX 20912 • PHOENIX, ARIZONA 85036 • A SUBSIDIARY OF MOTOROLA INC.

4567-1 PRINTED IN USA 6-69 IMPERIAL LITHO B11766   15M

DS 8529

449

MOTOROLA Semiconductors
BOX 20912 • PHOENIX, ARIZONA 85036

MLED50
MLED55

VISIBLE RED LIGHT-EMITTING DIODES

... designed for applications requiring high visibility, low-drive power and high reliability. These devices can be used as circuit status indicators, panel indicators in large matrix displays, and for film annotation. The MLED50 is a high intensity point source in a clear plastic package. The MLED55, because of its diffusing red plastic package appears as a large area light source with wide viewing angle.

- High Luminous Intensity – MLED50 – 1.0 mcd (Typ)
  MLED55 – 0.6 mcd (Typ)
- Solid State Reliability
- Compatible with IC's – Low Drive Current
- Economical Plastic Package – Clear or Diffusing Red
- Resistant to Shock and Vibration
- Wide Viewing Angle
- Easy Cathode Indentification – Wider Lead
- Visible Red Emission – 660 nM (Typ)

LIGHT-EMITTING DIODE
VISIBLE RED

GALLIUM
ARSENIDE PHOSPHIDE

120 MILLIWATTS

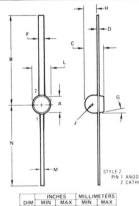

MLED50 – Clear Plastic
MLED55 – Diffusing Red Plastic

MAXIMUM RATINGS

Rating	Symbol	Value	Unit
Reverse Voltage	V_R	3.0	Volts
Forward Current-Continuous	I_F	50	mA
Total Device Dissipation @ T_A = 25°C Derate above 25°C	P_D(1)	120 2.0	mW mW/°C
Operating and Storage Junction Temperature Range	T_J, T_{stg} (2)	-40 to +85	°C

THERMAL CHARACTERISTICS

Characteristic	Symbol	Max	Unit
Thermal Resistance, Junction to Ambient	θ_{JA}(1)	500	°C/W
Solder Temperature	260°C for 3 sec. – 1/16" from Case		

(1) Printed Circuit Board Mounting
(2) Heat Sink should be applied to leads during soldering to prevent Case Temperature exceeding 85°C.

FIGURE 1 – TYPICAL NORMALIZED LIGHT OUTPUT versus INSTANTANEOUS FORWARD CURRENT

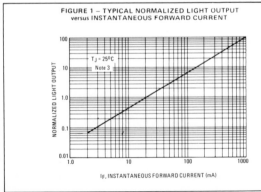

T_J = 25°C
Note 3

NORMALIZED LIGHT OUTPUT

I_F, INSTANTANEOUS FORWARD CURRENT (mA)

DIM	INCHES		MILLIMETERS	
	MIN	MAX	MIN	MAX
A	0.083	0.093	2.11	2.36
B	0.445	0.450	11.30	11.43
C	0.094	0.104	2.39	2.64
D	0.008	0.012	0.203	0.305
F	0.026	0.028	0.660	0.711
G	9°	11°	9°	11°
H	0.062	0.072	1.57	1.83
J	0.040	0.045	1.02	1.14
L	0.092	0.102	2.34	2.59
M	0.019	0.021	0.483	0.533
N	0.405	0.410	10.29	10.41

STYLE 2
PIN 1. ANODE
2 CATHODE

CASE 234 02

DS 2624 R2
(Replaces DS 2624 R1

ELECTRICAL CHARACTERISTICS (T_A = 25°C unless otherwise noted)

Characteristic	Fig. No.	Symbol	Min	Typ	Max	Unit
Reverse Leakage Current (V_R = 3.0 V, R_L = 1.0 Megohm)	–	I_R	–	100	–	nA
Reverse Breakdown Voltage (I_R = 100 μA)	–	BV_R	3.0	–	–	Volts
Forward Voltage (I_F = 20 mA)	2	V_F	–	1.6	2.0	Volts
Total Capacitance (V_R = 0 V, f = 1.0 MHz)	–	C_T	–	150	–	pF

OPTICAL CHARACTERISTICS (T_A = 25°C unless otherwise noted)

Characteristic		Fig. No.	Symbol	Min	Typ	Max	Unit
Axial Instantaneous Luminous Intensity (I_F = 20 mA) Note 1	MLED50 MLED55	1 1	I_o	0.5 0.3	1.0 0.6	– –	mcd
Brightness (I_F = 20 mA) Note 2	MLED50		B	–	750		fL
Peak Emission Wavelength			λ_p	–	660	–	nM
Spectral Line Half Width			$\Delta\lambda$	–	10	–	nM

TYPICAL CHARACTERISTICS

FIGURE 2 – FORWARD CHARACTERISTICS

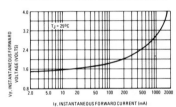

FIGURE 3 – AXIAL LUMINOUS INTENSITY
versus JUNCTION TEMPERATURE

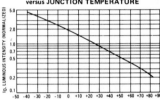

FIGURE 4 – AXIAL LUMINOUS INTENSITY
versus CONTINUOUS FORWARD CURRENT

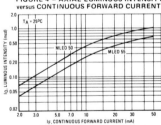

FIGURE 5 – SPATIAL RADIATION PATTERN

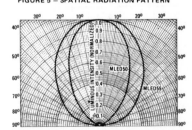

NOTES:

1. Axial Luminous Intensity (I_o) is measured using a Kerant K1100 Light-Emitting Diode (LED) Photometer incorporating a photometric sensor (detector and filter) matched to the CIE* standard observers eye response. I_o is defined as the ratio of the luminous flux emitted by a source to an incremental on axis solid angle subtended by a sensor; i.e., candela = lumens/steradian. Since I_o is a photometric measurement, it provides an accurate indication of the visibility of an LED that includes the physical characteristics of the package such as encapsulant and lens design. The spatial radiation pattern and I_o clearly define the light emitting characteristics of an LED.

As seen from the specification, the MLED50 has a much higher I_o than the MLED55 because of the diffusing nature of the encapsulant used for the MLED55. The result is a large uniform field of emitted light for the MLED55 and a sharp intense field for the MLED50 as shown in Figure 5.

2. Brightness (B) measured with a Photo Research Spectra Spot Brightness Meter Model UB 1/4° with Spectra L-175 lens.

3. To estimate output level under non continuous current drive at junction temperature other than 25°C, first the average junction temperature can be calculated from

$$T_{J(av)} = T_A + \theta_{JA} \times V_F \times I_F \times D$$

where D is the duty cycle of the applied current (I_F). Then the normalized luminous intensity at this junction temperature can be read from Figure 3. Use of the above method should be restricted to drive conditions employing pulses of less than 10 μs duration to avoid errors caused by high peak junction temperatures.

© MOTOROLA INC., 1972   *International Commission on Illumination

MOTOROLA Semiconductor Products Inc.

BOX 20912 • PHOENIX, ARIZONA 85036 • A SUBSIDIARY OF MOTOROLA INC.

6980-3 PRINTED IN USA 6-72 IMPERIAL LITHO 829847   10M   DS 2624 R2

MOTOROLA Semiconductors

BOX 20912 • PHOENIX, ARIZONA 85036

MLED930

INFRARED-EMITTING DIODE

. . . designed for applications requiring high power output, low drive power and very fast response time. This device is used in industrial processing and control, light modulators, shaft or position encoders, punched card readers, optical switching, and logic circuits. It is spectrally matched for use with silicon detectors.

- High-Power Output — 650, μW (Typ) @ I_F = 100 mA
- Infrared-Emission — 9000 Å (Typ)
- Low Drive Current — 10 mA for 70 μW (Typ)
- Popular TO-18 Type Package for Easy Handling and Mounting
- Hermetic Metal Package for Stability and Reliability

INFRARED-EMITTING DIODE
9000 Å
PN GALLIUM ARSENIDE

250 MILLIWATTS

CONVEX LENS

MAXIMUM RATINGS

Rating	Symbol	Value	Unit
Reverse Voltage	V_R	3.0.	Volts
Forward Current-Continuous	I_F	150	mA
Total Device Dissipation @ T_A = 25°C	$P_D(1)$	250	mW
Derate above 25°C		2.5	mW/°C
Operating and Storage Junction Temperature Range	T_J, T_{stg}	–65 to +125	°C

THERMAL CHARACTERISTICS

Characteristics	Symbol	Max	Unit
Thermal Resistance, Juntion to Ambient	θ_{JA}	400	°C/W

(1)Printed Circuit Board Mounting

FIGURE 1 — RELATIVE SPECTRAL OUTPUT

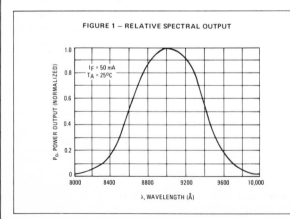

I_F = 50 mA
T_A = 25°C

P$_0$, POWER OUTPUT (NORMALIZED)

λ, WAVELENGTH (Å)

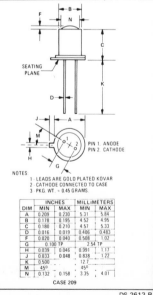

SEATING PLANE

PIN 1. ANODE
PIN 2. CATHODE

NOTES:
1 · LEADS ARE GOLD PLATED KOVAR
2 · CATHODE CONNECTED TO CASE
3 · PKG. WT. ≈ 0.45 GRAMS

DIM	INCHES MIN	INCHES MAX	MILLIMETERS MIN	MILLIMETERS MAX
A	0.209	0.230	5.31	5.84
B	0.178	0.195	4.52	4.95
C	0.180	0.210	4.57	5.33
D	0.016	0.019	0.406	0.483
F	0.020	0.040	0.508	1.02
G	0.100 TP		2.54 TP	
H	0.039	0.046	0.991	1.17
J	0.033	0.048	0.838	1.22
K	0.500		12.7	
M	45°		45°	
N	0.132	0.158	3.35	4.01

CASE 209

DS 2613 R1

ELECTRICAL CHARACTERISTICS (T_A = 25°C unless otherwise noted)

Characteristic	Fig. No.	Symbol	Min	Typ	Max	Unit
Reverse Leakage Current (V_R = 3.0 V, R_L = 1.0 Megohm)	–	I_R	–	50	–	nA
Reverse Breakdown Voltage (I_R = 100 μA)	–	BV_R	3.0	–	–	Volts
Forward Voltage (I_F = 50 mA)	2	V_F	–	1.2	1.5	Volts
Total Capacitance (V_R = 0 V, f = 1.0 MHz)	–	C_T	–	150	–	pF

OPTICAL CHARACTERISTICS (T_A = 25°C unless otherwise noted)

Characteristic	Fig. No.	Symbol	Min	Typ	Max	Unit
Total Power Output (Note 1) (I_F = 50 mA)	3, 4	P_O	200	650	–	μW
Radiant Intensity (Note 2) (I_F = 100 mA)		I_O	–	1.5	–	mW/steradian
Peak Emission Wavelength	1	λ_P	–	9000	–	Å
Spectral Line Half Width	1	$\Delta\lambda$	–	400	–	Å

NOTE:

1. Power Output, P_O, is the total power radiated by the device into a solid angle of 2π steradians. It is measured by directing all radiation leaving the device, within this solid angle, onto a calibrated silicon solar cell.

2. Irradiance from a Light Emitting Diode (LED) can be calculated by:

$$H = \frac{I_O}{d^2} \quad \text{where H is irradiance in mW/cm}^2; I_O \text{ is radiant intensity in mW/steradian;}$$
$$d \text{ is distance from LED to the detector in cm.}$$

FIGURE 2 – FORWARD CHARACTERISTICS

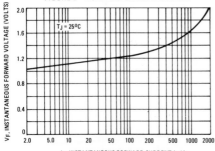

I_F, INSTANTANEOUS FORWARD CURRENT (mA)

FIGURE 3 – POWER OUTPUT versus JUNCTION TEMPERATURE

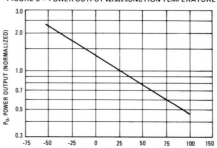

T_J, JUNCTION TEMPERATURE (°C)

FIGURE 4 – INSTANTANEOUS POWER OUTPUT versus FORWARD CURRENT

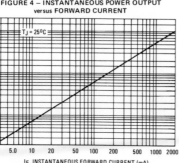

I_F, INSTANTANEOUS FORWARD CURRENT (mA)

FIGURE 5 – SPATIAL RADIATION PATTERN

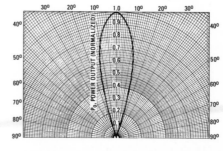

Output saturation effects are not evident at currents up to 2 A as shown on Figure 4. However, saturation does occur due to heating of the semiconductor as indicated by Figure 3. To estimate output level, average junction temperature may be calculated from:

$$T_{J(AV)} = T_A + \theta_{JA} \, V_F I_F D$$

where D is the duty cycle of the applied current, I_F. Use of the above method should be restricted to drive conditions employing pulses of less than 10 μs duration to avoid errors caused by high peak junction temperatures.

© MOTOROLA INC., 1972

 MOTOROLA Semiconductor Products Inc.

BOX 20912 • PHOENIX, ARIZONA 85036 • A SUBSIDIARY OF MOTOROLA INC.

6066-2 PRINTED IN USA 3-72 IMPERIAL LITHO B27778   14M   DS 2613 R1

PIN Silicon Photodiode MRD 500
Courtesy of Motorola Semiconductor Products, Inc.

MOTOROLA Semiconductors
BOX 20912 • PHOENIX, ARIZONA 85036

MRD500
MRD510

PIN SILICON PHOTO DIODE

**100 VOLT
PHOTO DIODE
PIN SILICON**

100 MILLIWATTS

... designed for application in laser detection, light demodulation, detection of visible and near infrared light-emitting diodes, shaft or position encoders, switching and logic circuits, or any design requiring radiation sensitivity, ultra high-speed, and stable characteristics.

- Ultra Fast Response — (<1.0 ns Typ)
- High Sensitivity — MRD500 (1.2 μA/mW/cm^2 Min)
  MRD510 (0.3 μA/mW/cm^2 Min)
- Available With Convex Lens (MRD500) or Flat Glass (MRD510) for Design Flexibility
- Popular TO-18 Type Package for Easy Handling and Mounting
- Sensitive Throughout Visible and Near Infrared Spectral Range for Wide Application
- Annular[†] Passivated Structure for Stability and Reliability

MRD500
(CONVEX LENS)
CASE 209-1

MRD510
(FLAT GLASS)
CASE 210-1

MAXIMUM RATINGS (T$_A$ = 25°C unless otherwise noted)

Rating	Symbol	Value	Unit
Reverse Voltage	V$_R$	100	Volts
Total Device Dissipation @ T$_A$ = 25°C Derate above 25°C	P$_D$	100 0.57	mW mW/°C
Operating and Storage Junction Temperature Range	T$_J$,T$_{stg}$	−65 to +200	°C

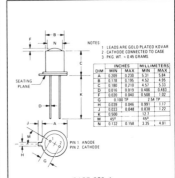

NOTES
1. LEADS ARE GOLD PLATED KOVAR
2. CATHODE CONNECTED TO CASE
3. PKG. WT. = 0.45 GRAMS.

DIM	INCHES		MILLIMETERS	
	MIN	MAX	MIN	MAX
A	0.209	0.230	5.31	5.84
B	0.178	0.195	4.52	4.95
C	0.180	0.210	4.57	5.33
D	0.016	0.019	0.406	0.483
F	0.020	0.040	0.508	1.02
G	0.100 TP		2.54 TP	
H	0.039	0.046	0.991	1.17
J	0.033	0.048	0.838	1.22
K	0.500		12.7	
M	45º		45º	
N	0.132	0.158	3.35	4.01

PIN 1 ANODE
PIN 2 CATHODE

CASE 209-1

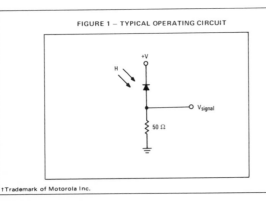

FIGURE 1 — TYPICAL OPERATING CIRCUIT

+V

H

V$_{signal}$

50 Ω

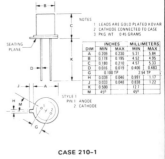

NOTES
1. LEADS ARE GOLD PLATED KOVAR
2. CATHODE CONNECTED TO CASE
3. PKG. WT. 0.45 GRAMS

DIM	INCHES		MILLIMETERS	
	MIN	MAX	MIN	MAX
A	0.209	0.230	5.31	5.84
B	0.178	0.195	4.52	4.95
C	0.180	0.210	4.57	5.33
D	0.016	0.019	0.406	0.483
G	0.100 TP		2.54 TP	
H	0.039	0.046	0.991	1.17
J	0.033	0.048	0.838	1.22
K	0.500		12.7	
M	45º		45º	

STYLE 1
PIN 1 ANODE
2 CATHODE

CASE 210-1

†Trademark of Motorola Inc.

DS 2608 R1

MRD500 ● MRD510

STATIC ELECTRICAL CHARACTERISTICS (T_A = 25°C unless otherwise noted)

Characteristic	Fig. No.	Symbol	Min	Typ	Max	Unit

Dark Current
(V_R = 20 V, R_L = 1.0 megohm; Note 2)
          T_A = 25°C
          T_A = 100°C
	4 and 5	I_D	
			
			–
			–
				
				–
				14
					
					2.0
					–
						nA

Reverse Breakdown Voltage (I_R = 10 µA)	–	BV_R	100	200	–	Volts
Forward Voltage (I_F = 50 mA)	–	V_F	–	–	1.1	Volts
Series Resistance (I_F = 50 mA)	–	R_s	–	–	10	ohms
Total Capacitance (V_R = 20 V; f = 1.0 MHz)	6	C_T	–	–	4	pF

OPTICAL CHARACTERISTICS (T_A = 25°C)

Characteristic		Fig. No.	Symbol	Min	Typ	Max	Unit
Radiation Sensitivity (V_R = 20 V, Note 1)	MRD500 MRD510	2 and 3	S_R	1.2 0.3	1.8 0.42	– –	µA/mW/cm^2
Sensitivity at 0.8 µm (V_R = 20 V, Note 3)	MRD500 MRD510	– –	$S_{(\lambda = 0.8\,\mu m)}$	– 	6.6 1.5	– –	µA/mW/cm^2
Response Time (V_R = 20 V, R_L = 50 ohms)		– –	$t_{(resp)}$		1.0		ns
Wavelength of Peak Spectral Response		7	λ_s	–	0.8	–	µm

NOTES:

1. Radiation Flux Density (H) equal to 5.0 mW/cm^2 emitted from a tungsten source at a color temperature of 2870 K.

2. Measured under dark conditions. (H ≈ 0).

3. Radiation Flux Density (H) equal to 0.5 mW/cm^2 at 0.8 µm.

MOTOROLA *Semiconductor Products Inc.*

TYPICAL ELECTRICAL CHARACTERISTICS

FIGURE 2 – IRRADIATED VOLTAGE – CURRENT
CHARACTERISTIC FOR MRD500

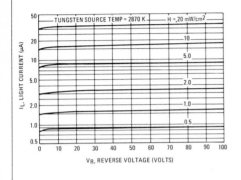

FIGURE 3 – IRRADIATED VOLTAGE – CURRENT
CHARACTERISTIC FOR MRD 510

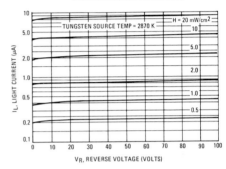

FIGURE 4 – DARK CURRENT versus TEMPERATURE

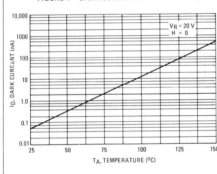

FIGURE 5 – DARK CURRENT versus REVERSE VOLTAGE

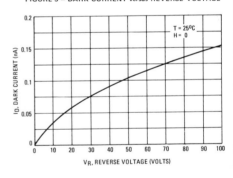

FIGURE 6 – CAPACITANCE versus VOLTAGE

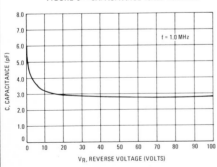

FIGURE 7 – RELATIVE SPECTRAL RESPONSE

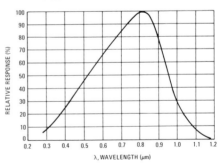

 MOTOROLA *Semiconductor Products Inc.*

MRD500 ● MRD510

MRD500 AND MRD510
OPTOELECTRONIC DEFINITIONS, CHARACTERISTICS, AND RATINGS

BV_R Reverse Breakdown Voltage – The minimum dc reverse breakdown voltage at stated diode current and ambient temperature.

C_T Total Capacitance

H Radiation Flux Density (Irradiance) $[mW/cm^2]$ – The total incident radiation energy measured in power per unit area.

I_D Dark Current – The maximum reverse leakage current through the device measured under dark conditions, ($H\approx0$), with a stated reverse voltage, load resistance, and ambient temperature.

P_D Power Dissipation

R_s Series Resistance – The maximum dynamic series resistance measured at stated forward current and ambient temperature.

S_R Radiation Sensitivity ($\mu A/mW/cm^2$) – The ratio of photo-induced current to the incident radiant energy measured at the plane of the lens of the photo device under stated conditions of radiation flux density (H), reverse voltage, load resistance, and ambient temperature.

T_A Ambient Temperature

T_J Junction Temperature

T_{stg} Storage Temperature

V_F Forward Voltage – The maximum forward voltage drop across the diode at stated diode current and ambient temperature.

V_R Reverse Voltage – The maximum allowable value of dc reverse voltage which can be applied to the device at the rated temperature.

$\lambda_s(\mu m)$ Wavelength of peak spectral response in micro – meters.

OPTO DEVICES

AN-440 – THEORY AND CHARACTERISTICS OF PHOTO TRANSISTORS

A brief history of the photoelectric effect is discussed, followed by a comprehensive analysis of the effect in bulk semiconductors, pn junctions and phototransistors. A model is presented for the phototransistor. Static and transient data for the MRD300 provide typical phototransistor characteristics. Appendices provide a discussion of the relationship of irradiation and illumination and define terms specifically related to phototransistors.

AN-508 Applications of Phototransistors in Electro-Optic Systems

This note reviews phototransistor theory, characteristics and terminology, then discusses the design of electro-optic systems using device information and geometric considerations. It also includes several circuit designs that are suited to dc, low-frequency and high-frequency applications.

To obtain a copy of this note list the AN number on your company letterhead and send your request to:

Technical Information Center
Motorola Semiconductor Products Inc.
P. O. Box 20924
Phoenix, Arizona 85036

 MOTOROLA *Semiconductor Products Inc.*

BOX 20912 ● PHOENIX, ARIZONA 85036 ● A SUBSIDIARY OF MOTOROLA INC.
5206-4 PRINTED IN USA 3-72 IMPERIAL LITHO B27738   13M   DS 2608 R1

MOTOROLA
Semiconductors
BOX 20912 • PHOENIX, ARIZONA 85036

MRD300
MRD310

NPN SILICON HIGH SENSITIVITY
PHOTO TRANSISTOR

. . . designed for application in industrial inspection, processing and control, counters, sorters, switching and logic circuits or any design requiring radiation sensitivity, and stable characteristics.

- Popular TO-18 Type Package for Easy Handling and Mounting
- Sensitive Throughout Visible and Near Infra-Red Spectral Range for Wider Application
- Minimum Light Current 4 mA at H = 5 mW/cm^2 (MRD 300)
- External Base for Added Control
- Annular ◆ Passivated Structure for Stability and Reliability

50 VOLT
PHOTO TRANSISTOR
NPN SILICON

250 MILLIWATTS

MAXIMUM RATINGS (T_A = 25°C unless otherwise noted)

Rating (Note 1)	Symbol	Value	Unit
Collector-Emitter Voltage	V_{CEO}	50	Volts
Emitter-Collector Voltage	V_{ECO}	7.0	Volts
Collector-Base Voltage	V_{CBO}	80	Volts
Total Device Dissipation @ T_A = 25°C	P_D	250	mW
Derate above 25°C		1.43	mW/^{o}C
Operating Junction and Storage Temperature Range	T_J, T_{stg}	–65 to +200	^{o}C

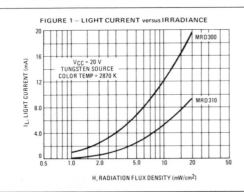

FIGURE 1 – LIGHT CURRENT versus IRRADIANCE

MRD300

V_{CC} = 20 V
TUNGSTEN SOURCE
COLOR TEMP = 2870 K

MRD310

I_L, LIGHT CURRENT (mA)

H, RADIATION FLUX DENSITY (mW/cm^2)

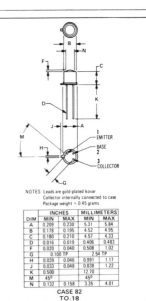

NOTES: Leads are gold-plated kovar
Collector internally connected to case
Package weight = 0.45 grams

DIM	INCHES		MILLIMETERS	
	MIN	MAX	MIN	MAX
A	0.209	0.230	5.31	5.84
B	0.178	0.195	4.52	4.95
C	0.180	0.210	4.57	4.33
D	0.016	0.019	0.406	0.483
F	0.020	0.040	0.508	1.02
G	0.100 TP		2.54 TP	
H	0.039	0.046	0.991	1.17
J	0.033	0.048	0.838	1.22
K	0.500		12.70	
M	45^0		45^0	
N	0.132	0.158	3.35	4.01

CASE 82
TO-18

1 EMITTER
2 BASE
3 COLLECTOR

◆Annular Semiconductors Patented by Motorola Inc.

DS 2601 R2

MRD300 ● MRD310

STATIC ELECTRICAL CHARACTERISTICS ($T_A = 25^{\circ}C$ unless otherwise noted)

Characteristic	Symbol	Min	Typ	Max	Unit
Collector Dark Current (V_{CC} = 20 V, H≈0) $T_A = 25^{\circ}C$ $T_A = 100^{\circ}C$	I_{CEO}	– –	– 4.0	25 –	na µA
Collector-Base Breakdown Voltage (I_C = 100 µA)	BV_{CBO}	80	–	–	Volts
Collector-Emitter Breakdown Voltage (I_C = 100 µA)	BV_{CEO}	50	–	–	Volts
Emitter-Collector Breakdown Voltage (I_E = 100 µA)	BV_{ECO}	7.0	–	–	Volts

OPTICAL CHARACTERISTICS ($T_A = 25^{\circ}C$ unless otherwise noted)

Characteristic	Device Type	Symbol	Min	Typ	Max	Unit
Light Current (V_{CC} = 20 V, R_L = 100 ohms) Note 1	MRD300 MRD310	I_L	4.0 1.0	7.5 2.5	– –	mA
Light Current (V_{CC} = 20 V, R_L = 100 ohms) Note 2	MRD300 MRD310	I_L	– –	2.5 0.8	– –	mA
Photo Current Rise Time (Note 3) (R_L = 100 ohms I_L = 1.0 mA peak)		tr	–	–	2.5	µs
Photo Current Fall Time (Note 3) (R_L = 100 ohms I_L = 1.0 mA peak)		tf	–	–	4.0	µs

NOTES:

1. Radiation flux density (H) equal to 5.0 mW/cm^2 emitted from a tungsten source at a color temperature of 2870 K.
2. Radiation flux density (H) equal to 0.5 mW/cm^2 (pulsed) from a GaAs (gallium-arsenide) source at λ ≈ 0.9 µm.
3. For unsaturated response time measurements, radiation is provided by pulsed GaAs (gallium-arsenide) light-emitting diode (λ ≈ 0.9 µm) with a pulse width equal to or greater than 10 microseconds (see Figure 6) I_L = 1.0 mA peak.

MOTOROLA *Semiconductor Products Inc.*

TYPICAL ELECTRICAL CHARACTERISTICS

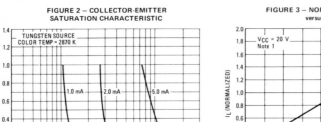

FIGURE 2 – COLLECTOR-EMITTER
SATURATION CHARACTERISTIC

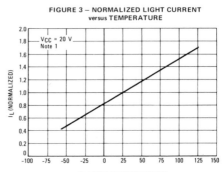

FIGURE 3 – NORMALIZED LIGHT CURRENT
versus TEMPERATURE

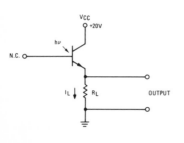

FIGURE 4 – RISE TIME versus
LIGHT CURRENT

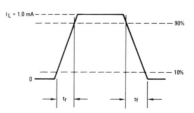

FIGURE 5 – FALL TIME versus
LIGHT CURRENT

FIGURE 6 – PULSE RESPONSE TEST CIRCUIT AND WAVEFORM

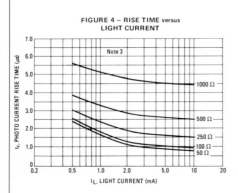

MRD300 ● MRD310

FIGURE 7 – DARK CURRENT versus TEMPERATURE

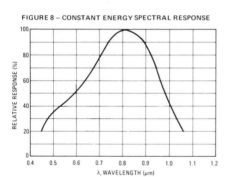

I_{CEO}, COLLECTOR, DARK CURRENT (μA)

$H = 0$
$V_{CE} = 20$ V

T_A, AMBIENT TEMPERATURE (^{0}C)

FIGURE 8 – CONSTANT ENERGY SPECTRAL RESPONSE

RELATIVE RESPONSE (%)

λ, WAVELENGTH (μm)

FIGURE 9 – ANGULAR RESPONSE

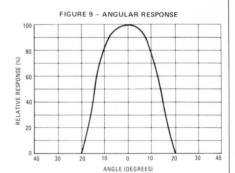

RELATIVE RESPONSE (%)

ANGLE (DEGREES)

MOTOROLA Semiconductor Products Inc.

BOX 20912 ● PHOENIX, ARIZONA 85036 ● A SUBSIDIARY OF MOTOROLA INC.

2004-8 PRINTED IN USA 3–72 IMPERIAL LITHO B27801   7M   DS 2601 R2

bibliography

AHRONS, RICHARD W. "Gallium Phosphide Diodes Offer Brighter Lamps and Displays." *Computer Design*, January 1971.

BAASCH, THOMAS L. "Light Emitting Diodes, How to Select Them, How to Use Them." *Electronic Products Magazine*, 20 September 1971.

BLISS, JOHN. *Applications of Phototransistors in Electric Systems.* Phoenix, Arizona: Motorola Semiconductor Products, Inc., AN508, 1970.

———. "Build Reliable Optoelectronic Circuits." *Electronic Design 3*, 3 February 1972.

———. *Theory and Characteristics of Photo Transistors.* Phoenix, Arizona: Motorola Semiconductor Products, Inc., AN440, 1968.

BOTOS, R. *Low Frequency Applications of Field Effect Transistors.* Phoenix, Arizona: Motorola Semiconductor Products, Inc., AN511, 1969.

BOTTINI, MIKE. "Don't Forget the Optical Isolator When Solving Your Coupling Problems." *Electronic Design News*, 15 April 1972.

BRANCH, JAMES K. "Don't Let LED Output Measurements Throw You." *Electronic Design News*, 1 January 1972.

CLAIREX ELECTRONICS, INC. *Photoconductive Cell Design Manual.* New York: New York, 1966.

COUGHLIN, ROBERT F. *Principles and Applications of Semiconductors and Circuits.* Englewood Cliffs, N. J.: Prentice-Hall, Inc., 1971.

——— et al. *Laboratory Manual in Transistors and Semiconductor Devices.* Englewood Cliffs, N. J.: Prentice-Hall, Inc., 1970.

CUSHMAN, ROBERT. "Where to Use LEDs." *Electronic Design News*, October 1968.

DEBOO, GORDON J., and CLIFFORD N. BURROUS. "Integrated Circuits and Semiconductor Devices: Theory and Application." McGraw-Hill Book Company, New York: 1971.

DRISCOLL, FREDERICK F. *Analysis of Electric Circuits.* Englewood Cliffs, N. J.: Prentice-Hall, Inc., 1973.

FAIRCHILD SEMICONDUCTOR. *Linear Integrated Circuits Applications Handbook.* Mountain View, Cal.: 1967.

———. *Linear Integrated Circuits Data Catalog.* Mountain View, Cal.: 1971.

GENERAL ELECTRIC CO. *Miniature Lamp Catalog 3-6253.* Nela Park, Cleveland, Ohio: 1971.

————. *SCR Manual*, 5th ed. Syracuse, N. Y.: 1972.

GIENGER, M., and D. KESNER. *Voltage and Current Boost Techniques Using the MC 1560-61*. Phoenix, Arizona: Motorola Semiconductor Products, Inc., AN498, 1969.

HAVER, R. J., and B. C. SHINER. *Theory, Characteristics and Applications of the Programmable Unijunction Transistor*. Phoenix, Arizona: Motorola Semiconductor Products, Inc., AN527, 1971.

HERTZ, L. M. *Solid State Lamps Application Manual 3-0121*. Nela Park, Cleveland, Ohio: General Electric, 1970.

————. *Solid State Lamps Theory and Characteristics Manual 3-8270R*. Nela Park, Cleveland, Ohio: General Electric, 1970.

HEWLETT, H., and A. S. VAUSE. *Lamps and Lighting*. New York: American Elsevier Publishing Co., 1966.

HEWLETT PACKARD. *An Attenuator Design Using PIN Diodes, Application Note 912*. Palo Alto, Cal.: 1970

————. *DC Power Supply*. New Jersey: 1970.

————. *HP PIN Photodiode, Application Note 917*. Palo Alto, Cal.: October 1970.

————. *Threshold Detection of Visible and Infrared Radiation with PIN Photodiodes, Application Note 915*. Palo Alto, Cal.: 1970.

HOWELL, E. K. *The Light Activated SCR, Application Note 200.34*. Neva Park, Cleveland, Ohio: General Electric, 1965.

INTERNATIONAL RECTIFIER. "Photovoltaic Light Sensors." *Rectifier News*, El Segundo, California, 1969.

KEPCO CO. *Power Supply Handbook*. Flushing, N. Y.: 1965.

KORN, SEBALD R. *How to Evaluate Light Emitters and Optical Systems for Light Sensitive Silicon Devices, Application Note 200.59*. Syracuse, N. Y.: General Electric, 1971.

M 7 INC. *Silicon Photovoltaic Converters*. Arlington Heights, Ill.: 1971.

MAJOR, LARRY D., and RONALD D. GEROTTI. *Efficient High-Power GaAs Emitters*, Dallas, Texas: Texas Instruments, Inc., Bulletin CA-131, July 1969.

MCDERMOTT, JIM. "Solid-State Optoelectronic Components Put Imagination in Engineering." *Electronic Design* 11, 27 May, 1971.

MIMS, FORREST M. "Light Emitting Diodes." *Popular Electronics*, November 1970.

MONSANTO COMMERCIAL PRODUCTS CO. *GaAs Lite*. Cupertino, Cal.: 1972.

NATIONAL SEMICONDUCTOR CORP. *Linear Applications*. Santa Clara, Cal.: 1972.

NEWMIRE, LEWIS J. *Theory, Characteristics and Applications of Silicon Unilateral, and Bilateral Switches*. Phoenix, Arizona: Motorola Semiconductor Products, Inc., AN526, 1970.

RCA CORPORATION. *Solid-State Power Circuits.* Somerville, N. J.: 1971.

SIGNALITE, INC. *Neon Glow Lamps.* Neptune, N. J. 1967.

SMITH, GEORGE. "Applications of Opto-isolators." Litronix, Appnote 2. 1971

———. "LEDs and Photometry." Litronix, Appnote 1. 1971

SMITH, WARREN J. *Modern Optical Engineering.* New York: McGraw-Hill Book Company, 1966.

SOLAR SYSTEMS, INC. *Light Sensitive Devices.* Skokie, Ill.: 1969

TEXAS INSTRUMENTS, INC. *FET Design Ideas.* Dallas, Texas: 1971.

———. *Optically Coupled Isolators in Circuits.* Dallas, Texas: Bulletin CA-156, September 1970.

VILLANUCCI, ROBERT S., ALEXANDER W. AVTGIS, and WILLIAM F. MEGOW. *Electronic Techniques: Shop Practices and Construction.* Englewood Cliffs, N. J.: Prentice-Hall, Inc., 1974.

WETTERAU, LIN. "Put Optoelectronic Components to Work in Systems Design." *Electronic Design 12,* 10 June 1971.

ZINDER, DAVID A. *Electronic Speed Control for Appliance Motors.* Phoenix, Arizona: Motorola Semiconductor Products, Inc., AN482, 1969.

———. *Unijunction Trigger Circuits for Grated Thyristors.* Phoenix, Arizona: Motorola Semiconductor Products, Inc., AN413, 1969.

index

A

Acceptor atoms, 12
Ac load line, 80–84
Ac, small signal models, 64–66, 114
α_F, 41–42
Air flow measurement, 9–10
Alarm system, 419–420
Alpha cutoff, 146
Ambient, 148, 150
Amplifiers
  class A, 159–167
  complementary, 167–172
  differential, 217–220
  error, 218
  integrated circuit, 172, 216–220
  operational (*see* Operational amplifier)
  power, 159–167, 173, 238
  regulator, 241–243
And gate, 28
Angstrom, 355, 356, 380
Anode
  diode, 15
  programmable UJT, 286
  two transistor switch, 281
Anode characteristic, SCR, 313
Anode program voltage, 295
Application notes, 251
Arcing, 327
ASBS, 349–350
Asymmetrical silicon bilateral switch, 301
Audio level control, 130
Avalanche breakdown, 31
Average current, SCR, 316–327
Average light intensity, 381
Average voltage, 317–327

B

β_o, 46–47, 65–68
β_F, 42–47, 51–52
Ballasting resistor, 368

Base, 38
Base bulk resistance, 144
Battery charger, 333–334, 404
Beta, (*see* β_F, β_o)
Biasing
  common-base, 91
  common-collector, 86
  common-emitter, 52–63
  complementary amplifier, 168
  FET, 104–112
  Zener diode, 220–221
Bidirectional, 305
Bidirectional trigger diode, 296–297
Bilateral, 305
Bipolar device, 37
Bipolar power supply, 190–192
Bipolar junction transistors (BJT)
  action, 40–41
  common-base, 41, 42
  common-collector, 42, 43
  common-emitter, 42
  operating modes, 39
BJT (*see* Bipolar junction transistor)
Body resistance, 15
Bootstrapping, 172, 173
Break-frequency, 414
Breakover voltage, 297–298
Bridge, 8
Brightness
  apparent, 378
  physical, 378
Bulk resistance, 15
Bulk type photodetectors, 384, 385

C

Cadmium selenide, 387, 389
Cadmium sulfide, 386, 387, 389
Cadmium sulphide cell, 273
Calibrated bias lines, 106–108
Candle, 380
Candle power, 359, 360–362, 390
Capacitance, BJT, 144–145